AF780881

Nanomaterial-Enabled Environmental Remediation and Removal of Emerging Pollutants

Nanomaterial-Enabled Environmental Remediation and Removal of Emerging Pollutants

Guest Editors

Chuanjia Jiang
Shengwei Liu
Tanapon Phenrat
Qian Sui

MDPI

Basel • Beijing • Wuhan • Barcelona • Belgrade • Novi Sad • Cluj • Manchester

Guest Editors
Chuanjia Jiang
College of Environmental Science and Engineering
Nankai University
Tianjin
China

Shengwei Liu
School of Environmental Science and Engineering
Sun Yat-sen University
Guangzhou
China

Tanapon Phenrat
Department of Civil Engineering
Naresuan University
Phitsanulok
Thailand

Qian Sui
School of Resources and Environmental Engineering
East China University of Science & Technology
Shanghai
China

Editorial Office
MDPI AG
Grosspeteranlage 5
4052 Basel, Switzerland

This is a reprint of the Special Issue, published open access by the journal *Toxics* (ISSN 2305-6304), freely accessible at: https://www.mdpi.com/journal/toxics/special_issues/AIA76D9Y0G.

For citation purposes, cite each article independently as indicated on the article page online and as indicated below:

Lastname, A.A.; Lastname, B.B. Article Title. *Journal Name* **Year**, *Volume Number*, Page Range.

ISBN 978-3-7258-5689-3 (Hbk)
ISBN 978-3-7258-5690-9 (PDF)
https://doi.org/10.3390/books978-3-7258-5690-9

© 2025 by the authors. Articles in this book are Open Access and distributed under the Creative Commons Attribution (CC BY) license. The book as a whole is distributed by MDPI under the terms and conditions of the Creative Commons Attribution-NonCommercial-NoDerivs (CC BY-NC-ND) license (https://creativecommons.org/licenses/by-nc-nd/4.0/).

Contents

toxics

MDPI

Editorial

Nanomaterial-Enabled Environmental Remediation and Removal of Emerging Pollutants

Chuanjia Jiang [1,*], Shengwei Liu [2,*], Tanapon Phenrat [3,*] and Qian Sui [4,*]

1 Ministry of Education Key Laboratory of Pollution Processes and Environmental Criteria, Tianjin Key Laboratory of Environmental Remediation and Pollution Control, College of Environmental Science and Engineering, Nankai University, Tianjin 300350, China
2 Guangdong Provincial Key Laboratory of Environmental Pollution Control and Remediation Technology, School of Environmental Science and Engineering, Sun Yat-sen University, Guangzhou 510006, China
3 Research Unit for Integrated Natural Resources Remediation and Reclamation (IN3R), Department of Civil Engineering, Faculty of Engineering, Naresuan University, Phitsanulok 65000, Thailand
4 State Environmental Protection Key Laboratory of Environmental Risk Assessment and Control on Chemical Process, School of Resources and Environmental Engineering, East China University of Science and Technology, Shanghai 200237, China
* Correspondence: jiangcj@nankai.edu.cn (C.J.); liushw6@mail.sysu.edu.cn (S.L.); tanaponph@nu.ac.th (T.P.); suiqian@ecust.edu.cn (Q.S.)

Rapid industrialization and urbanization in recent decades have benefited human society unprecedentedly. However, the concomitant release of various toxic and harmful substances into the environment has caused considerable ecological and human health risks [1–3]. In particular, emerging contaminants—including perfluorinated and polyfluoroalkyl substances (PFASs), pharmaceuticals and personal care products, organophosphate esters, and micro-/nanoplastics—have been detected across diverse environmental matrices and caused global concerns [4–7]. Therefore, enormous efforts have been devoted to the development of new technologies for removing environmental pollutants. However, there remain obstacles to high-efficiency and low-carbon environmental remediation [8], and the effective removal of emerging contaminants (e.g., PFASs and microplastics) poses significant new challenges [9,10].

The rapid development of nanotechnology has opened up new opportunities for more efficient and cost-effective pollution control and environmental remediation. A myriad of novel nanomaterials with large specific surface area and abundant surface reactive sites have been explored for the enhanced removal of various legacy and emerging pollutants via adsorption, membrane separation, catalytic oxidation/reduction/hydrolysis, and photocatalysis, etc. [11,12]. Nanomaterials can act as efficient adsorbents, not only owing to their high surface areas and well-developed pore network, but also to their nano-specific surface structures [13,14]. Synthetic membranes with confined nanostructures have demonstrated selective ion separation from complex aqueous matrices, achieving ultrahigh selectivity for a range of monovalent and divalent ions [15]. Meanwhile, the catalytic efficiency of nanomaterials for degrading pollutants can be regulated by modulating their surface atomic arrangement via facet and defect engineering [16,17]. Moreover, nanomaterials have shown tremendous potential for in situ remediation of contaminated soil and groundwater [18].

The applications and potential environmental implications of nanomaterials have emerged as one of the most active and productive research frontiers in the field of environmental science and engineering. Nevertheless, nanomaterials still face a number of challenges in practical applications, including relatively high cost, material instability, and potential environmental impact. Structural degradation, aggregation, and the loss of active sites can significantly compromise their performance [19,20]. Moreover, current applications of nanomaterials remain largely confined to laboratory or pilot-scale studies. An understanding of their realistic performance as well as environmental and health impacts remains inadequate. These uncertainties hinder the large-scale deployment of nanotechnology in environmental applications, and call for safety- and sustainability-by-design strategies [21]. Addressing these challenges requires robust interdisciplinary collaboration, spanning materials science, environmental chemistry, engineering, toxicology, ecology, and policy. Notably, emerging tools such as theoretical computation and machine learning are critical for predicting the pollutant removal performances and environmental behaviors of nanomaterials and guiding the rational design of next-generation environmental functional materials with minimized risk profiles [22,23].

This Special Issue comprises two comprehensive reviews and eight original research articles, with international authorship from six countries (Contributions 1–10). These papers highlight recent progress in the development of nanomaterials, including metal-based, carbon-based, and composite nanomaterials, for the removal of both organic pollutants (e.g., pesticides, dyes, and formaldehyde) and heavy metals. In addition to experimental investigations, we have included works to emphasize the role of computational simulation in predicting and optimizing the performance of nanomaterials. Moreover, we identify critical knowledge gaps and propose future research directions to advance the field of environmental nanotechnology. We hope that the advances and insights presented herein will inspire further innovation and foster interdisciplinary collaboration, ultimately contributing to safer, more efficient, and sustainable nanotechnology for environmental protection applications.

Author Contributions: Conceptualization, C.J., S.L., T.P. and Q.S.; writing—original draft preparation, C.J.; writing—review and editing, C.J., S.L., T.P. and Q.S. All authors have read and agreed to the published version of the manuscript.

Data Availability Statement: No new data were created or analyzed in this study. Data sharing is not applicable to this article.

Conflicts of Interest: The authors declare no conflict of interest.

List of Contributions:

1. Wongrerkdee, S.; Wongrerkdee, S.; Boonruang, C.; Sujinnapram, S. Enhanced Photocatalytic Degradation of Methylene Blue Using Ti-Doped ZnO Nanoparticles Synthesized by Rapid Combustion. *Toxics* **2023**, *11*, 33. https://doi.org/10.3390/toxics11010033
2. Yi, Y.; Guan, Q.; Wang, W.; Jian, S.; Li, H.; Wu, L.; Zhang, H.; Jiang, C. Recyclable Carbon Cloth-Supported ZnO@Ag_3PO_4 Core–Shell Structure for Photocatalytic Degradation of Organic Dye. *Toxics* **2023**, *11*, 70. https://doi.org/10.3390/toxics11010070
3. Ezzeddine, Z.; Batonneau-Gener, I.; Pouilloux, Y. Zinc Removal from Water via EDTA-Modified Mesoporous SBA-16 and SBA-15. *Toxics* **2023**, *11*, 205. https://doi.org/10.3390/toxics11030205
4. Krobthong, S.; Rungsawang, T.; Wongrerkdee, S. Comparison of ZnO Nanoparticles Prepared by Precipitation and Combustion for UV and Sunlight-Driven Photocatalytic Degradation of Methylene Blue. *Toxics* **2023**, *11*, 266. https://doi.org/10.3390/toxics11030266
5. Neelgund, G.M.; Jimenez, E.A.; Ray, R.L.; Kurkuri, M.D. Facilitated Adsorption of Mercury(II) and Chromium(VI) Ions over Functionalized Carbon Nanotubes. *Toxics* **2023**, *11*, 545. https://doi.org/10.3390/toxics11060545
6. Krobthong, S.; Rungsawang, T.; Khaodara, N.; Kaewtrakulchai, N.; Manatura, K.; Sukiam, K.; Wathinputthiporn, D.; Wongrerkdee, S.; Boonruang, C.; Wongrerkdee, S. Sustainable Development of ZnO Nanostructure Doping with Water Hyacinth-Derived Activated Carbon for Visible-Light Photocatalysis. *Toxics* **2024**, *12*, 165. https://doi.org/10.3390/toxics12030165

7. Lu, Z.; Wang, Z. Complete Photooxidation of Formaldehyde to CO_2 via Ni-Dual-Atom Decorated Crystalline Triazine Frameworks: A DFT Study. *Toxics* **2024**, *12*, 242. https://doi.org/10.3390/toxics12040242
8. Tu, W.; Cai, W. Selective Adsorption of Hazardous Substances from Wastewater by Hierarchical Oxide Composites: A Review. *Toxics* **2024**, *12*, 447. https://doi.org/10.3390/toxics12070447
9. He, F.; Xu, L.; Wang, H.; Jiang, C. Recent Progress in Molecular Oxygen Activation by Iron-Based Materials: Prospects for Nano-Enabled In Situ Remediation of Organic-Contaminated Sites. *Toxics* **2024**, *12*, 773. https://doi.org/10.3390/toxics12110773
10. Yu, S.; Wang, Z. Density Functional Theory Insight in Photocatalytic Degradation of Dichlorvos Using Covalent Triazine Frameworks Modified by Various Oxygen-Containing Acid Groups. *Toxics* **2024**, *12*, 928. https://doi.org/10.3390/toxics12120928

References

1. Driscoll, C.T.; Mason, R.P.; Chan, H.M.; Jacob, D.J.; Pirrone, N. Mercury as a Global Pollutant: Sources, Pathways, and Effects. *Environ. Sci. Technol.* **2013**, *47*, 4967–4983. [CrossRef] [PubMed]
2. Alengebawy, A.; Abdelkhalek, S.T.; Qureshi, S.R.; Wang, M.Q. Heavy Metals and Pesticides Toxicity in Agricultural Soil and Plants: Ecological Risks and Human Health Implications. *Toxics* **2021**, *9*, 42. [CrossRef]
3. Arp, H.P.H.; Gredelj, A.; Glüge, J.; Scheringer, M.; Cousins, I.T. The Global Threat from the Irreversible Accumulation of Trifluoroacetic Acid (TFA). *Environ. Sci. Technol.* **2024**, *58*, 19925–19935. [CrossRef]
4. Qu, R.; Wang, J.X.; Li, X.J.; Zhang, Y.; Yin, T.L.; Yang, P. Per- and Polyfluoroalkyl Substances (PFAS) Affect Female Reproductive Health: Epidemiological Evidence and Underlying Mechanisms. *Toxics* **2024**, *12*, 678. [CrossRef]
5. Boxall, A.B.A.; Brooks, B.W. Pharmaceuticals and Personal Care Products in the Environment: What Progress Has Been Made in Addressing the Big Research Questions? *Environ. Toxicol. Chem.* **2024**, *43*, 481–487. [CrossRef] [PubMed]
6. Yang, Y.; Meng, Y.; Liu, S.; Wei, L.; Huang, Q.H. Insights into organophosphate esters (OPEs) in aquatic ecosystems: Occurrence, environmental behavior, and ecological risk. *Crit. Rev. Environ. Sci. Technol.* **2024**, *54*, 641–675. [CrossRef]
7. Wang, C.H.; Zhao, J.; Xing, B.S. Environmental source, fate, and toxicity of microplastics. *J. Hazard. Mater.* **2021**, *407*, 124357. [CrossRef]
8. Stroo, H.F.; Leeson, A.; Marqusee, J.A.; Johnson, P.C.; Ward, C.H.; Kavanaugh, M.C.; Sale, T.C.; Newell, C.J.; Pennell, K.D.; Lebrón, C.A.; et al. Chlorinated Ethene Source Remediation: Lessons Learned. *Environ. Sci. Technol.* **2012**, *46*, 6438–6447. [CrossRef]
9. Wang, B.; Yu, G. Emerging contaminant control: From science to action. *Front. Environ. Sci. Eng.* **2022**, *16*, 81. [CrossRef]
10. Liu, W.Y.; Zhang, J.L.; Liu, H.; Guo, X.N.; Zhang, X.Y.; Yao, X.L.; Cao, Z.G.; Zhang, T.T. A review of the removal of microplastics in global wastewater treatment plants: Characteristics and mechanisms. *Environ. Int.* **2021**, *146*, 106277. [CrossRef]
11. Lu, F.; Astruc, D. Nanocatalysts and other nanomaterials for water remediation from organic pollutants. *Coordin. Chem. Rev.* **2020**, *408*, 213180. [CrossRef]
12. Wen, M.C.; Li, G.Y.; Liu, H.L.; Chen, J.Y.; An, T.C.; Yamashita, H. Metal-organic framework-based nanomaterials for adsorption and photocatalytic degradation of gaseous pollutants: Recent progress and challenges. *Environ. Sci-Nano* **2019**, *6*, 1006–1025. [CrossRef]
13. Auffan, M.; Rose, J.; Proux, O.; Borschneck, D.; Masion, A.; Chaurand, P.; Hazemann, J.L.; Chaneac, C.; Jolivet, J.P.; Wiesner, M.R.; et al. Enhanced adsorption of arsenic onto maghemites nanoparticles: As(III) as a probe of the surface structure and heterogeneity. *Langmuir* **2008**, *24*, 3215–3222. [CrossRef]
14. Auffan, M.; Rose, J.; Bottero, J.Y.; Lowry, G.V.; Jolivet, J.P.; Wiesner, M.R. Towards a definition of inorganic nanoparticles from an environmental, health and safety perspective. *Nat. Nanotechnol.* **2009**, *4*, 634–641. [CrossRef] [PubMed]
15. Liu, K.R.; Epsztein, R.; Lin, S.H.; Qu, J.H.; Sun, M. Ion-Ion Selectivity of Synthetic Membranes with Confined Nanostructures. *ACS Nano* **2024**, *18*, 21633–21650. [CrossRef]
16. Jin, X.; Wu, D.D.; Liu, C.; Huang, S.H.; Zhou, Z.Y.; Wu, H.; Chen, X.R.; Huang, M.Y.; Zhou, S.D.; Gu, C. Facet effect of hematite on the hydrolysis of phthalate esters under ambient humidity conditions. *Nat. Commun.* **2022**, *13*, 6125. [CrossRef] [PubMed]
17. Wang, Y.X.; Duan, X.G.; Xie, Y.B.; Sun, H.Q.; Wang, S.B. Nanocarbon-Based Catalytic Ozonation for Aqueous Oxidation: Engineering Defects for Active Sites and Tunable Reaction Pathways. *ACS Catal.* **2020**, *10*, 13383–13414. [CrossRef]
18. Zhang, T.; Lowry, G.V.; Capiro, N.L.; Chen, J.M.; Chen, W.; Chen, Y.S.; Dionysiou, D.D.; Elliott, D.W.; Ghoshal, S.; Hofmann, T.; et al. In situ remediation of subsurface contamination: Opportunities and challenges for nanotechnology and advanced materials. *Environ. Sci-Nano* **2019**, *6*, 1283–1302. [CrossRef]
19. Fu, H.B.; Xu, T.G.; Zhu, S.B.; Zhu, Y.F. Photocorrosion Inhibition and Enhancement of Photocatalytic Activity for ZnO via Hybridization with C_{60}. *Environ. Sci. Technol.* **2008**, *42*, 8064–8069. [CrossRef]

20. Fanourakis, S.K.; Peña-Bahamonde, J.; Rodrigues, D.F. Inorganic salts and organic matter effects on nanorod, nanowire, and nanoplate MoO_3 aggregation, dissolution, and photocatalysis. *Environ. Sci-Nano* **2020**, *7*, 3794–3804. [CrossRef]
21. Wu, S.; Peng, J.; Lee, S.L.J.; Niu, X.; Jiang, Y.; Lin, S. Let the two sides of the same coin meet-Environmental health and safety-oriented development of functional nanomaterials for environmental remediations. *Eco-Environ. Health* **2024**, *3*, 494–504. [CrossRef] [PubMed]
22. Dangayach, R.; Jeong, N.; Demirel, E.; Uzal, N.; Fung, V.; Chen, Y.S. Machine Learning-Aided Inverse Design and Discovery of Novel Polymeric Materials for Membrane Separation. *Environ. Sci. Technol.* **2024**, *59*, 993–1012. [CrossRef] [PubMed]
23. Mou, T.Y.; Pillai, H.S.; Wang, S.W.; Wan, M.Y.; Han, X.; Schweitzer, N.M.; Che, F.L.; Xin, H.L. Bridging the complexity gap in computational heterogeneous catalysis with machine learning. *Nat. Catal.* **2023**, *6*, 122–136. [CrossRef]

Disclaimer/Publisher's Note: The statements, opinions and data contained in all publications are solely those of the individual author(s) and contributor(s) and not of MDPI and/or the editor(s). MDPI and/or the editor(s) disclaim responsibility for any injury to people or property resulting from any ideas, methods, instructions or products referred to in the content.

toxics

MDPI

Article

Enhanced Photocatalytic Degradation of Methylene Blue Using Ti-Doped ZnO Nanoparticles Synthesized by Rapid Combustion

Sutthipoj Wongrerkdee [1], Sawitree Wongrerkdee [2], Chatdanai Boonruang [3,4] and Supphadate Sujinnapram [1,*]

1 Department of Physics, Faculty of Liberal Arts and Science, Kasetsart University Kamphaeng Saen Campus, Kamphaeng Saen, Nakhon Pathom 73140, Thailand
2 Faculty of Engineering, Rajamangala University of Technology Lanna Tak, Muang, Tak 63000, Thailand
3 Department of Physics and Materials Science, Faculty of Science, Chiang Mai University, Chiang Mai 50200, Thailand
4 Center of Excellence in Materials Science and Technology, Chiang Mai University, Chiang Mai 50200, Thailand
* Correspondence: supphadate.s@ku.ac.th

Abstract: ZnO and Ti-doped ZnO (Ti-ZnO) nanoparticles were synthesized using rapid combustion. The morphology of ZnO and Ti-ZnO featured nanoparticles within cluster-like structures. The ZnO and Ti-ZnO structures exhibited similar hexagonal wurtzite structures and crystal sizes. This behavior occurred because Zn^{2+} sites of the ZnO lattice were substituted by Ti^{4+} ions. The chemical structure characterization implied the major vibration of the ZnO structure. The physisorption analysis showed similar mesoporous and non-rigid aggregation structures for ZnO and Ti-ZnO using N_2 adsorption–desorption. However, Ti-ZnO demonstrated a specific surface area two times higher than that of ZnO. This was a major factor in improving the photocatalytic degradation of methylene blue (MB). The photocatalytic degradation analysis showed a kinetic degradation rate constant of 2.54 $\times 10^{-3}$ min^{-1} for Ti-ZnO, which was almost 80% higher than that of ZnO (1.40 $\times 10^{-3}$ min^{-1}). The transformation mechanism of MB molecules into other products, including carbon dioxide, aldehyde, and sulfate ions, was also examined.

Keywords: ZnO; titanium; combustion; photocatalyst; surface area; methylene blue

1. Introduction

Due to the expansion of the industrial sector and an increase in pollution over the past few decades, a wide range of pollutants have been released and accumulated in environmental systems, especially water. Wastewater and the treatment of complex contaminations in wastewater are major environmental concerns. Moreover, synthetic dyes are modern contaminants in wastewater. Dye molecules are typically designed to be environmentally stable, complex, and non-biodegradable, making them difficult to degrade or remove [1,2]. One of the interesting wastewater treatment strategies is photocatalytic degradation, a low-cost and effective method for pollutant transformation [3–7]. The photocatalytic degradation process requires semiconducting materials to act as a photocatalyst, which can be activated by absorbing the irradiated photons with an appropriate wavelength. After absorbing the photon energy by the photocatalyst, the electrons in the valence band (VB) are excited and promoted to the conduction band (CB), leaving holes at VB [8]. The generated electron-hole (e^--h^+) carriers continuously migrate to the surface to interact with H_2O/O_2, inducing $OH^{\cdot-}/O_2^{\cdot-}$ radicals. The radicals then interact and transform dye molecules into other phases, such as minerals and CO_2 [9,10].

ZnO is one of the popular semiconductors used as an alternative photocatalyst for effective wastewater treatment [11]. Moreover, ZnO is non-toxic, chemically and optically stable, biocompatible, and cheap [12]. However, the low surface area of ZnO, compared with TiO_2 [2,8], is an important issue because the surface area is an essential factor directly affecting photocatalytic performance [10], as chemical reactions of pollutant molecules occur at the photocatalyst's surfaces. Nevertheless, the low surface area also decreases the charge transfer affecting the recombination, which diminishes the photocatalytic performance [13]. To overcome this issue, several studies have presented various kinds of strategies to increase the surface area and reduce the recombination of ZnO photocatalysts [14–16]. The application of ZnO doped with an alkaline earth metal, such as barium (Ba), was reported as the photocatalyst [14]. The synthesis of Ba-doped ZnO (BaZnO) was performed by facile chemical precipitation using zinc nitrate hexahydrate and barium nitrate. The particle and crystal sizes of BaZnO were almost halved compared with those of pure ZnO. This caused a small increase in the surface area from 13.35 to 15.00 m^2/g and a larger induced dislocation density from 1.52×10^{-4} to 5.88×10^{-4} nm^{-2}. The photocatalytic activity of BaZnO to degrade carbaryl solution using UV irradiation was also evaluated. BaZnO exhibited a better photocatalytic performance than that of pure ZnO by around 1.4 times. The incorporation of Pd-metal plasmons into ZnO was presented as an effective way to improve the ZnO surface area [15]. The Pd-ZnO nanoparticles (NPs) were synthesized using a green microwave-assisted method from orange peel extract as a capping agent with a Pd addition. The optimum Pd addition resulted in obtaining the maximum surface area of 16.9 m^2/g for Pd-ZnO NPs in comparison with that of 9.2 m^2/g for pure ZnO. The increased surface area was correlated to the decreased crystal size, suggesting the role of aggregation prevention for Pd in ZnO crystallization and the decrease in particle sizes. The Pd-ZnO NPs were then used in a photocatalytic process of degrading the reactive yellow 15 organic dye under sunlight illumination. The Pd-ZnO NPs showed excellent photocatalytic degradation performance, better than that of pure ZnO by around 1.9 times due to the increased surface area. Another interesting photocatalyst structure of ZnO/TiO_2 composites was synthesized using a hydrothermal process with different capping agents, including cetyltrimethylammonium bromide (CTAB), acetylacetone (ACAC), and poly (4-styrene sulfonic acid) (PS) [16]. The composites prepared with ACAC showed a considerable surface area of 217.08 m^2/g in comparison with that of 14.05 m^2/g for pure ZnO. This was caused by the small crystal size, similar to the abovementioned study [14,15]. The increased surface area of the ZnO/TiO_2 structure caused the enhanced photocatalytic degradation of Tartrazine dye aqueous solution, which was almost two times higher than pure ZnO. The enhancement was attributed to the high surface area of ZnO/TiO_2 composites, due to the synergy of ZnO and TiO_2 NPs. However, it is a complex method for synthesizing composite structures and requires many initial materials, which leads to high-cost production. Therefore, the modification of ZnO material using a simple method was considered in this study. Pure ZnO with a low specific surface area was enhanced using Ti-doping. It was aimed at improving the surface areas of Ti-doped ZnO (Ti-ZnO). Furthermore, the doping effect also caused varieties in the mechanical, electronic, and optical properties of the ZnO host material. Generally, there are several methods for synthesizing ZnO, such as precipitation, sol-gel, hydrothermal method, electrochemical process, and combustion [17,18]. In this study, rapid combustion was used to synthesize ZnO and Ti-ZnO because of its simplicity and short time process. The synthesized ZnO and Ti-ZnO were then characterized by several techniques and utilized as photocatalysts for degrading methylene blue (MB) in water.

2. Materials and Methods

ZnO was synthesized by the rapid combustion technique. For the preparation of the Zn source, zinc nitrate hexahydrate ($Zn(NO_3)_2{\cdot}6H_2O$) was dissolved in 20 mL of 2-ethoxyethanol ($C_4H_{10}O_2$) at the initial concentration of 1 M under continuous stirring at room temperature for 30 min to form a homogenous zinc complex solution. Before combustion, the solution was heated to 350 °C for 10 min. Then, the solution was dropped on heated-glass substrates (350 °C) for the smoldering combustion process. After 20 min, the substrate was left to cool down to room temperature. Next, the obtained white product of ZnO was scratched from the glass substrate, ground for 1 h, and stocked in a brown bottle. A similar process was carried out to synthesize Ti-ZnO using a zinc–titanium complex solution instead of a zinc complex solution. The zinc–titanium complex solution was prepared by co-dissolving zinc nitrate hexahydrate and 1 mol.% titanium chloride tetrahydrofuran complex ($TiCl_4 \cdot 2THF$) in 20 mL of 2-ethoxyethanol. The preparation flow diagram was illustrated in Figure 1.

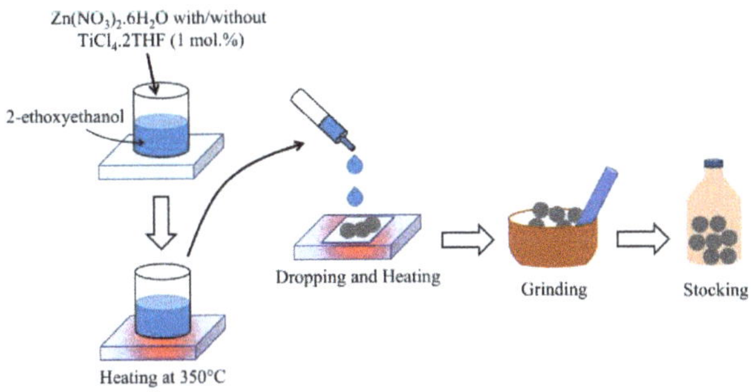

Figure 1. The illustrated flow diagrams of ZnO and Ti-ZnO NPs synthesized by rapid combustion.

The synthesized combustion products of ZnO and Ti-ZnO were characterized using several methods. The morphology was observed using a scanning electron microscope (SEM, JSM-6610 LV, JOEL) equipped with an energy dispersive X-ray spectrometer (EDS, INCAx-act, OXFORD) for elemental detection. The high-resolution morphology and selected area electron diffraction (SAED) were investigated using a transmission electron microscope (TEM, TECNAI G^2 20 S-TWIN, FEI). The crystal structure was determined by X-ray diffraction (XRD) patterns recorded on an X-ray diffractometer (D8 Advance, Bruker). The crystal sizes of ZnO and Ti-ZnO were calculated using the Debye–Scherrer equation (Equation (1)) [14,19]:

$$D = k\lambda/\beta\cos\theta \quad (1)$$

where k and λ are the shape factor of the XRD measurement (0.89) and the wavelength of the X-ray source (1.5406 Å), respectively. The β and θ are the full width at half maximum (FWHM) and the diffraction angle for each plane, respectively. A surface functional group analysis was conducted using Fourier transform infrared spectroscopy (FTIR, Spectrum Two, PerkinElmer). A pore analysis was performed by physisorption of N_2 adsorption–desorption at 77 K using a Tristar II3020, Micrometrics. The adsorption–desorption isotherm was plotted to identify the pore structure. The pore width and specific surface area were analyzed using the Barrett–Joyner–Halenda (BJH) and Brunauer–Emmett–Teller (BET) methods. The band gap energy (E_g) was investigated from the diffuse reflectance absorption spectra for ZnO and Ti-ZnO using an ultraviolet-visible-near infrared (UV–Vis-NIR) spectrophotometer (UV-3101 PC, Shimadzu). The recorded spectra were then analyzed according to the Tauc's relation (Equation (2)) [20–22]:

$$(\alpha h\nu)^{1/n} - A(h\nu - E_g) \quad (2)$$

where α, A, h, and υ are the optical absorbance coefficient, constant, Planck's constant, and frequency, respectively. For a direct electron transition material, including ZnO, the parameter n is 1/2. The synthesized ZnO and Ti-ZnO were utilized to degrade the MB solution under UV irradiation and demonstrate their potential as an applicable photocatalyst. The MB solution was prepared by dissolving 5 mg of MB in 1000 mL of DI water. It was then stirred for 1 h in the dark. The amount of 0.1 g of ZnO or Ti-ZnO was added into a separate beaker containing 100 mL of the MB solution. It was then stirred for 15 min to disperse the photocatalyst. Then, the photocatalyst-contained MB solution was left for 30 min in the dark condition to reach an adsorption–desorption equilibrium. The photocatalytic process was then activated by the irradiation of a UV lamp (λ~365 nm, 4.5 W/m^2) for 150 min. To evaluate the photocatalytic degradation performance, a portion of the degraded MB solution was collected at every 30 min interval. The collected solution was examined by measuring the absorbance to monitor the remaining MB molecules. To analyze the MB degradation performance, the degradation rate constant (k) was estimated using the pseudo-first-order kinetics (Equation (3)) [21,23]:

$$\ln(A/A_0) = -kt \quad (3)$$

where A_0 and A are the absorbances of the MB solution at the initial and interval irradiation times, respectively. The degraded MB solution was further investigated using FTIR to monitor the transformed products.

3. Results and Discussion

SEM images of ZnO and Ti-ZnO (Figure 2) were observed for the morphological investigation. The morphology showed ZnO and Ti-ZnO nanoparticles (NPs). It seems that numerous accumulated and dense NPs formed cluster-like structures, which is consistent with the TEM images in Figure 3. It is believed that the rapid combustion process provides for fast and disordered crystallization. The process might limit the reaction between zinc and hydroxide ions during the rapid combustion for a short duration, resulting in a randomized reaction that results in clustered NPs [24]. The particle size was estimated using image-J software, as depicted in Figure 4. The maximum particle size distribution for ZnO and Ti-ZnO was found to be 10 nm and 18 nm, respectively, indicating their small size. To examine the chemical elements, EDS results were tabulated in Table 1, indicating the detection of Zn, O, and Ti. The detected elements revealed only the presence of Zn and O for ZnO. In the case of Ti-ZnO, adding the Ti element indicated the presence of Ti in Ti-ZnO structures.

Figure 2. SEM images of (**a**) ZnO and (**b**) Ti-ZnO NPs.

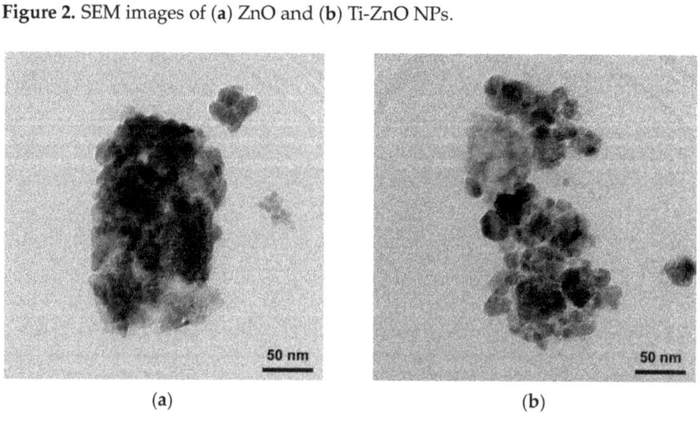

Figure 3. TEM images of (**a**) ZnO and (**b**) Ti-ZnO NPs.

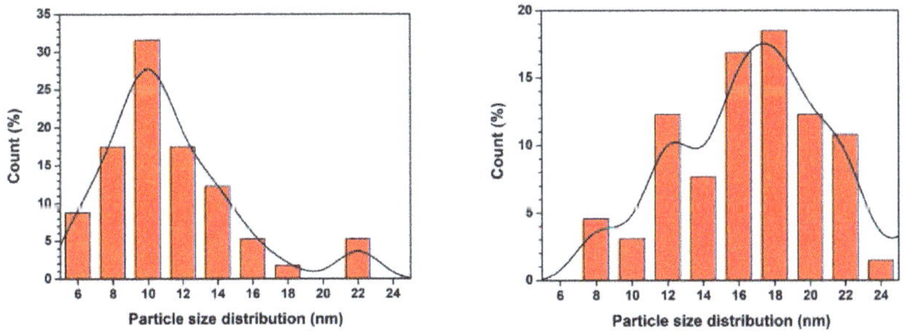

(**a**) (**b**)

Figure 4. Particle size distribution of (**a**) ZnO and (**b**) Ti-ZnO NPs.

Table 1. Detected chemical composition of ZnO and Ti-ZnO using EDS.

Sample	Element (Atomic%)		
	Zn	O	Ti
ZnO	42.97	57.03	-
Ti-ZnO	33.43	66.05	0.51

The XRD patterns of ZnO and Ti-ZnO in Figure 5 show the diffraction angles at 2θ of 31.7°, 34.5°, 36.2°, 47.5°, 56.6°, 62.9°, and 68.0° indexed to the crystal planes of (100), (002), (101), (102), (110), (103), and (122), respectively [22,25]. The patterns corresponded well with the hexagonal wurtzite structure of ZnO, according to the standard JCPDS no. 36-1451 [2]. The peak at 2θ~23.5° might be related to the orthorhombic $Zn(OH)_2$ phase (JCPDS no. 01-071-2215). To estimate the crystal sizes of ZnO and Ti-ZnO, the three major peaks, including the (100), (002), and (101) planes, were considered for the calculation using the Debye–Scherrer equation (Equation (1)). The calculated crystal sizes were 12.88 ± 2.33 and 12.81 ± 2.88 nm for ZnO and Ti-ZnO, respectively. The results presented no differences in the average crystal size for ZnO and Ti-ZnO. This can indicate that Ti^{4+} ions can substitute into Zn^{2+} sites, because the ionic radius of 0.61 Å for Ti^{4+} ions is lower than that of 0.74 Å for Zn^{2+} ions [26]. However, the change in crystal size was not significant for ZnO and Ti-ZnO due to employing the rapid combustion synthesis method. The SAED patterns (Figure 6) displayed the ring patterns for both ZnO and Ti-ZnO implying the polycrystalline structure. This was caused by the limited crystallization due to the short combustion reaction, which caused low crystallization rates and an uncontrollable crystal growth mechanism.

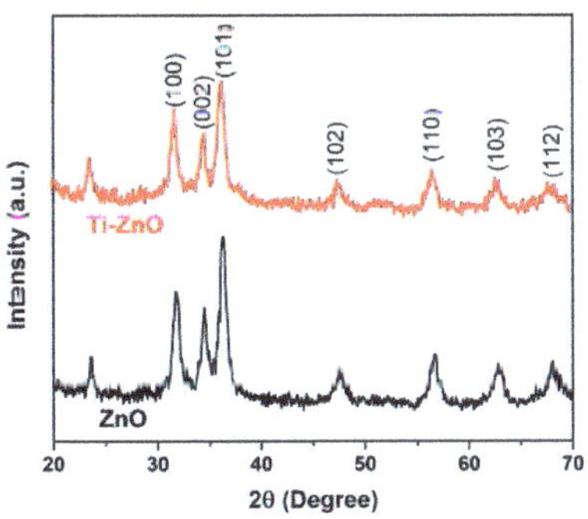

Figure 5. XRD patterns of ZnO and Ti-ZnO NPs.

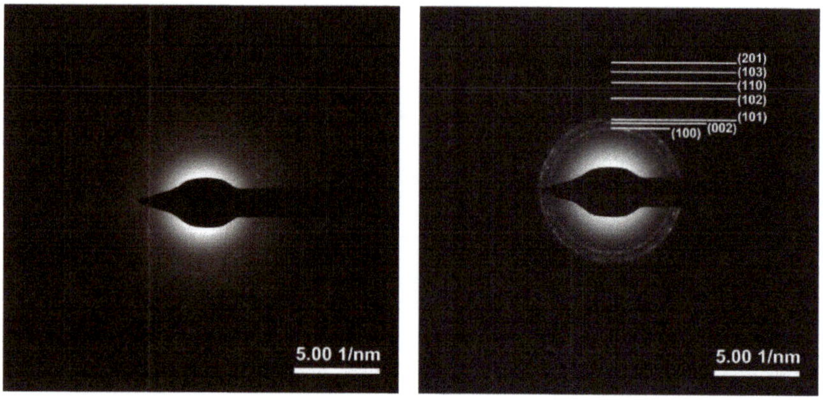

(**a**) (**b**)

Figure 6. SAED patterns of (**a**) ZnO and (**b**) Ti-ZnO NPs.

Figure 7 shows the FTIR spectra of ZnO and Ti-ZnO in the range of 4000–450 cm^{-1} for the examination of the surface chemical structures. The peak at around 461 cm^{-1}, referring to the Zn−O vibration, confirmed the ZnO structure [26,27]. The other peaks were marked as residuals, including organic compounds and water. The peaks at 1321 and 1644 cm^{-1} were assigned to C−O stretching and C=O stretching vibrations, respectively [22]. The peak at 820 cm^{-1} was attributed to the OH libration, caused by the $Zn(OH)_2$ phase, and it was consistent with the impurity peak in the XRD results. The peak at 3395 cm^{-1} was assigned to the stretching vibration of the OH groups.

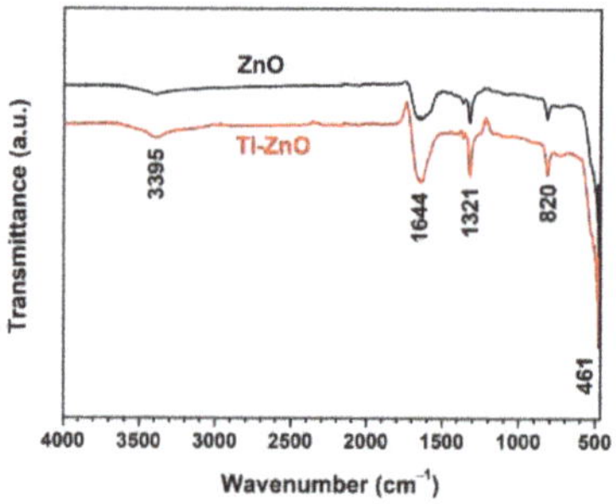

Figure 7. FTIR spectra of ZnO and Ti-ZnO NPs for surface chemical structure analysis.

The band gap energy (E_g) for the combustion-synthesized ZnO and Ti-ZnO samples was assessed using the Tauc's plot relation, as shown in Figure 8. For this purpose, a linear line with the maximum slope was extrapolated to the hν-axis. Then, the E_g value was estimated from the intercept on the hν-axis. It was found that the E_g values for ZnO and Ti-ZnO were 3.06 and 3.10 eV, respectively. However, the E_g values exhibited small differences; the obtained values in this study were lower than those in the literature [2,9,12,28,29], which is due to the low crystallization and the defects [20]. The estimated E_g values were correlated with the UV energy, indicating that ZnO and Ti-ZnO can be used as photocatalysts and activated by UV light.

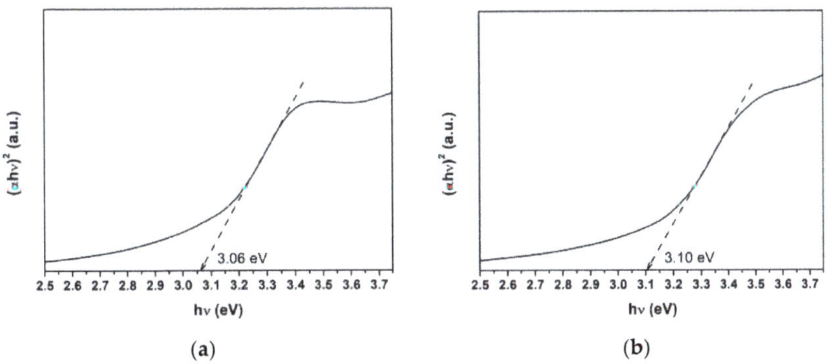

Figure 8. Band gap energy estimation using the Tauc plot for (**a**) ZnO and (**b**) Ti-ZnO.

The N_2 adsorption–desorption was recorded (Figure 9a) to examine the pore characteristics. The type IV isotherm was similarly observed for both ZnO and Ti-ZnO indicating the mesoporous structure. The H3 hysteresis loop was observed, implying the non-rigid aggregation structures that confirm the cluster-like structure of ZnO and Ti-ZnO. In Figure 9b, the pore width shows a similar pore width distribution. However, the pore widths of 14.75 and 9.12 nm for ZnO and Ti-ZnO, respectively, were estimated using the BJH analysis. S_{BET} was then calculated for the type IV isotherms to estimate the active surface areas of ZnO and Ti-ZnO photocatalysts. The calculated S_{BET} values of 27.84 and 63.85 m^2/g for ZnO and Ti-ZnO, respectively, were achieved. The results indicate that Ti-ZnO has a greater active surface area than that of ZnO by over two times, which confirms its suitability for photocatalytic applications.

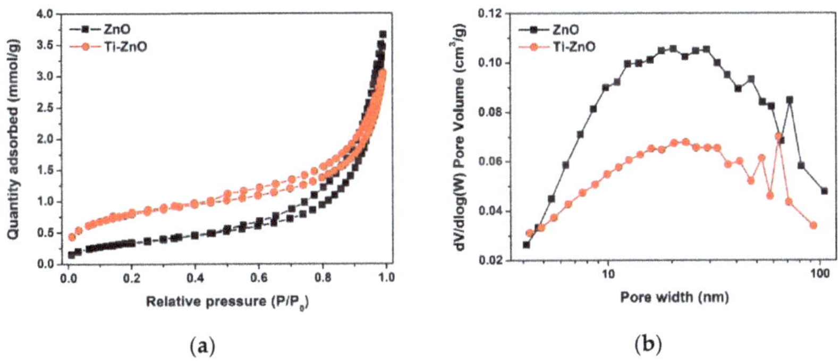

Figure 9. (**a**) N_2 adsorption–desorption and (**b**) pore width distribution for ZnO and Ti-ZnO NPs.

The synthesized ZnO and Ti-ZnO were used as the photocatalysts for photocatalytic degradation of MB under UV activation. The degradation of MB using only a photocatalyst was examined, as shown in Figure S1 (Supplementary Materials). The negligible change in absorbance of MB demonstrates that the photocatalyst had no significant effect on MB without UV light. To further investigate the influence of photocatalytic activity on MB, the experiment was repeated using ZnO and Ti-ZnO photocatalysts under UV light. The absorbance of MB was measured to investigate the degradation performance, as shown in Figure 10. It was observed that the absorbance of MB for both ZnO and Ti-ZnO photocatalysts decreased as a function of the UV irradiation time. The absorbance of MB for Ti-ZnO showed stronger decreasing trends than that for ZnO. This indicates a better photocatalytic activity for Ti-ZnO to degrade MB molecules compared with ZnO. This is correlated with the higher S_{BET} value for Ti-ZnO. It is believed that the high surface areas can provide interaction between electron-hole carriers and surrounding molecules [21,30], whereas, the unchanged absorbance of MB after UV irradiation in the absence of a photocatalyst confirms that the UV light had no effect on MB. In the analysis of MB degradation performance, the degradation rate constant was estimated using pseudo-first-order kinetics. From Equation (3), the ln (A/A_0) curve was plotted versus the irradiation time (Figure 11), and the degradation rate constant was calculated by tracking the slopes of the curve. The degradation rate constant values of MB degradation were 1.40×10^{-3} and 2.54×10^{-3} min^{-1} for ZnO and Ti-ZnO photocatalysts, respectively. The results confirm that Ti-ZnO prepared by combustion has efficient photocatalytic activity for MB degradation, which is consistent with Ti-ZnO prepared by chemical precipitation as described elsewhere [30]. Remarkably, the degradation rate constant for Ti-ZnO increased by over 80% in comparison with ZnO. This result implies that Ti-ZnO improved the kinetic activity of MB degradation under UV illumination. The improved activity is believed to be due to the increased surface area and the reduced pore width. The mechanism of photocatalytic degradation can be briefly described in Equations (4)–(7) [2,31,32]:

$$ZnO + h\nu \rightarrow e^- + h^+ \quad (4)$$

$$e^- + O_2 \rightarrow O_2^{\cdot -} \quad (5)$$

$$h^+ + H_2O \rightarrow OH^{\cdot -} + H^+ \quad (6)$$

$$Dyes + O_2^{\cdot -} \text{ or } OH^{\cdot -} \rightarrow intermediates \rightarrow CO_2 + H_2O + NH_4^+ + SO_4^{2-} \quad (7)$$

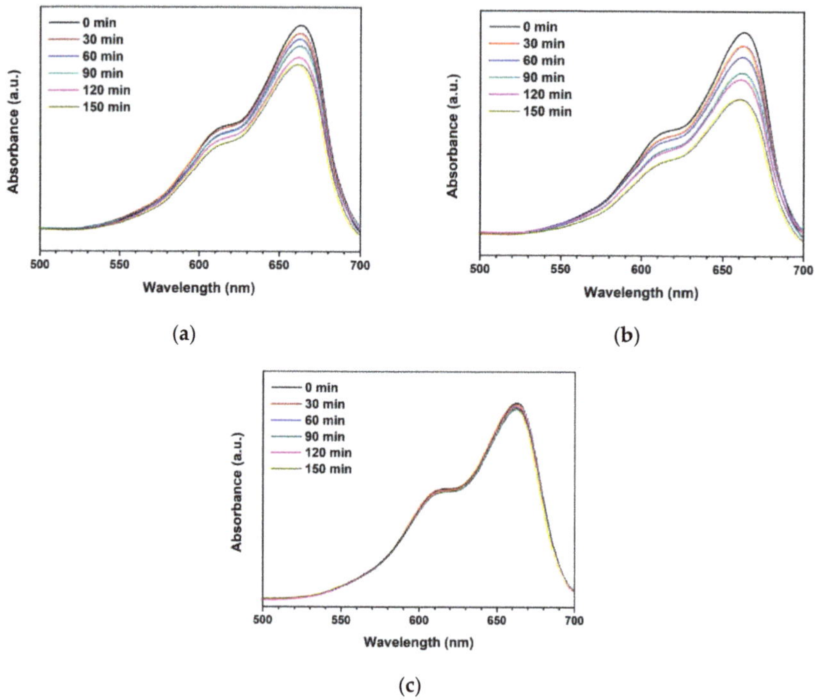

Figure 10. Absorbances of MB under photocatalytic degradation for different UV irradiation time using (**a**) ZnO photocatalyst, (**b**) Ti-ZnO photocatalyst, and (**c**) no photocatalyst.

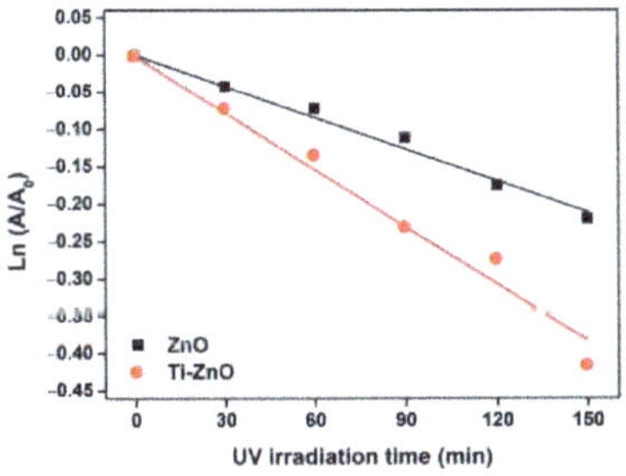

Figure 11. Plots of ln (A/A_0) versus UV irradiation time for analysis of degradation rate constant.

The electron-hole (e^--h^+) pairs are simultaneously generated as the photocatalyst is activated by absorbing photons. They also continuously diffuse onto the photocatalyst surface [33]. The reaction between the electrons and dissolved oxygen (O_2) in the MB solution induces superoxide radicals ($O_2^{\cdot-}$). In another reaction, the holes react with water to produce hydroxyl radicals ($OH^{\cdot-}$). These reactive radicals react efficiently to degrade MB dyes into other products. To examine the products, in this study, the MB solution was monitored before and after the photocatalytic degradation for 150 min using ZnO and Ti-ZnO (Figure 12). The initial MB solution peaks were observed at the wavenumber of 1615, 1417, and 1355 cm^{-1} corresponding to the C=N stretching of the central ring, the multiple rings stretching, and C−N stretching of the side aromatic ring and a nitrogen atom, respectively [30]. The intensity of the peaks decreased after photocatalytic degradation. This was attributed to the transformation of MB molecules into intermediates. Moreover, the presence of peaks at 2360, 1738, and 1220 cm^{-1} was assigned to the C=O asymmetric stretching of CO_2, C=O stretching of aldehyde, and S−O stretching of sulfate ions (SO_4^{2-}) [34,35], respectively. This indicates the conversion of MB into other products. Furthermore, the intensity of the CO_2 peak for the Ti-ZnO photocatalyst was higher than that of the ZnO. It can be concluded that the high surface area of Ti-ZnO plays a role in offering active sites for the reaction between electron-hole pairs and MB molecules. This effect leads to accelerating the degradation mechanism and raising the degradation performance.

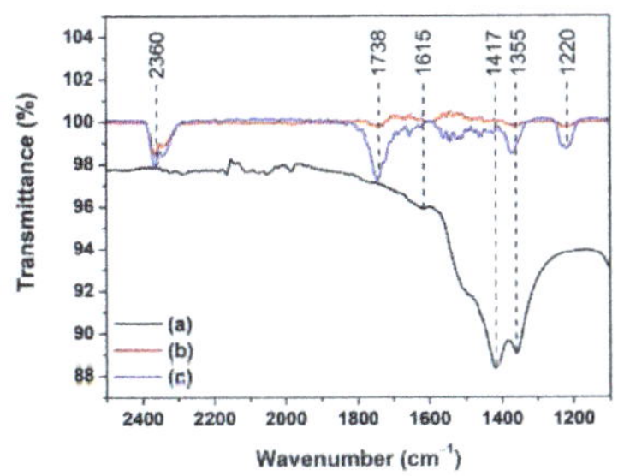

Figure 12. FTIR spectra of (**a**) initial MB solution; and MB solution after photocatalytic degradation for 150 min using (**b**) ZnO and (**c**) Ti-ZnO photocatalysts.

4. Conclusions

ZnO and Ti-ZnO NPs were synthesized using rapid combustion. The ZnO and Ti-ZnO structures exhibited similar structures and sizes of polycrystalline hexagonal wurtzite structures, which was due to the substitution of Zn^{2+} sites in the ZnO host lattice by Ti^{4+}. The chemical structure examination implied the major vibrations of the ZnO structure. However, some residuals, including organic compounds and water, were also observed. The N_2 adsorption–desorption analysis confirmed the mesoporous and non-rigid aggregation structures for both ZnO and Ti-ZnO. However, Ti-ZnO showed a higher surface area than that of ZnO by more than double. This is the major factor in improving the photocatalytic activity of MB degradation. The photocatalytic degradation analysis showed the kinetic degradation rate constant of 2.54×10^{-3} min^{-1} for Ti-ZnO, which was higher than that of 1.40×10^{-3} min^{-1} for ZnO by over 80%. The transformation mechanism of MB molecules into other products, including carbon dioxide, aldehyde, and sulfate ions, was also interpreted in this study.

Supplementary Materials: The following supporting information can be downloaded at: https://www.mdpi.com/article/10.3390/toxics11010033/s1, Figure S1: Absorbances of MB in the dark condition with the dispersion of (a) ZnO and (b) Ti-ZnO photocatalysts.

Author Contributions: Conceptualization, S.W. (Sutthipoj Wongrerkdee), S.W. (Sawitree Wongrerkdee) and S.S.; Methodology, S.W. (Sutthipoj Wongrerkdee) and S.W. (Sawitree Wongrerkdee); Investigation, S.W. (Sutthipoj Wongrerkdee), S.W. (Sawitree Wongrerkdee), C.B. and S.S.; Resources, S.W. (Sutthipoj Wongrerkdee), S.W. (Sawitree Wongrerkdee), C.B. and S.S.; Writing—original draft, S.W. (Sutthipoj Wongrerkdee) and S.W. (Sawitree Wongrerkdee); Writing—review and editing preparation, S.W. (Sutthipoj Wongrerkdee) and S.S.; Supervision, S.W. (Sutthipoj Wongrerkdee) and S.S.; Funding acquisition, S.W. (Sutthipoj Wongrerkdee) and S.S. All authors have read and agreed to the published version of the manuscript.

Funding: This research was funded by the Research Promotion and Technology Transfer Center (RPTTC), Faculty of Liberal Arts and Science, Kasetsart University Kamphaeng Saen Campus (grant no. 202/2565); and the Department of Physics, Faculty of Liberal Arts and Science, Kasetsart University Kamphaeng Saen Campus (grant no. PHY-2021/06 and PHY-2021/09).

Institutional Review Board Statement: Not applicable.

Informed Consent Statement: Not applicable.

Data Availability Statement: Not applicable.

Acknowledgments: This work was also partially supported by the Center of Excellence in Materials Science and Technology, Chiang Mai University. The authors acknowledge the Research Promotion and Technology Transfer Center (RPTTC), Faculty of Liberal Arts and Science, Kasetsart University Kamphaeng Saen Campus for FTIR measurement; and the Center of Excellence in Glass Technology and Materials Science (CEGM), Nakhon Pathom Rajabhat University for XRD measurement.

Conflicts of Interest: The authors declare no conflict of interest.

References

1. Hosseini, A.; Karimi, K.; Foroughi, J.; Sabzehmeidani, M.M.; Ghaedi, M. Heterogeneous photoelectro-Fenton using ZnO and TiO2 thin film as photocatalyst for photocatalytic degradation Malachite Green. *Appl. Surf. Sci. Adv.* **2021**, *6*, 100126. [CrossRef]
2. Yudasari, N.; Anugrahwidya, R.; Tahir, D.; Suliyanti, M.M.; Herbani, Y.; Imawan, C.; Khalil, M.; Djuhana, D. Enhanced photocatalytic degradation of rhodamine 6G (R6G) using ZnO–Ag nanoparticles synthesized by pulsed laser ablation in liquid (PLAL). *J. Alloys Compd.* **2021**, *886*, 161291. [CrossRef]
3. Ding, C.; Zhu, Q.; Yang, B.; Petropoulos, E.; Xue, L.; Feng, Y.; He, S.; Yang, L. Efficient photocatalysis of tetracycline hydrochloride (TC-HCl) from pharmaceutical wastewater using AgCl/ZnO/g-C3N4 composite under visible light: Process and mechanisms. *J. Environ. Sci.* **2023**, *126*, 249–262. [CrossRef] [PubMed]
4. Lincho, J.; Zaleska-Medynska, A.; Martins, R.C.; Gomes, J. Nanostructured photocatalysts for the abatement of contaminants by photocatalysis and photocatalytic ozonation: An overview. *Sci. Total Environ.* **2022**, *837*, 155776. [CrossRef] [PubMed]
5. Liu, C.; Mao, S.; Shi, M.; Hong, X.; Wang, D.; Wang, F.; Xia, M.; Chen, Q. Enhanced photocatalytic degradation performance of BiVO4/BiOBr through combining Fermi level alteration and oxygen defect engineering. *Chem. Eng. J.* **2022**, *449*, 137757. [CrossRef]
6. Liu, C.; Mao, S.; Wang, H.; Wu, Y.; Wang, F.; Xia, M.; Chen, Q. Peroxymonosulfate-assisted for facilitating photocatalytic degradation performance of 2D/2D WO3/BiOBr S-scheme heterojunction. *Chem. Eng. J.* **2022**, *430*, 132806. [CrossRef]
7. Liu, C.; Mao, S.; Shi, M.; Wang, F.; Xia, M.; Chen, Q.; Ju, X. Peroxymonosulfate activation through 2D/2D Z-scheme CoAl-LDH/BiOBr photocatalyst under visible light for ciprofloxacin degradation. *J. Hazard. Mater.* **2021**, *420*, 126613. [CrossRef]
8. Qumar, U.; Hassan, J.Z.; Bhatti, R.A.; Raza, A.; Nazir, G.; Nabgan, W.; Ikram, M. Photocatalysis vs adsorption by metal oxide nanoparticles. *J. Mater. Sci. Technol.* **2022**, *131*, 122–166. [CrossRef]
9. Ishchenko, O.; Rogé, V.; Lamblin, G.; Lenoble, D.; Fechete, I. TiO2, ZnO, and SnO2-based metal oxides for photocatalytic applications: Principles and development. *C. R. Chim.* **2021**, *24*, 103–124. [CrossRef]

10. Hezam, A.; Drmosh, Q.A.; Ponnamma, D.; Bajiri, M.A.; Qamar, M.; Namratha, K.; Zare, M.; Nayan, M.B.; Onaizi, S.A.; Byrappa, K. Strategies to Enhance ZnO Photocatalyst's Performance for Water Treatment: A Comprehensive Review. *Chem. Rec.* **2022**, *22*, e202100299. [CrossRef]
11. Sanakousar, F.M.; Vidyasagar, C.C.; Jimenez-Perez, V.M.; Prakash, K. Recent progress on visible-light-driven metal and non-metal doped ZnO nanostructures for photocatalytic degradation of organic pollutants. *Mater. Sci. Semicond.* **2022**, *140*, 106390. [CrossRef]
12. Wang, X.; Li, J. Facile Liquid-Phase Synthesis of a High-Performance Cd-Doped ZnO-Quantum-Dot-Based Photocatalyst. *ECS J. Solid State Sci. Technol.* **2021**, *10*, 124003. [CrossRef]
13. Abdullah, F.H.; Abu Bakar, N.H.H.; Abu Bakar, M. Current advancements on the fabrication, modification, and industrial application of zinc oxide as photocatalyst in the removal of organic and inorganic contaminants in aquatic systems. *J. Hazard. Mater.* **2022**, *424*, 127416. [CrossRef] [PubMed]
14. Wongrerkdee, S.; Krobthong, S. Synthesis, Characterization, and Photocatalytic Property of Ba-Doped ZnO Nanoparticles Synthesized Using Facile Precipitation. *Integr. Ferroelectr.* **2022**, *224*, 192–204. [CrossRef]
15. Mbrouk, O.A.; Fawzy, M.; Elshafey, H.M.; Saif, M.; Hafez, H.; Abdel Mottaleb, M.S.A. Green synthesized plasmonic Pd–ZnO nanomaterials for visible light-induced photobiogas production from industrial wastewater. *Appl. Organomet. Chem.* **2022**, *36*, e6807. [CrossRef]
16. Mbrouk, O.; Hafez, H.; Mozia, S.; Othman, A.M.; Abdel Mottaleb, M.S.A. Stimulated generation of photobiogas by morphologically tuned nanostructured ZnO and ZnO/TiO2. *BMC Chem.* **2022**, *16*, 74. [CrossRef] [PubMed]
17. Tangcharoen, T.; Klysubun, W.; Kongmark, C. Synthesis and characterization of nanocrystalline CuO/ZnO composite powders with enhanced photodegradation performance under sunlight irradiation. *J. Mater. Sci. Mater. Electron.* **2022**, *31*, 12807–12822. [CrossRef]
18. Al-Namshah, K.S.; Adewinbi, S.A.; Shkir, M.; Ashraf, I.M.; Hamdy, M.S. Enhancement of photodegradation activity of ZnO nanoparticles prepared by one-pot flash combustion route: An effect of Ag doping. *Opt. Quant. Electron.* **2022**, *54*, 838. [CrossRef]
19. Nethravathi, P.C.; Suresh, D. Silver-doped ZnO embedded reduced graphene oxide hybrid nanostructured composites for superior photocatalytic hydrogen generation, dye degradation, nitrite sensing and antioxidant activities. *Inorg. Chem. Commun.* **2021**, *134*, 109051. [CrossRef]
20. Moungsrijun, S.; Wongrerkdee, S. Investigation of structural, optical, and electrical properties of zno thin films for electro-optical devices. *Suranaree J. Sci. Technol.* **2022**, *29*, 030085.
21. Sujinnapram, S.; Wongrerkdee, S. Synergistic effects of structural, crystalline, and chemical defects on the photocatalytic performance of Y-doped ZnO for carbaryl degradation. *J. Environ. Sci.* **2023**, *124*, 667–677. [CrossRef] [PubMed]
22. Algarni, T.S.; Abduh, N.A.Y.; Kahtani, A.A.; Aouissi, A. Photocatalytic degradation of some dyes under solar light irradiation using ZnO nanoparticles synthesized from Rosmarinus officinalis extract. *Green. Chem. Lett. Rev.* **2022**, *15*, 460–473. [CrossRef]
23. Hakimi-Tehrani, M.J.; Hassanzadeh-Tabrizi, S.A.; Koupaei, N.; Saffar-Teluri, A.; Rafiei, M. Facile thermal synthesis of g–C3N4/ZnO nanocomposite with antibacterial properties for photodegradation of methylene blue. *Mater. Res. Express* **2021**, *8*, 125002. [CrossRef]
24. Weerathunga, H.; Tang, C.; Brock, A.J.; Sarina, S.; Wang, T.; Liu, Q.; Zhu, H.Y.; Du, A.; Waclawik, E.R. Nanostructure Shape-Effects in ZnO heterogeneous photocatalysis. *J. Colloid Interface Sci.* **2022**, *606*, 588–599. [CrossRef] [PubMed]
25. Długosz, O.; Banach, M. Continuous synthesis of photocatalytic nanoparticles of pure ZnO and ZnO modified with metal nanoparticles. *J. Nanostructure Chem.* **2021**, *11*, 601–617. [CrossRef]
26. Saxena, P.; Choudhary, P.; Yadav, A.; Dewangan, B.; Rai, V.N.; Mishra, A. Improved structural and dielectric properties of Cd and Ti dual doped ZnO nanoparticles. *Appl. Phys. A* **2020**, *126*, 765. [CrossRef]
27. Winiarski, J.; Tylus, W.; Winiarska, K.; Szczygieł, I.; Szczygieł, B. XPS and FT-IR characterization of selected synthetic corrosion products of zinc expected in neutral environment containing chloride ions. *J. Spectrosc.* **2018**, *2018*, 2079278. [CrossRef]
28. Rungsawang, T.; Sujinnapram, S.; Nilphai, S.; Wongrerkdee, S. Influence of yttrium doping on ZnO nanoparticles for enhanced photocatalytic degradation of methylene blue. *Dig. J. Nanomater. Biostructures* **2021**, *16*, 1209–1217.
29. Buasakun, J.; Srilaoong, P.; Rattanakam, R.; Duangthongyou, T. Synthesis of heterostructure of ZnO@MOF-46(Zn) to improve the photocatalytic performance in methylene blue degradation. *Crystals* **2021**, *11*, 1379. [CrossRef]
30. Siva, N.; Kannadasan, N.; Shanmugam, N.; Ragupathy, S.; Sakthi, D.; Arun, V. Effect of Ti-doping on photocatalytic activity of ZnO nanocatalyst under sunlight irradiation. *Inorg. Chem. Commun.* **2022**, *146*, 110097. [CrossRef]
31. Moungsrijun, S.; Sujinnapram, S.; Sutthana, S. Synthesis and characterization of zinc oxide prepared with ammonium hydroxide and photocatalytic application of organic dye under ultraviolet illumination. *Monatsh Chem.* **2017**, *148*, 1177–1183. [CrossRef]
32. Pannak, P.; Songsasen, A.; Foytong, W.; Kidkhunthod, P.; Sirisaksoontorn, W. Homogeneous distribution of nanosized ZnO in montmorillonite clay sheets for the photocatalytic enhancement in degradation of Rhodamine B. *Res. Chem. Intermed.* **2018**, *44*, 6861–6875. [CrossRef]
33. Buasakun, J.; Srilaoong, P.; Chaloeipote, G.; Rattanakram, R.; Wongchoosuk, C.; Duangthongyou, T. Synergistic effect of ZnO/ZIF8 heterostructure material in photodegradation of methylene blue and volatile organic compounds with sensor operating at room temperature. *J. Solid State Chem.* **2020**, *289*, 121494. [CrossRef]

34. Yu, Z.; Chuang, S.S.C. Probing Methylene Blue Photocatalytic Degradation by Adsorbed Ethanol with In Situ IR. *J. Phys. Chem. C* **2007**, *111*, 13813–13820. [CrossRef]
35. Schott, J.A.; Do-Thanh, C.-L.; Shan, W.; Puskar, N.G.; Dai, S.; Mahurin, S.M. FTIR investigation of the interfacial properties and mechanisms of CO2 sorption in porous ionic liquids. *Green Chem. Eng.* **2021**, *2*, 392–401. [CrossRef]

Disclaimer/Publisher's Note: The statements, opinions and data contained in all publications are solely those of the individual author(s) and contributor(s) and not of MDPI and/or the editor(s). MDPI and/or the editor(s) disclaim responsibility for any injury to people or property resulting from any ideas, methods, instructions or products referred to in the content.

toxics

MDPI

Article

Recyclable Carbon Cloth-Supported ZnO@Ag_3PO_4 Core–Shell Structure for Photocatalytic Degradation of Organic Dye

Yuan Yi [1], Qifang Guan [1], Wenguang Wang [1,*], Siyuan Jian [1], Hengchao Li [1], Liangpeng Wu [2], Haiyan Zhang [1] and Chuanjia Jiang [3,*]

1 School of Materials and Energy, Guangdong University of Technology, Guangzhou Higher Education Mega Center 100#, Guangzhou 510006, China
2 Advanced Energy Science and Technology Guangdong Laboratory, Huizhou 516000, China
3 College of Environmental Science and Engineering, Nankai University, 38 Tongyan Rd., Tianjin 300350, China
* Correspondence: wenguangwang2005@163.com (W.W.); jiangcj@nankai.edu.cn (C.J.)

Abstract: The extensive use of organic dyes in industry has caused serious environmental problems, and photocatalysis is a potential solution to water pollution by organic dyes. The practical application of powdery photocatalysts is usually limited by the rapid recombination of charge carriers and difficulty in recycling. In this study, recyclable carbon cloth-supported ZnO@Ag_3PO_4 composite with a core–shell structure was successfully prepared by solvothermal treatment and subsequent impregnation–deposition. The as-prepared carbon cloth-supported ZnO@Ag_3PO_4 composite showed an improved photocatalytic activity and stability for the degradation of rhodamine B (RhB), a model organic dye, under visible light irradiation. The decomposition ratio of RhB reached 87.1% after exposure to visible light for 100 min, corresponding to a reaction rate constant that was 4.8 and 15.9 times that of carbon cloth-supported Ag_3PO_4 or ZnO alone. The enhanced performance of the composite can be attributed to the effectively inhibited recombination of photoinduced electron–hole pairs by the S-scheme heterojunction. The carbon fibers further promoted the transfer of charges. Moreover, the carbon cloth-supported ZnO@Ag_3PO_4 can be easily separated from the solution and repeatedly used, demonstrating a fair recyclability and potential in practical applications.

Keywords: carbon cloth; silver phosphate; zinc oxide; S-scheme heterojunction; wastewater treatment

1. Introduction

With the rapid development of the textile and dyeing industry, large quantities of organic dye-laden wastewater have been discharged into the natural waters, causing a range of environmental problems. Many of the organic dyes are toxic to human health or aquatic life [1,2]. Over the past decades, various destructive and nondestructive techniques have been explored for removing organic dyes from wastewater, including adsorption, filtration, biodegradation, advanced oxidation processes, photocatalysis, etc. [3–6]. Amongst these approaches, semiconductor-based photocatalysis can degrade and detoxify organic pollutants, including dyes, utilizing inexhaustible solar energy, which is expected to solve the problems of increasingly serious water pollution by organic dyes [7–10].

Zinc oxide (ZnO) is a versatile semiconductor photocatalysts with high ultraviolet (UV) shielding and refractive index, high electron mobility, and strong luminescence at room temperature [11], and ZnO-based photocatalysis has been widely explored for removing wastewater pollutants, including organic dyes [12–14]. However, as a typical broad-bandgap semiconductor, ZnO can be mainly excited by UV light, resulting in a low utilization rate of the visible light fraction of solar radiation, which, together with its vulnerability to corrosion under both irradiated and dark conditions [15,16], hinders the large-scale application of pure ZnO-based materials in the field of environmental photocatalysis [17]. The coupling of two band-matched semiconductors to form a heterojunction is an effective way to reduce the recombination of photoelectron–hole pairs and improve the performance of photocatalytic materials [18,19]. Moreover, the coating of a material less prone to corrosion onto the surface of ZnO can dramatically enhance the stability of ZnO-based photocatalysts [15,20]. Since Ye et al. reported the use of silver phosphate (Ag_3PO_4) in water oxidization and organic contaminant decomposition, Ag_3PO_4, a visible light-responsive semiconductor, has been widely considered as a promising material for solar photocatalysis [21–23]. Despite its low solubility in water, however, we have previously observed that, during extended photocatalytic tests, photoelectrons can induce the precipitation of metallic Ag films on the surface of Ag_3PO_4, which shield light and lead to the deactivation of Ag_3PO_4 [24]. It has been demonstrated that the combination of ZnO with Ag_3PO_4 could improve the photocatalytic activity of ZnO/Ag_3PO_4 [25–27], though an improvement in performance stability is less straightforward, with contradicting results reported [25,28].

Up to now, a majority of the reported photocatalysts exist in the state of powder, which causes difficulty in separation and recycling when it comes to practical applications [29,30], highlighting the need for preparing immobilized photocatalysts with both high activity and recyclability [31]. Carbon cloth has unique physical and chemical properties such as a large specific surface area, excellent electrical conductivity, high strength, good thermal stability, and good corrosion resistance, which has been widely used in supercapacitors, lithium–ion batteries, electrocatalysis, microwave absorption, and other fields [32–34]. It was reported in our previous work that using carbon cloth as the catalyst support can effectively improve the photocatalytic performance and recyclability of the photocatalyst [35]. Thus, it is hypothesized that loading composites of ZnO and Ag_3PO_4 with well-designed structures onto carbon cloth may yield a recyclable photocatalyst with high activity.

In this work, we reported the synthesis of a hybrid ZnO@Ag_3PO_4 core–shell structure grown on carbon cloth via a calcination treatment followed by a two-step solution route (including a solvothermal treatment and subsequent impregnation–deposition) and evaluated its performance for the photocatalytic degradation of rhodamine B (RhB), a widely used organic dye known to be toxic to both humans and aquatic life [1,36]. The ZnO@Ag_3PO_4 heterostructure is conductive to photocatalytic activity improvement under visible light irradiation, whereas the carbon fibers, as high-speed electron transfer channels, are also beneficial to charge a separation. As a result, the carbon cloth-supported ZnO@Ag_3PO_4 composite material exhibits a dramatic visible light photocatalytic activity compared with that of carbon cloth-supported ZnO and Ag_3PO_4. Additionally, the free-standing carbon cloth-supported ZnO@Ag_3PO_4 can be easily withdrawn from a simulated wastewater solution and reused, which makes it a recyclable photocatalyst with potential for practical applications.

2. Materials and Methods

The procedures for synthesizing the materials are illustrated in Figure 1.

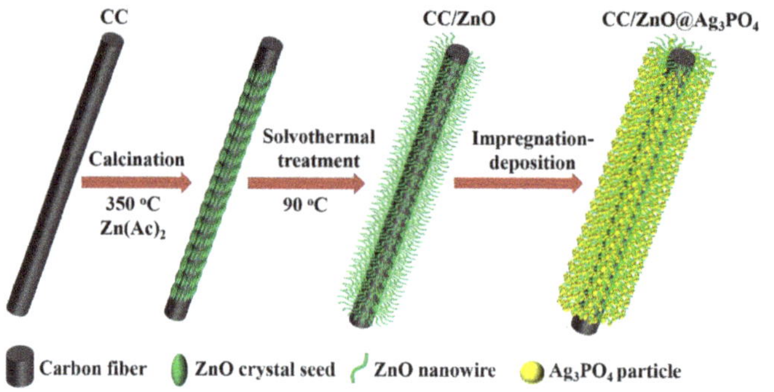

Figure 1. Schematic illustration for the synthesis procedure of the carbon cloth-supported ZnO and ZnO@Ag_3PO_4 composite.

2.1. Deposition of ZnO Crystal Seeds on Carbon Cloth

The carbon cloth (WOS: 1009-17070602, average thickness: 0.33 mm, and nominal basic weight: 115.0 g/m^2) with dimensions of 2 cm × 2 cm was successively dipped into acetone, anhydrous ethanol, and deionized water for 20 min in ultrasonic bath to remove the surface impurities. Then, the carbon cloth was soaked in nitric acid (65%) for 24 h to improve the surface hydrophilicity. After that, the carbon cloth was cleaned with deionized water several times and dried in an oven at 60 °C for 12 h.

ZnO crystal seeds were deposited on the carbon cloth by the calcination method. Specifically, 0.18 mmol of anhydrous zinc acetate was completely dissolved in 30 mL of anhydrous ethanol under continuous magnetic stirring. Afterwards, the mixed solution was slowly dripped onto the pretreated carbon cloth, and the obtained wet carbon cloth was further heated in a muffle furnace at 350 °C for 20 min in air. After cooling to room temperature, white ZnO crystal seeds could be found to adhere to the surfaces of carbon fibers.

2.2. Growth of ZnO Nanowires on Carbon Cloth

ZnO nanowires were grown on carbon cloth by a simple solvothermal method. Firstly, 3 mmol of zinc nitrate hexahydrate (Zn $(NO_3)_2 \cdot 6H_2O$) and 2 mmol of hexamethylenetetramine (HMTA) were dissolved in a certain volume of mixed solution containing aqueous ammonia (3 mL) and deionized water (70 mL), followed by being poured into a 100 mL autoclave with Teflon liner. Subsequently, the carbon cloth with ZnO crystal seeds was vertically soaked into the solution, and then, the autoclave was sealed and heated at 90 °C for 24 h. After the reaction, the autoclave naturally cooled down to room temperature. Finally, the carbon cloth with ZnO nanowires (denoted as CC/ZnO) was removed from the autoclave and washed with deionized water and ethanol several times and dried in an oven at 60 °C overnight.

2.3. Preparation of Carbon Cloth-Supported ZnO@Ag_3PO_4 Composite

The carbon cloth-supported ZnO@Ag_3PO_4 composite was prepared by a simple impregnation–deposition method. The whole experiment process was carried out in dark conditions to avoid undesired photo corrosion. A certain amount of $AgNO_3$ and $(NH_4)_2HPO_4$ were separately dissolved in deionized water to form solutions with concentrations of 0.5 and 0.17 mol/L, respectively. Then, the as-prepared carbon cloth-supported ZnO was soaked in $AgNO_3$ solution for 6 h. After that, the wet $AgNO_3$-infiltrated CC/ZnO was taken out and immersed in $(NH_4)_2HPO_4$ solution for 1 min and then was again dipped back into the $AgNO_3$ solution. This soaking process was repeated 20 times. Finally, the obtained carbon cloth-supported ZnO@Ag_3PO_4 composite (denoted as CC/ZnO@Ag_3PO_4) was washed with deionized water and dried at 60 °C for 12 h. For comparison, carbon cloth-supported Ag_3PO_4 (denoted as CC/Ag_3PO_4) was also prepared by directly soaking pure carbon cloth into the precursor solutions in the same way as described above.

2.4. Characterization

The morphological observation of the carbon cloth-supported photocatalysts was accomplished on a SU8010 field emission scanning electron microscope (FESEM, Hitachi, Japan) and a JEM2010F transmission electron microscope (TEM, JEOL, Japan). The crystal structure and phase composition of the samples were analyzed by a D/MAX-Ultima IV X-ray diffractometer (XRD, Rigaku, Japan) using Cu Kα radiation with a scanning rate of 8°/min. X-ray photoelectron spectroscopy (XPS) was measured on an EscaLab 250Xi electron spectrometer (Thermo, Waltham, MA, USA) excited by using Al Kα radiation. UV–Vis diffuse reflectance spectra (DRS) of the samples were recorded on a UV9000 UV–Vis spectrophotometer (Metash, Shanghai, China), with $BaSO_4$ used as a reflectance standard.

2.5. Photocatalytic Activity Measurement

The photocatalytic activities of the carbon cloth-supported photocatalysts were evaluated by the degradation of RhB (1.0×10^{-5} mol/L) as the target organic pollutant at room temperature, and a LED lamp with a fixed wavelength of 420 nm was used as the visible light source to trigger the reaction. To be more specific, a piece of carbon cloth-supported photocatalyst (2 cm × 2 cm) was vertically placed into a beaker containing 60 mL of RhB solution. Before turning on the LED light, the catalytic reaction system was magnetically stirred in dark for 30 min to reach adsorption equilibrium. During the process of light irradiation, aliquots of the supernatant were removed with a disposable dropper at regular time intervals and transfused into a colorimeter cell. The absorbance of the solution was measured by a UV–visible spectrophotometer, and the activities of the photocatalysts were estimated by comparing the apparent reaction rate constant, which can be calculated by $\ln(C_0/C_t) = kt$. In this equation, C_0 and C_t represent the concentration of RhB solution initially at the adsorption equilibrium (before the light was turned on) and at the reaction time of t, respectively, where k represents the reaction rate constant, and t represents the irradiation time.

3. Results and Discussion

3.1. Phase Structure

The crystal structure and phase composition of the samples were characterized by XRD, and the related patterns are shown in Figure 2. The pure carbon cloth (CC) exhibits a relatively wide diffraction peak at 26°, corresponding to the (002) plane of the carbon fiber [37]. In the case of CC/ZnO, besides the weak peak of carbon cloth, many new diffraction peaks (labeled with ♦) appear in the XRD pattern, including three strong peaks at 31.7°, 34.4°, and 36.3°, which can be assigned to the (100), (002), and (101) planes of the hexagonal ZnO (JCPDS card No. 89-0510), respectively [38]. This indicates that ZnO has successfully grown on the carbon cloth. The XRD pattern of CC/Ag_3PO_4 shows diffraction peaks corresponding to both carbon fiber and Ag_3PO_4 (JCPDS card No. 74-1876, labeled with ▼) [39]. As for the CC/ZnO@Ag_3PO_4 composite, its XRD pattern displays strong diffraction peaks of Ag_3PO_4 and extremely weak diffraction peaks of ZnO, which is probably due to the full coverage of ZnO by Ag_3PO_4. Meanwhile, the characteristic peak of carbon fiber at 26° is practically unobservable, mainly attributed to the fact that the surface of the carbon fibers was entirely coated by ZnO and Ag_3PO_4. In addition, the peak intensity of Ag_3PO_4 in CC/ZnO@Ag_3PO_4 is stronger than that in CC/Ag_3PO_4. It can be inferred that more Ag_3PO_4 was deposited on the surface of CC/ZnO than directly on the pure carbon cloth.

3.2. Morphology

The morphology of the bare carbon cloth, CC/ZnO, CC/Ag_3PO_4, and CC/ZnO@Ag_3PO_4 was investigated by SEM (Figure 3). The microstructure of the bare carbon cloth consisted of carbon fibers with an average diameter of about 10 μm (Figure 3a). A high-magnification SEM image showed that the surface of the carbon fiber was relatively smooth, with wrinkles parallel to the direction of the fibers (Figure 3b). After solvothermal treatment in the precursor solution of ZnO, the carbon fibers became brushy, and many nanowires were found to grow evenly on the surface (Figure 3c). More detailed observation revealed that these ZnO nanowires were crisscross-distributed on the carbon fibers with diameters of 60–160 nm and lengths of more than 2 μm (Figure 3d). The ZnO nanowires with a crisscross structure tended to have high specific surface areas that were available for the deposition of more catalyst particles or adsorption of target molecules. As expected, the surface of CC/ZnO was completely covered by a large number of particles after the deposition of Ag_3PO_4 (Figure 3e). Further observation of the CC/ZnO@Ag_3PO_4 composite showed that these Ag_3PO_4 particles have a polyhedral shape with sizes of 300 nm to 1 μm (Figure 3f). Meanwhile, the ZnO nanowires can be hardly seen in Figure 3f, probably because they were entirely encased by the Ag_3PO_4 particles. In contrast, it is evident from Figure 3g,h that only a small quantity of Ag_3PO_4 particles are sparsely distributed on the surface of the carbon fibers in CC/Ag_3PO_4, owing to the relatively smooth surfaces of carbon fibers. The mass loading of ZnO@Ag_3PO_4 and pure Ag_3PO_4 on the carbon cloth was 7.8×10^{-3} and 1.8×10^{-3} g/cm^2, respectively. The above results are in good correspondence with the XRD analysis.

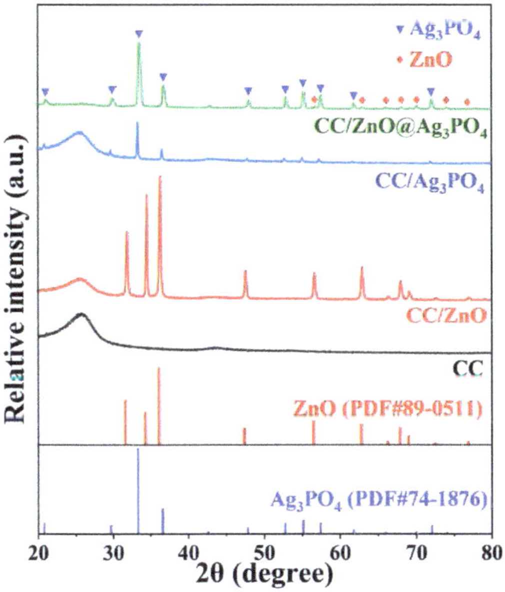

Figure 2. XRD patterns of the CC, CC/ZnO, CC/Ag_3PO_4, and CC/ZnO@Ag_3PO_4 composites.

In order to further investigate the presence of ZnO nanowires in the CC/ZnO@Ag_3PO_4 composite, a control experiment was carried out by shortening the repeated soaking times of CC/ZnO in $AgNO_3$ and $(NH_4)_2HPO_4$ solutions to 15 times, and the SEM image of the as-obtained CC/ZnO@Ag_3PO_4 composite is shown in Figure 4a. Obviously, many nanoparticles aggregated on the surfaces of ZnO nanowires (marked by the dotted circle). The diameter of these ZnO@Ag_3PO_4 nanowires was about 200 nm. In addition, the ZnO@Ag_3PO_4 nanowires were scraped off from the surfaces of carbon fibers for TEM characterization to further confirm the existence of ZnO nanowires. As shown in Figure 4b, the nanowires (dark) were completely covered by many nanoparticles (light) with sizes of about 75 nm, forming a core–shell structure. Additionally, bigger polyhedral particles were found to be adjacent to these nanoparticles. The TEM image is well consistent with the SEM observations in Figure 4a. Furthermore, the elemental mappings of the Ag, Zn, P, and O elements also support the core–shell structure of the CC/ZnO@Ag_3PO_4 composite (Figure 4c). The Zn element only exists in the core nanowires, while the Ag, P, and O elements are distributed near the ZnO nanowires but cover a larger extent. Such tight contact between ZnO and Ag_3PO_4 is conductive to the transfer of charge carriers.

3.3. Surface Chemistry

The surface composition and chemical states of CC/ZnO, CC/Ag_3PO_4, and CC/ZnO@Ag_3PO_4 were further determined by XPS, and the obtained spectra are shown in Figure 5. The full-scanned spectrum of CC/ZnO@Ag_3PO_4 indicates the presence of the Zn, Ag, O, P, and C elements (Figure 5a), which originated from the ternary composites containing carbon fiber, ZnO, and Ag_3PO_4. A close look at the full spectra reveals that the C 1s peak intensity of CC/Ag_3PO_4 is much stronger than that of CC/ZnO and CC/ZnO@Ag_3PO_4, owing to the low loading of Ag_3PO_4 on the pure carbon cloth, which results in a larger fraction of surface-exposed carbon fibers. This result is in good agreement with the XRD analysis. The high-resolution XPS spectrum of Ag 3d in CC/Ag_3PO_4 centered at 374.0 and 368.0 eV (Figure 5b), corresponding to the Ag $3d_{3/2}$ and Ag $3d_{5/2}$ core levels of Ag^+, respectively [34]. Meanwhile, the Ag 3d peaks in $CC/ZnO@Ag_3PO_4$ exhibited a positive shift by 0.6 eV compared with those for CC/Ag_3PO_4, reflecting an increase in the electron density around the Ag ions (corresponding to a lower formal charge) [40,41], which likely came from the transfer of electrons from ZnO to Ag_3PO_4. The characteristic peaks of Zn 2p appeared at 1023.0 and 1046.1 eV (Figure 5c), which correspond to Zn $2p_{3/2}$ and Zn $2p_{1/2}$, indicating the existence of Zn^{2+} from ZnO [42]. The O 1s peak of CC/ZnO could be fitted into three peaks, as shown in Figure 5d. The dominant peak at 530.2 eV was from the Zn–O bonding of ZnO, and the other two peaks at 531.1 and 532.8 eV were associated with dissociatively adsorbed water (Zn–OH) and physically adsorbed H_2O molecules, respectively [43,44]. For CC/Ag_3PO_4, the O 1s XPS spectrum could also be deconvoluted into three peaks at 530.0, 531.6, and 532.6 eV, which are related to the lattice oxygen of Ag_3PO_4, chemisorbed oxygen of surface –OH group, and physically adsorbed H_2O molecules [45–47]. The high-resolution XPS spectrum of O 1s in $CC/ZnO@Ag_3PO_4$ could be disintegrated into four peaks. The two peaks at lower binding energy (529.9 and 530.7 eV) are assigned to the lattice oxygen of Ag_3PO_4 and Zn–O bonding of ZnO [48,49], respectively. The other two peaks at higher binding energy (531.5 and 532.4 eV) are related to the surface –OH group and H_2O molecules absorbed on the surface of the $CC/ZnO@Ag_3PO_4$ composite [50]. The slight shift of O 1s peaks related to Zn–O and Ag–O in the $CC/ZnO@Ag_3PO_4$ composite was observed, compared with O 1s peaks in CC/ZnO and CC/Ag_3PO_4, implying the formation of chemical bonding between ZnO and Ag_3PO_4. The above XPS results further confirmed the successful synthesis of a $CC/ZnO@Ag_3PO_4$ composite photocatalyst.

3.4. Photocatalytic Dye Removal Performance

The photocatalytic activities of the prepared samples for the removal of RhB from an aqueous solution were performed under visible light from LED lamp irradiation. As can be seen from Figure 7a, only a small amount of RhB was adsorbed for all samples in the dark conditions. The CC/ZnO sample exhibited weak photocatalytic activity, with only about 15.4% removal achieved after 100 min of light illumination. The CC/Ag_3PO_4 sample also exhibited a modest photocatalytic performance, with 35.3% RhB removal. By contrast, 87.1% of RhB was degraded by the $CC/ZnO@Ag_3PO_4$ composite, which showed the highest photocatalytic activity with a reaction rate constant of 0.0191 min^{-1}, exceeding that of CC/Ag_3PO_4 by 4.8 times and that of CC/ZnO by a factor of 15.9 (Figure 7b). The stability and reusability are essential to the practical application of the photocatalysts. Herein, three cycles of photocatalytic experiments were carried out to investigate the photocatalytic stability of $CC/ZnO@Ag_3PO_4$ and CC/Ag_3PO_4. The RhB removal ratio for the $CC/ZnO@Ag_3PO_4$ composite was 30.9% after three cycle tests (Figure 7c), showing a 64.5% activity loss compared with its performance in the first test. However, only 4.7% of the decomposition ratio of RhB remained for CC/Ag_3PO_4 after three cycles of photocatalytic degradation tests (Figure 7d), which exhibited an 88.0% activity loss compared with itself. These results indicated that the heterostructure formed between ZnO and Ag_3PO_4 is beneficial to the improvement of photocatalytic activity and stability of the $CC/ZnO@Ag_3PO_4$ composite.

3.5. Optical Absorption Property

The UV–Visible DRS test was carried out to explore the optical absorption of CC, CC/ZnO, CC/Ag_3PO_4, $CC/ZnO@Ag_3PO_4$, and pure Ag_3PO_4 powder, and the results are shown in Figure 6. It can be found that the dark grey CC has a strong light absorption within the entire UV–visible light region examined. The CC/ZnO exhibited strong absorption in the ultraviolet light region of 200–400 nm, and its relatively strong absorption in the visible light region was brought about by the carbon cloth. The intrinsic absorption edge of pure

Ag_3PO_4 powder is about 536.5 nm, corresponding to a bandgap energy of 2.31 eV. The CC/ZnO@Ag_3PO_4 also displayed a similar light absorption curve to that of pure Ag_3PO_4, except for an enhanced absorption in the 470–800 nm region, possibly because the unique structure of the ZnO nanowires grown on carbon fibers enhanced light scattering. This can also be verified from the relatively strong absorption of CC/ZnO in the visible light region. In addition, the light absorbance of CC/Ag_3PO_4 is higher than that of CC/ZnO@Ag_3PO_4, which can be attributed to the fact that the quantity of Ag_3PO_4 particles directly deposited on the surface of the carbon cloth was much less than that deposited on CC/ZnO, leading to more exposed carbon fibers in CC/Ag_3PO_4 than in CC/ZnO@Ag_3PO_4. It is well known that the dark grey carbon fibers can almost absorb all the light. The above results also match well with the XRD and SEM analyses.

Figure 3. SEM images of CC (**a**,**b**), CC/ZnO (**c**,**d**), CC/ZnO@Ag_3PO_4 (**e**,**f**), and CC/Ag_3PO_4 (**g**,**h**).

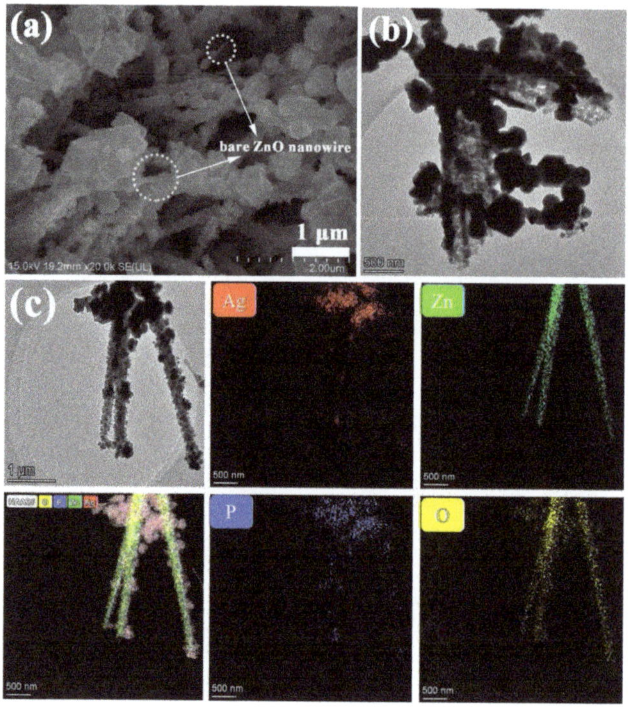

Figure 4. SEM image (**a**), TEM image, (**b**) and corresponding elemental mappings (**c**) of the CC/ZnO@Ag_3PO_4 composite, which show distribution of the Ag, Zn, P, and O elements.

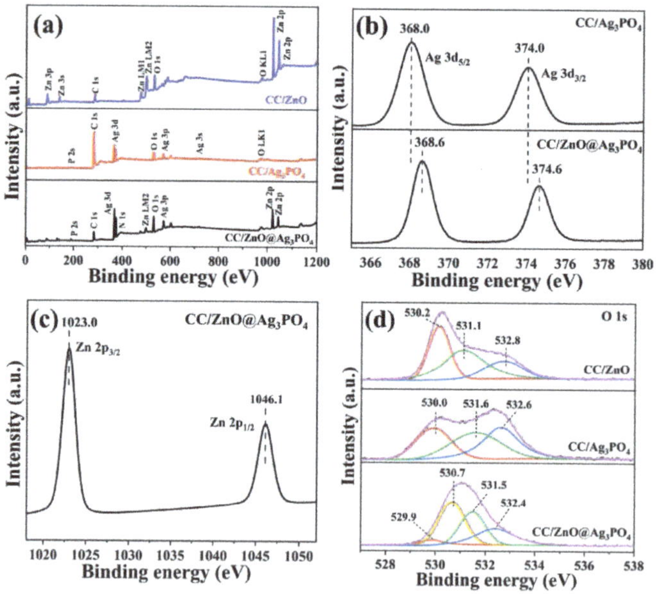

Figure 5. XPS survey spectrum (**a**) and high-resolution XPS spectra of Ag 3d (**b**), Zn 2p (**c**), O 1s (**d**) of CC/ZnO, CC/Ag_3PO_4, and the CC/ZnO@Ag_3PO_4 composite.

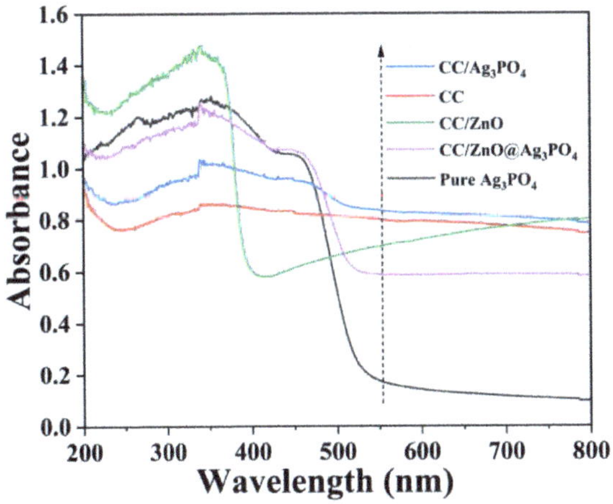

Figure 6. UV–Vis diffuse reflectance spectra of the as-prepared samples.

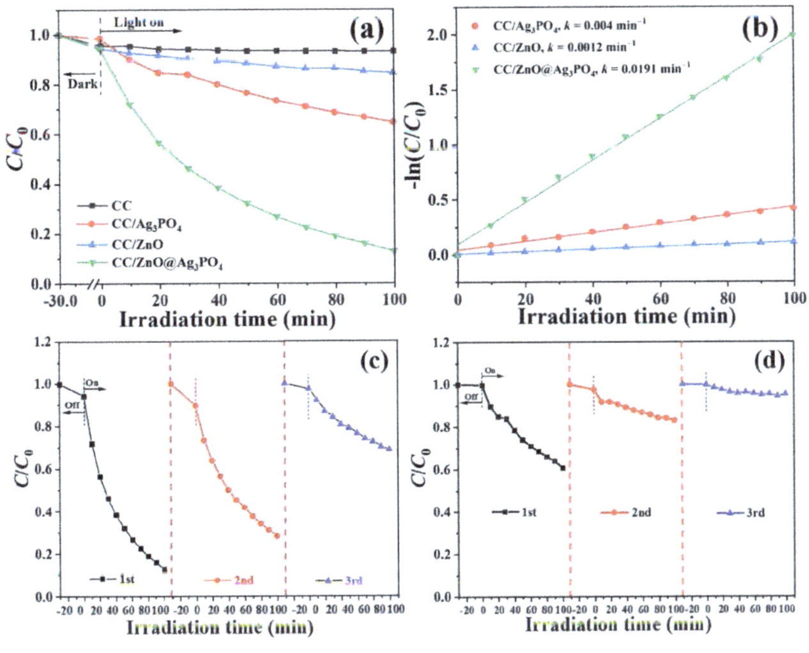

Figure 7. Comparison of the photocatalytic activities (**a**) and apparent reaction rate constants (**b**) of CC, CC/ZnO, CC/ZnO@Ag_3PO_4, and CC/Ag_3PO_4 for the degradation of RhB under visible light irradiation, and photocatalytic recyclability of the CC/ZnO@Ag_3PO_4 composite (**c**) and CC/Ag_3PO_4 (**d**).

4. Possible Photocatalytic Mechanism

On the basis of the above results and the literature, a possible charge separation and transfer mechanism for the enhanced activity and stability of the CC/ZnO@Ag_3PO_4 composite was proposed (Figure 8). According to the literature, ZnO possesses a more negative valence band (VB) and conduction band (CB) edge than that of Ag_3PO_4 [51]. When they come into contact with each other, electrons transfer from ZnO to Ag_3PO_4, and an internal electric field (IEF) is created at the interface of ZnO and Ag_3PO_4 [52]. The direction of the IEF points from ZnO to Ag_3PO_4, as marked by the arrow in Figure 8. Under visible light excitation, ZnO shows weak photocatalytic activity, probably due to its intrinsic defect in the crystal and inefficient light absorption [53]. Simultaneously, both ZnO and Ag_3PO_4 can be excited, and the electrons in the VB leap into the CB, while the positively charged holes (h^+) are left in the VB. Subsequently, the photogenerated electrons in the CB of Ag_3PO_4 shift towards ZnO, as driven by the interfacial IEF, and recombine with the h^+ in the VB of ZnO. Apparently, an S-scheme mechanism applies well to the heterojunction between Ag_3PO_4 and ZnO, and the transfer of electrons and holes in the IEF region follows a slide-like pathway. Benefitting from this ZnO@Ag_3PO_4 S-scheme heterojunction, the photogenerated charges efficiently separate. Consequently, the holes are enriched in the VB of Ag_3PO_4, while the electrons are enriched in the CB of ZnO. It has been reported that the h^+ and superoxide anion radicals ($\cdot O_2^-$) are the main active species in the ZnO/Ag_3PO_4 reaction system, according to the free radical capture experiment [54,55]. The holes reserving high oxidation ability in the VB of Ag_3PO_4 can directly participate in the degradation of RhB. In addition, it is well known that the carbon fibers with a one-dimensional linear structure possess excellent electrical conductivity, which rapidly receive electrons from the CB of ZnO. Then, the dissolved oxygen molecules are captured by the electrons enriched in the carbon fibers and reduced into $\cdot O_2^-$ radicals, which further degrade RhB molecules into CO_2, H_2O, and small molecular compounds [35]. Moreover, the ZnO nanowires directly grown on the conductive carbon fibers ensure good adhesion of the ZnO@Ag_3PO_4 core–shell structure with the carbon cloth substrate.

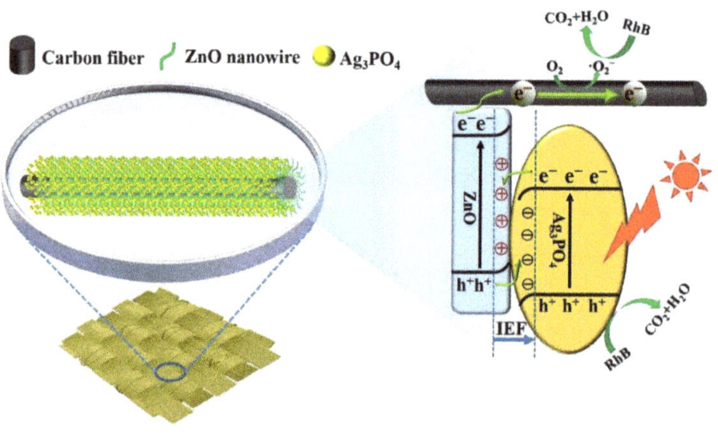

Figure 8. Schematic diagram for the improved charge separation of the CC/ZnO@Ag_3PO_4 composite.

5. Conclusions

In summary, the carbon cloth-supported ZnO@Ag_3PO_4 composite with a core–shell structure was successfully prepared by a two-step process including a solvothermal method and a succeeding impregnation–deposition method. The as-prepared CC/ZnO@Ag_3PO_4 displayed an enhanced photocatalytic activity for degrading RhB under visible light irradiation compared with that of CC/ZnO and CC/Ag_3PO_4, mainly attributed to the synergistic effect of the carbon cloth, ZnO, and Ag_3PO_4. The construction of a S-scheme heterojunction between ZnO and Ag_3PO_4 was mainly responsible for the enhanced activity of the composite, which suppressed the recombination of the charge carriers within ZnO or Ag_3PO_4 itself by transferring electrons from the CB of Ag_3PO_4 to the VB of ZnO through the internal electric field. Additionally, carbon fibers further accelerated the transmission of electrons; therefore, the charge separation efficiency was further improved. The carbon cloth-supported ZnO@Ag_3PO_4 can be easily separated from the solution and repeatedly used, demonstrating a fair recyclability and potential in practical applications. Nevertheless, the decreased activity of CC/ZnO@Ag_3PO_4 after many cycles of photocatalytic tests still needs to be further explored.

Author Contributions: Conceptualization, W.W. and C.J.; sample preparation, Y.Y., Q.G., S.J. and H.L.; sample characterization, Y.Y. and Q.G.; photocatalytic test, Y.Y., Q.G. and S.J.; writing—original draft preparation, Y.Y.; writing—review and editing, W.W. and C.J.; supervision, W.W.; project administration, L.W. and H.Z.; and funding acquisition, W.W., C.J. and L.W. All authors have read and agreed to the published version of the manuscript.

Funding: This research was funded by the National Natural Science Foundation of China (grant numbers 52072079 and 51802305), the Natural Science Foundation of Guangdong Province, China (No. 2021A1515010445), and the Fundamental Research Funds for the Central Universities of Nankai University.

Institutional Review Board Statement: Not applicable.

Informed Consent Statement: Not applicable.

Data Availability Statement: All the data are available within the manuscript.

Conflicts of Interest: The authors declare no conflict of interest. The funders had no role in the design of the study; in the collection, analyses, or interpretation of the data; in the writing of the manuscript; or in the decision to publish the results.

References

1. Dire, D.J.; Wilkinson, J.A. Acute exposure to rhodamine B. *J. Toxicol. Clin. Toxicol.* **1987**, *25*, 603–607. [CrossRef]
2. Verma, Y. Acute toxicity assessment of textile dyes and textile and dye industrial effluents using Daphnia magna bioassay. *Toxicol. Ind. Health* **2008**, *24*, 491–500.
3. Fernández, C.; Larrechi, M.S.; Callao, M.P. An analytical overview of processes for removing organic dyes from wastewater effluents. *TrAC Trends Anal. Chem.* **2010**, *29*, 1202–1211. [CrossRef]
4. Shanker, U.; Rani, M.; Jassal, V. Degradation of hazardous organic dyes in water by nanomaterials. *Environ. Chem. Lett.* **2017**, *15*, 623–642. [CrossRef]
5. Loc, N.X.; Tuyen, P.T.T.; Mai, L.C.; Phuong, D.T.M. Chitosan-modified biochar and unmodified biochar for methyl orange: Adsorption characteristics and mechanism exploration. *Toxics* **2022**, *10*, 500. [CrossRef]
6. Gomaa, H.; El-Monaem, E.M.A.; Eltaweil, A.S.; Omer, A.M. Efficient removal of noxious methylene blue and crystal violet dyes at neutral conditions by reusable montmorillonite/$NiFe_2O_4$@amine-functionalized chitosan composite. *Sci. Rep.* **2022**, *12*, 15499. [CrossRef]
7. Jaleel, R.M.; Uc, J.R.; Pinheiro, D.; NK, R.; Devi KR, S.; Park, J.; Manickam, S.; Choi, M.Y. Architecture of visible-light induced Z-scheme MoS_2/g-C_3N_4/ZnO ternary photocatalysts for malachite green dye degradation. *Environ. Res.* **2022**, *214*, 113742.
8. Chu, M.N.; Nguyen, L.T.H.; Truong, M.X.; Do, H.T.; Duong, T.T.A.; Nguyen, L.T.T.; Pham, M.A.; Tran, T.K.N.; Ngo, T.C.Q.; Pham, V.H. Ce^{3+}/Ce^{4+}-Doped ZrO_2/CuO nanocomposite for enhanced photocatalytic degradation of methylene blue under visible light. *Toxics* **2022**, *10*, 463. [CrossRef]
9. Gomaa, H.; Hussein, M.A.T.; Motawea, M.M.; Aboraia, A.M.; Cheira, M.F.; Alotaibi, M.T.; El-Bahy, S.M.; Ali, H.M. A hybrid mesoporous CuO@barley straw-derived SiO_2 nanocomposite for adsorption and photocatalytic degradation of methylene blue from real wastewater. *Colloids Surf. A* **2022**, *644*, 128811. [CrossRef]
10. Motawea, M.M.; Hussein, M.A.T.; Elsenety, M.M.; Ali, H.M.; El Nasr, T.A.S.; Gomaa, H. Mesoporous hierarchical ZrO_2@rice straw-derived SiO_2 nanocomposite for rapid adsorption and sunlight-driven photocatalytic degradation of methylene blue. *J. Photochem. Photobiol. A* **2022**, *426*, 113758. [CrossRef]
11. Chang, J.S.; Strunk, J.; Chong, M.N.; Poh, P.E.; Ocon, J.D. Multi-dimensional zinc oxide (ZnO) nanoarchitectures as efficient photocatalysts: What is the fundamental factor that determines photoactivity in ZnO. *J. Hazard. Mater.* **2020**, *381*, 120958. [CrossRef]
12. Lam, S.M.; Sin, J.C.; Abdullah, A.Z.; Mohamed, A.R. Degradation of wastewaters containing organic dyes photocatalysed by zinc oxide: A review. *Desalin. Water Treat.* **2012**, *41*, 131–169. [CrossRef]
13. Biglar, F.; Talaiekhozani, A.; Aminsharei, F.; Park, J.; Baighi, A.; Rezania, S. Application of ZnO-Nd nano-photocatalyst for the reactive red 198 dye decolorization in the falling-film photocatalytic reactor. *Toxics* **2021**, *9*, 254. [CrossRef]
14. Kassem, K.O.; Hussein, M.A.T.; Motawea, M.M.; Gomaa, H.; Alrowaili, Z.A.; Ezzeldien, M. Design of mesoporous ZnO @ silica fume-derived SiO_2 nanocomposite as photocatalyst for efficient crystal violet removal: Effective route to recycle industrial waste. *J. Clean. Prod.* **2021**, *326*, 129416. [CrossRef]
15. Zhang, L.W.; Cheng, H.Y.; Zong, R.L.; Zhu, Y.F. Photocorrosion suppression of ZnO nanoparticles via hybridization with graphite-like carbon and enhanced photocatalytic activity. *J. Phys. Chem. C* **2009**, *113*, 2368–2374. [CrossRef]

16. Jiang, C.J.; Aiken, G.R.; Hsu-Kim, H. Effects of natural organic matter properties on the dissolution kinetics of zinc oxide nanoparticles. *Environ. Sci. Technol.* **2015**, *49*, 11476–11484. [CrossRef]
17. Kumar, S.G.; Rao, K.S.R.K. Zinc oxide based photocatalysis: Tailoring surface-bulk structure and related interfacial charge carrier dynamics for better environmental applications. *RSC Adv.* **2015**, *5*, 3306–3351. [CrossRef]
18. Zhu, B.C.; Xia, P.F.; Li, Y.; Ho, W.K.; Yu, J.G. Fabrication and photocatalytic activity enhanced mechanism of direct Z-scheme g-C_3N_4/Ag_2WO_4 photocatalyst. *Appl. Surf. Sci.* **2017**, *391*, 175–183. [CrossRef]
19. Xiao, T.T.; Tang, Z.; Yang, Y.; Tang, L.P.; Zhou, Y.; Zou, Z.G. In situ construction of hierarchical WO_3/g-C_3N_4 composite hollow microspheres as a Z-scheme photocatalyst for the degradation of antibiotics. *Appl. Catal. B* **2018**, *220*, 417–428. [CrossRef]
20. Yang, X.B.; Hu, J.P.; Pan, J.J.; Shen, Y.B.; Cheng, K.J. Fabrication of Ag/ZnO@N-carbon core@shell photocatalyst for efficient photocatalytic degradation of rhodamine B. *Front. Chem.* **2022**, *10*, 950007. [CrossRef]
21. Yi, Z.G.; Ye, J.H.; Kikugawa, N.; Kako, T.; Ouyang, S.X.; Williams, H.S.; Yang, H.; Cao, J.Y.; Luo, W.J.; Li, Z.S.; et al. An orthophosphate semiconductor with photooxidation properties under visible-light irradiation. *Nat. Mater.* **2010**, *9*, 559–564. [CrossRef]
22. Liu, Y.; Wang, W.G.; Si, M.Z.; Yu, Y.F.; Zhang, H.Y. (Yb^{3+},Er^{3+}) co-doped TiO_2/Ag_3PO_4 hybrid photocatalyst with enhanced activity for photodegradation of phenol. *Appl. Surf. Sci.* **2019**, *463*, 159–168. [CrossRef]
23. Dong, P.Y.; Huo, G.H.; Liu, C.; Zhang, X.J.; Tian, H.; Xu, F.H.; Xi, X.G.; Shao, R. Origin of activity and stability enhancement for Ag_3PO_4 photocatalyst after calcination. *Materials* **2016**, *9*, 968. [CrossRef]
24. Wang, W.G.; Cheng, B.; Yu, J.G.; Liu, G.; Fan, W.H. Visible-light photocatalytic activity and deactivation mechanism of Ag_3PO_4 spherical particles. *Chem.-Asian J.* **2012**, *7*, 1902–1908. [CrossRef]
25. Dong, C.; Wu, K.L.; Li, M.R.; Liu, L.; Wei, X.W. Synthesis of Ag_3PO_4-ZnO nanorod composites with high visible-light photocatalytic activity. *Catal. Commun.* **2014**, *46*, 32–35. [CrossRef]
26. Liu, W.; Wang, M.L.; Xu, C.X.; Chen, S.F.; Fu, X.L. Ag_3PO_4/ZnO: An efficient visible-light-sensitized composite with its application in photocatalytic degradation of rhodamine B. *Mater. Res. Bull.* **2013**, *48*, 106–113. [CrossRef]
27. Yi, Z.; Li, X.; Wu, H.; Chen, X.F.; Yang, H.; Tang, Y.J.; Yi, Y.G.; Wang, J.Q.; Wu, P.H. Fabrication of ZnO@Ag_3PO_4 core-shell nanocomposite arrays as photoanodes and their photoelectric properties. *Nanomaterials* **2019**, *9*, 1254. [CrossRef]
28. Zhong, J.B.; Li, J.Z.; Wang, T.; Zeng, J.; Si, Y.J.; Cheng, C.Z.; Li, M.J.; Wang, P.; Ding, J. Improved solar-driven photocatalytic performance of Ag3PO4/ZnO composites benefiting from enhanced charge separation with a typical Z-scheme mechanism. *Appl. Phys. A Mater. Sci. Process.* **2016**, *122*, 4. [CrossRef]
29. Si, M.Z.; Wang, W.G.; Guan, Q.F.; Zhang, H.Y.; Puttaswamy, M. Facile fabrication of highly catalytic-active Ag_2CO_3/AgBr/graphene oxide ternary composites towards the photocatalytic wastewater treatment. *Environ. Sci. Pollut. Res.* **2021**, *28*, 4173–4183. [CrossRef]
30. Xie, Y.; Luo, S.Q.; Huang, H.W.; Huang, Z.H.; Liu, Y.G.; Fang, M.H.; Wu, X.W.; Min, X. Construction of an Ag_3PO_4 morphological homojunction for enhanced photocatalytic performance and mechanism investigation. *Colloids Surf. A* **2018**, *546*, 99–106. [CrossRef]
31. Hussein, M.A.T.; Motawea, M.M.; Elsenety, M.M.; El-Bahy, S.M.; Gomaa, H. Mesoporous spongy Ni-Co oxides@wheat straw-derived SiO_2 for adsorption and photocatalytic degradation of methylene blue pollutants. *Appl. Nanosci.* **2022**, *12*, 1519–1536. [CrossRef]
32. Yao, Y.; Dong, W.B.; Zhao, Z.; Cui, H.N.; Liao, G.L. Vertically aligned 1T-WS_2 nanosheets supported on carbon cloth as a high-performance flexible photocatalyst. *Colloids Surf. A* **2022**, *649*, 129533. [CrossRef]
33. Chen, H.H.; Deng, L.B.; Luo, S.; Ren, X.Z.; Li, Y.L.; Sun, L.N.; Zhang, P.X.; Chen, G.Q.; Gao, Y. Flexible three-dimensional heterostructured ZnO-Co_3O_4 on carbon cloth as free-standing anode with outstanding Li/Na storage performance. *J. Electrochem. Soc.* **2018**, *165*, A3932–A3942. [CrossRef]
34. Meng, S.J.; Hong, Y.; Dai, Z.Y.; Huang, W.; Dong, X.C. Simultaneous detection of dihydroxybenzene isomers with ZnO nanorod/carbon cloth electrodes. *ACS Appl. Mater. Interfaces* **2017**, *9*, 12453–12460. [CrossRef]
35. Liu, Y.; Wang, W.G.; Si, M.Z.; Zhang, H.Y. Carbon cloth-supported MoS2/Ag2S/Ag3PO4 composite with high photocatalytic activity and recyclability. *ChemCatChem* **2019**, *11*, 1017–1025.
36. Skjolding, L.M.; Jorgensen, L.vG.; Dyhr, K.S.; Koppl, C.J.; McKnight, U.S.; Bauer-Gottwein, P.; Mayer, P.; Bjerg, P.L.; Baun, A. Assessing the aquatic toxicity and environmental safety of tracer compounds Rhodamine B and Rhodamine WT. *Water Res.* **2021**, *197*, 117109. [CrossRef]
37. Yan, Y.P.; Wang, B.; Yan, C.Z.; Kang, D.J. Decorating ZnO nanoflakes on carbon cloth: Free-standing, highly stable lithium-ion battery anodes. *Ceram. Int.* **2019**, *45*, 15906–15912. [CrossRef]
38. Ashkarran, A.A.; Iraji-zad, A.; Mahdavi, S.M.; Ahadian, M.M. ZnO nanoparticles prepared by electrical arc discharge method in water. *Mater. Chem. Phys.* **2009**, *118*, 6–8. [CrossRef]
39. Li, W.; Qin, L.; Wang, Z.H.; Xu, G.C.; Zheng, H.C.; Zhou, L.M.; Chen, Z.Q. Efficient porous carbon nitride/Ag_3PO_4 photocatalyst for selective oxidation of amines to imines: Z-scheme heterojunction and interfacial adsorption. *Colloids Surf. A* **2022**, *652*, 129806. [CrossRef]
40. Liu, Y.P.; Fang, L.; Lu, H.D.; Liu, L.J.; Wang, H.; Hu, C.Z. Highly efficient and stable Ag/Ag_3PO_4 plasmonic photocatalyst in visible light. *Catal. Commun.* **2012**, *17*, 200–204. [CrossRef]

41. Cao, J.; Luo, B.D.; Lin, H.L.; Xu, B.Y.; Chen, S.F. Visible light photocatalytic activity enhancement and mechanism of AgBr/Ag_3PO_4 hybrids for degradation of methyl orange. *J. Hazard. Mater.* **2012**, *217–218*, 107–115. [CrossRef] [PubMed]
42. Saravanan, R.; Thirumal, E.; Gupta, V.K.; Narayanan, V.; Stephen, A. The photocatalytic activity of ZnO prepared by simple thermal decomposition method at various temperatures. *J. Mol. Liq.* **2013**, *177*, 394–401. [CrossRef]
43. Al-Gaashani, R.; Ramiman, S.; Daud, A.R.; Tabet, N.; Al-Douri, Y. XPS and optical studies of different morphologies of ZnO nanostructures prepared by microwave methods. *Ceram. Int.* **2013**, *39*, 2283–2292. [CrossRef]
44. Tian, Y.Y.; Fang, Q.H.; Li, J. Molecular dynamics simulations for nanoindentation response of nanotwinned FeNiCrCoCu high entropy alloy. *Nanotechnology* **2020**, *31*, 465701. [CrossRef] [PubMed]
45. Jin, B.; Zhou, X.S.; Luo, J.; Xu, X.Y.; Ma, L.; Huang, D.Y.; Shao, Z.L.; Luo, Z.H. Fabrication and characterization of high efficiency and stable Ag_3PO_4/TiO_2 nanowire array heterostructure photoelectrodes for the degradation of methyl orange under visible light irradiation. *RSC Adv.* **2015**, *59*, 48118–48123. [CrossRef]
46. Chai, B.; Zou, F.Y.; Chen, W.J. Facile synthesis of Ag_3PO_4/C_3N_4 composites with improved visible light photocatalytic activity. *J. Mater. Res.* **2015**, *30*, 1128–1136. [CrossRef]
47. Miao, X.L.; Yue, X.Y.; Ji, Z.Y.; Shen, X.P.; Zhou, H.; Liu, M.M.; Xu, K.Q.; Zhu, J.; Zhu, G.X.; Kong, L.R.; et al. Nitrogen-doped carbon dots decorated on g-C_3N_4/Ag_3PO_4 photocatalyst with improved visible light photocatalytic activity and mechanism insight. *Appl. Catal. B* **2018**, *227*, 459–469. [CrossRef]
48. Naraginti, S.; Yu, Y.Y.; Fang, Z.; Yong, Y.C. Novel tetrahedral Ag_3PO_4@N-rGO for photocatalytic detoxification of sulfamethoxazole: Process optimization, transformation pathways and biotoxicity assessment. *Chem. Eng. J.* **2019**, *375*, 122035. [CrossRef]
49. Choi, J.S.; Jung, R. In-situ XPS study of core-levels of ZnO thin films at the interface with graphene/Cu. *J. Korean Phys. Soc.* **2018**, *73*, 1546–1549. [CrossRef]
50. Shen, Y.Z.; Zhu, Z.D.; Wang, X.G.; Khan, A.; Gong, J.Y.; Zhang, Y.R. Synthesis of Z-scheme g-C_3N_4/Ag/Ag_3PO_4 composite for enhanced photocatalytic degradation of phenol and selective oxidation of gaseous isopropanol. *Mater. Res. Bull.* **2018**, *107*, 407–415. [CrossRef]
51. Wang, J.; Xia, Y.; Dong, Y.; Chen, R.S.; Xiang, L.; Komarneni, S. Defect-rich ZnO nanosheets of high surface area as an efficient visible-light photocatalyst. *Appl. Catal. B* **2016**, *192*, 8–16. [CrossRef]
52. Xu, Q.L.; Zhang, L.Y.; Cheng, B.; Fan, J.J.; Yu, J.G. S-Scheme heterojunction photocatalyst. *Chem* **2020**, *6*, 1543–1559. [CrossRef]
53. Pan, J.Q.; Zhang, X.F.; Mei, J.; Wang, S.; You, M.Z.; Zheng, Y.Y.; Cui, C.; Li, C.R. The cotton cellulose nanofibers framework of Z-Scheme ZnO/Ag_3PO_4 heterojunction for visible-light photocatalysis. *J. Mater. Sci. Mater. Electron.* **2017**, *28*, 17744–17749. [CrossRef]
54. Zhu, P.F.; Duan, M.; Wang, R.X.; Xu, J.; Zou, P.; Jia, H.S. Facile synthesis of ZnO/GO/Ag_3PO_4 heterojunction photocatalyst with excellent photodegradation activity for tetracycline hydrochloride under visible light. *Colloids Surf. A* **2020**, *602*, 125118. [CrossRef]
55. Du, C.Y.; Song, J.H.; Tan, S.Y.; Yang, L.; Yu, G.L.; Chen, H.; Zhou, L.; Zhang, Z.; Zhang, Y.; Su, Y.H.; et al. Facile synthesis of Z-scheme ZnO/Ag/Ag_3PO_4 composite photocatalysts with enhanced performance for the degradation of ciprofloxacin. *Mater. Chem. Phys.* **2021**, *260*, 124136. [CrossRef]

Disclaimer/Publisher's Note: The statements, opinions and data contained in all publications are solely those of the individual author(s) and contributor(s) and not of MDPI and/or the editor(s). MDPI and/or the editor(s) disclaim responsibility for any injury to people or property resulting from any ideas, methods, instructions or products referred to in the content.

toxics

MDPI

Article

Zinc Removal from Water via EDTA-Modified Mesoporous SBA-16 and SBA-15

Zeinab Ezzeddine [1,2,*], Isabelle Batonneau-Gener [1] and Yannick Pouilloux [1]

[1] Institut de Chimie des Milieux et Matériaux de Poitiers (IC2MP)-UMR 7285, Poitiers University, 86073 Poitiers, France
[2] Department of Chemistry, Lebanese University, Beirut P.O. Box 6573, Lebanon
* Correspondence: zeinabezzeddine.87@hotmail.com

Abstract: The removal of zinc ions from water was investigated using two types of ordered mesoporous silica (SBA-15 and SBA-16). Both materials were functionalized with APTES (3-aminopropyltri ethoxy-silane) and EDTA (ethylenediaminetetraacetic acid) through post grafting methods. The modified adsorbents were characterized by scanning electron microscopy (SEM) and transmission electron microscopy (TEM), X-ray diffraction (XRD), nitrogen (N_2) adsorption–desorption analysis, Fourier transform infrared spectroscopy (FT-IR), and thermogravimetric analysis. The ordered structure of the adsorbents was conserved after modification. SBA-16 was found to be more efficient than SBA-15 owing to its structural characteristics. Different experimental conditions were examined (pH, contact time, and initial zinc concentration). The kinetic adsorption data followed the pseudo-second-order model indicating favorable adsorption conditions. The intra-particle diffusion model plot represented a two-stage adsorption process. The maximum adsorption capacities were calculated by the Langmuir model. The adsorbent can be regenerated and reused several times without a significant decline in adsorption efficiency.

Keywords: mesoporous silica; SBA-16; SBA-15; zinc; adsorption; EDTA

1. Introduction

In the past few decades, water contamination has been considered a major problem worldwide. Effluents containing heavy metals such as copper, lead, zinc, and cadmium have been discharged without any treatment into the environment from several industries, such as from steel production, electroplating, and tanneries [1]. The discharged heavy-metal-contaminated wastewater is considered a serious threat to both human health and the ecosystem. Heavy metals are non-biodegradable and may be carcinogenic, thus their presence in water in high amounts could result in critical health issues to living organisms [2–4]. Zinc is largely spread in nature and is an essential trace metal for both humans and aquatic organisms [5]. However, if the zinc dosage exceeds a certain quantity, it becomes harmful to organisms [6] as it could interact with biological macromolecules, resulting in a change in their activity and poisoning [7]. Currently, various methods are available for heavy metal removal from water, including membrane filtration, coagulation, precipitation, ion-exchange, and adsorption [8]. The latter is very effective in eliminating heavy metals and is an attractive technique because it does not require complex and expensive installations [9,10]. The adsorption mechanism is defined by the physicochemical properties of the adsorbent and heavy metals and operating conditions (i.e., temperature, adsorbent amount, pH value, adsorption time, and initial concentration of metal ions) [11]. The efficiency of several low-cost adsorbents has been studied, including clay, chitosan, fly ash, zeolites, and activated carbon [12]. Moreover, adsorption is the best method to use when the metal ions' concentrations are below 100 mg L^{-1} [13]. The adsorption capacity is dependent on the pore size of adsorbents along with the active sites found on their surface [14] and, in recent years, many researchers have focused on developing new adsorbents that meet these criteria. The drawbacks of adsorbents derived from natural materials are their low mechanic resistance to abrasive forces and weak interactions with metal ions [15,16]. One of the advantages of the adsorption process is the ability to modify adsorbents. The carbon surface charges can be enhanced by surface functional groups (such as carboxyl, phenyl, and lactone groups) in order to improve the heavy metal uptake [17]. Surface modification often reduces its surface area and, in turn, increases the content of surface functional groups. Consequently, more metal ions can be adsorbed [18]. The disadvantages of carbon adsorbents are as follows: their high hydrophobicity and the rapid aggregation in aqueous solution owing to large Van der Waals forces, which decreases the adsorption capacity [11]. Chitosan, which is a natural adsorptive polymer, is another adsorbent used, because it has an affinity toward pollutants in wastewaters owing to the presence of amino and hydroxyl groups. However, it suffers from low mechanical strength and poor stability [19]. Mineral adsorbents such as zeolite, silica, and clay are considered good candidates for water purification with low operating costs. Clay has an extraordinary cation exchange capacity (CEC), cation exchange selectivity, surface hydrophilicity, high swelling/expanding capacity, and surface electronegativity [20]. However, their removal efficiency decreases after several sorption cycles. They are also affected by experimental conditions such as pH, irradiation time, adsorbent concentration, wastewater temperature, and the initial dosage of pollutants [21]. In order to overcome these disadvantages, various studies focused on the use of mesoporous silica as adsorbents [22,23], such as MCM-41 [24], SBA-15 [25], and SBA-16 [26]. The ordered structure of such mesoporous silica together with their tailored pore size made them suitable applications for the elimination of pollutants from water [27]. In addition, SBA-16 possesses a three-dimensional structure including cubic Im3m space group with a large pore diameter; dual porosity framework; and large surface area, i.e., a micro porous and mesoporous framework. Such properties make these materials dominant for adsorption [28]. Nevertheless, the surface of these materials has only silanol groups, so it is essential to modify them so that specific binding sites are added. Such chemical modification is achieved by functionalization with different groups, such as $-NH_2$ [29]. In addition, mesoporous materials' modification with chelating was studied as well, because chelating agents increase the metal adsorption capacity. Salicylic acid [30] and cyanex 272 [31] are examples of some of the agents used. The chelating agent ethylene diamine tetraacetic acid (EDTA) is very effective and widely known for heavy metals' complexion. Moreover, the structure formed between chelating agents and metal ions is stable and they reserve their metal binding properties after chemical regeneration [14].

Herein, SBA-16 and SBA-15 were synthesized, fully characterized, and modified with APTES (3-aminopropyltriethoxy-silane) and then with EDTA. The adsorption capacity for Zn^{2+} removal was then investigated. Moreover, the effect of various parameters on adsorption were studied, including pH, contact time, and zinc concentration. The kinetics of adsorption were also reported using pseudo-first, pseudo-second-order, and intra-diffusion models. The equilibrium isotherms of the experimental data were well fitted by Langmuir isotherm. The regeneration and reuse of the modified adsorbents were performed as well. It is worth mentioning that this manuscript aims to compare the structural impact of both adsorbents on zinc removal, which has not been studied before, other than the effectiveness of these materials as heavy metal adsorbents.

2. Materials and Methods

2.1. Chemicals

For the template of SBA-16 and SBA-15, Pluronic F127 (EO106PO70EO106) and Pluronic P123 (EO20PO70EO20) triblock copolymers, respectively, were used. Tetraethylorthosilicate (TEOS 98%) was used as the source of silica. For modification, 3-aminopropyl trimethoxysilane (APTES 99%) and ethylenediaminetetraaceticacid disodium salt (EDTA-Na_2) were utilized. Hydrochloric acid (HCl, 37%), sodium hydroxide (NaOH), toluene, zinc nitrate, and sodium bicarbonate ($NaHCO_3$) were also used in this study. All of the reagents were purchased from Sigma Aldrich (St. Louis, MO, USA) and utilized as received without any further purification. Ultrapure water was used throughout.

2.2. Adsorbents Synthesis and Modification

Mesoporous SBA-16 and SBA-15 were synthesized as described elsewhere in the literature [32,33]. Adsorbents' modifications were performed through a two-step post synthesis process [34], as illustrated in Figure 1. After APTES and EDTA modification, the adsorbents were denoted as SBA-16-NH_2 /SBA-15-NH_2 and SBA-16-EDTA/SBA-15-EDTA, respectively.

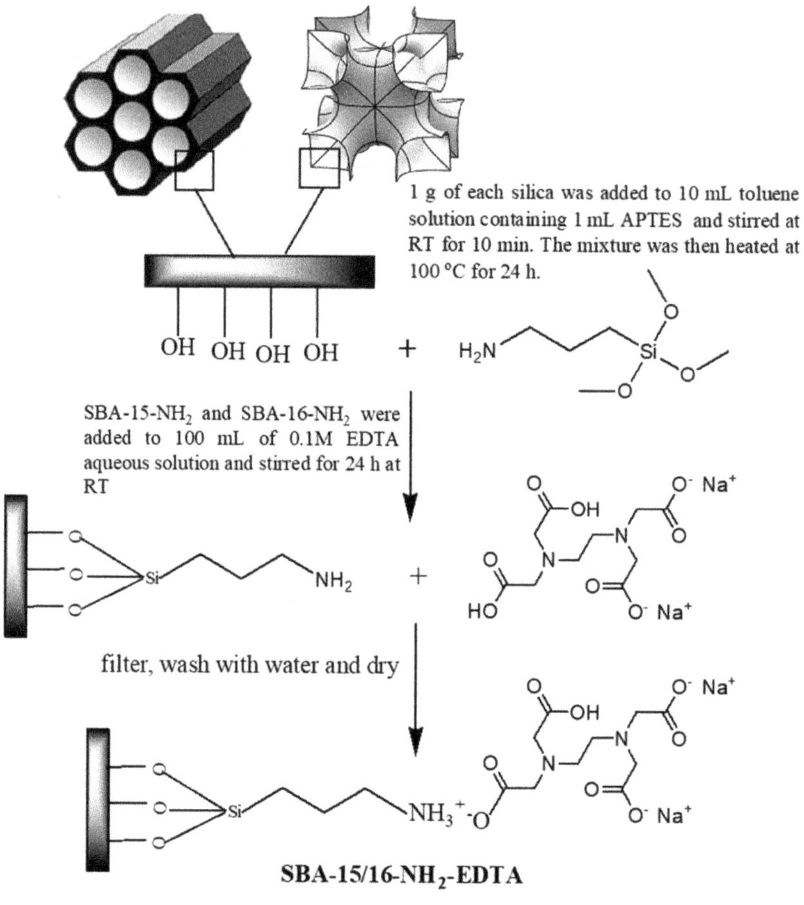

Figure 1. Modification of SBA-16 and SBA-15 with APTES and EDTA.

2.3. Adsorbents Characterization

For determining the textural properties, a Micromeritics TRISTAR sorptiometer (Micromeritics Instrument Corp., Norcross, GA, USA) was used and nitrogen adsorption–desorption isotherms were obtained at −196 °C. The as-synthesized samples were out gassed at 350 °C under vacuum for at least 5 h before measurement and overnight at 150 °C for the modified samples. Low angle X-ray diffraction (XRD) patterns were obtained with an Empyrean X-ray diffractometer (Malvern Panalytical Ltd., Royston, UK) with Cu Kα (λ = 1.54 Å) radiation and a 0.008° min^{-1} rate of scanning between 0.65° and 5° 2θ. SBA-16 and SBA-15 morphologies were determined by scanning electron microscopy (SEM, JEOL 7001 FEG, Tokyo, Japan) and transmission electron microscopy (TEM, JEOL 2100 UHR at

200 kV, Tokyo, Japan). Thermogravimetric analysis (TG) was conducted using SDT Q600 TA Instrument from 25 to 900 °C in air using SDT Q600 TA Instruments (New Castle, DE, USA). The functional groups were identified in the range of 4000–400 cm^{-1} by Fourier transform infrared spectroscopy (FT-IR–6300 JASCO, Oklahoma City, OK, USA) through mixing the samples with KBr and pressing them into pellets. The amount of immobilized carboxyl groups after EDTA modification was measured by back titration [35].

2.4. Batch Adsorption Experiments

The zinc nitrate salt was dissolved in ultrapure water and solutions with different zinc ion concentrations (between 10 and 500 ppm) were prepared. For batch adsorption studies, 20 mg of SBA-16-EDTA or SBA-15-EDTA was added to 20 mL of metal solution of concentration C and the flask was stirred at room temperature (RT) at 300 rpm for 180 min. At the end of each step, the zinc concentration was determined using an atomic adsorption spectrophotometer (AAS, Perkin Elmer AA200, Waltham, MA, USA). The removal efficiency was calculated by Equation (1) [36]:

$$R = \frac{C_0 - C_t}{C_0} \times \mathbf{100} \quad (1)$$

where C_0 and C_t are the heavy metal initial concentration and at concentration at time t, respectively. The adsorption capacity (mg g^{-1}) of the adsorbent at equilibrium was calculated by Equation (2) [36]:

$$q_e = \frac{(C_0 - C_e)V}{m} \quad (2)$$

where C_e is the concentration at equilibrium, (V) is the volume in L of metal solution, and m is the mass in g of the adsorbent.

The adsorption isotherms were established by varying the initial metal ion concentrations between 10 mg L^{-1} and 500 mg L^{-1}. The solutions were stirred for 180 min at RT and then filtered and the remaining metal ions were measured by AAS in order to calculate C_e and q_e. The pH effect was studied by varying the solution pH between 2 and 8 using 0.1 M HCl or 0.1 M NaOH. Adsorbents' regeneration was performed with 1 M HCl solution.

3. Results and Discussion

3.1. Adsorbents' Characterization

The surface morphologies of the adsorbents were obtained by SEM (Figure 2). SBA-16 appeared as fine cube particles while SBA-15 had rod-shaped particles. The crystallographic structure of both materials was investigated by TEM. SBA-16 images before and after EDTA modification (Figure 3A,C) showed arrays of highly ordered uniform cages demonstrating the 3D cubic structure of SBA-16 that remained unaffected after functionalization. As for SBA-15, Figure 3B,D revealed the highly ordered 2D hexagonal structure (honeycomb structure), which also remained intact after modification.

XRD patterns are shown in Figure 4. The peaks in SBA-16 diffractograms corresponding to the (110), (211), and (220) planes, which are indexed at $2\theta = 0.8$, 1.1, and 1.8, respectively, are characteristics of the cubic body-centered structure (Im3m) [37]. The three diffraction peaks in the SBA-15 pattern, indexed at (100), (110), and (200) planes, are characteristic of the two-dimensional hexagonal symmetry (P6mm) [38]. The three obvious characteristic peaks for SBA-15 are at $2\theta = 0.9$, 1.7, and 2.2, referring to the (100), (110), and (200) planes, respectively. After EDTA modification, there was a slight pattern shift due to the decrease in pore size, but the symmetrical structure was conserved. For SBA-16-EDTA, the (110) plane shifted to a lower diffraction angle (110) compared with unmodified mesoporous silica SBA-16. Such a shift is due to the decrease in lattice parameters a due to pore filling with amino and EDTA functional groups.

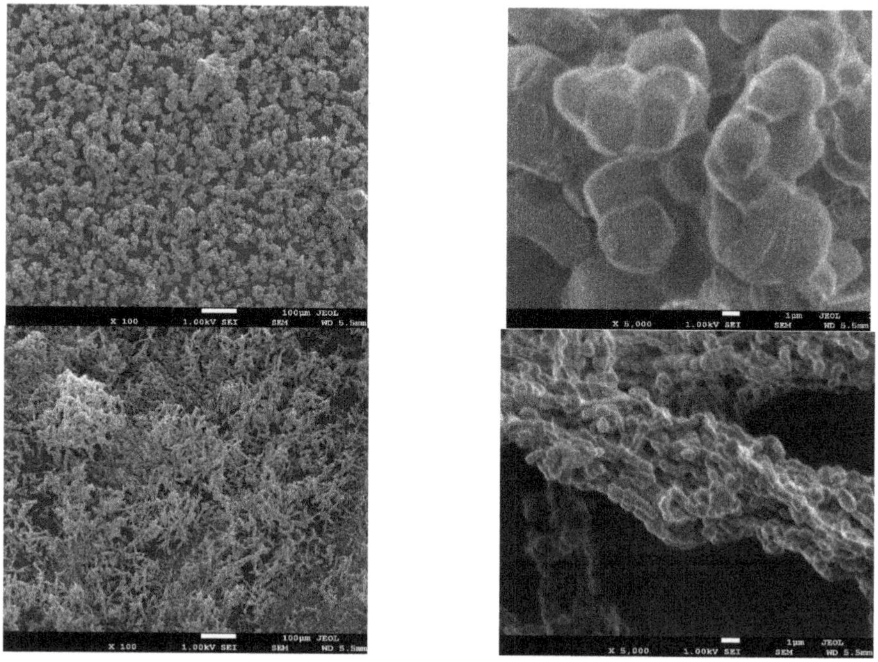

Figure 2. SEM images of of SBA-16 (**above**) and SBA-15 (**below**).

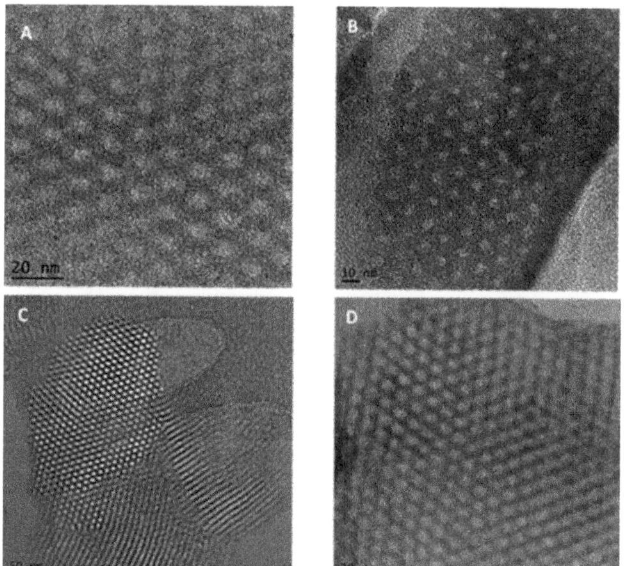

Figure 3. TEM images of SBA-16 (**above**) and SBA-15 (**below**) before (**A**,**C**) and after (**B**,**D**) EDTA modification.

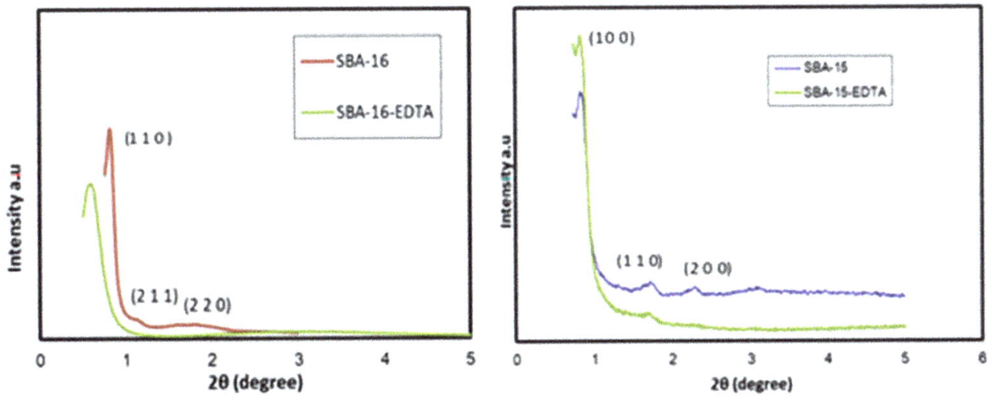

Figure 4. XRD patterns of SBA-16 and SBA-15 before and after modification.

Figure 5 illustrates the nitrogen adsorption–desorption isotherms, and the textural properties and pore size are included in Table 1. According to the IUPAC classification, classical type IV isotherms were made of both materials. SBA-15 exhibited an H1 hysteresis loop, affirming the presence of well-defined and cylindrical mesopores. On the other hand, SBA-16 exhibited an H2 hysteresis loop, which is characteristic of a cage-like mesoporous structure with narrower entrances than the cage itself [39]. For SBA-15, the capillary condensation occurred at a higher relative pressure than for SBA-16, showing that the mesopores of SBA-15 are larger than those of SBA-16. After modification with functional groups, the size of the mesopores along with the surface area and the mesoporous volume decreased for all of the samples. The large decrease in the surface area after modification is mainly the result of the micropores' blockage by amino propyl groups after modification. As for the shape of the mesopores, no important change was observed after modification and the structure remained intact, which was also proved by XRD and TEM.

Table 1. Textural properties of SBA-16 and SBA-15 before and after modification.

Sample	S_{BET} [a] ($m^2\ g^{-1}$)	Pore Size [b] (nm)	Mesopore Volume [c] ($cm^3\ g^{-1}$)	Micropore Volume ($cm^3\ g^{-1}$)
SBA-16	954.8	4.43	0.485	0.122
SBA-16-NH$_2$	567.8	4.14	0.223	0
SBA-16-EDTA	330.5	3.98	0.189	0
SBA-15	860.3	6.81	0.782	0.104
SBA-15-NH$_2$	422.2	6.01	0.587	0
SBA-15-EDTA	229.03	4.76	0.336	0

[a] S_{BET} is the BET surface area evaluated in the range of relative pressures p/p_o of 0.05–0.2. [b] Pore diameter calculated using the BJH method. [c] Total pore volumes were calculated by converting the amount adsorbed at p/p_o ~0.99.

Thermogravimetric analyses under air were also conducted for the samples before and after modification (Figure 6). The weight loss that occurred below 200 °C is due to the desorption of water. Before modification, the weight losses between 200 and 900 °C are attributed to the silicate networks' dehydroxylation. For the modified samples, significant weight losses were observed between 200 and 900 °C. Aminopropyl groups were thermally degraded between 100 and 550 °C and the decomposition of EDTA occurred in the same temperature range as well.

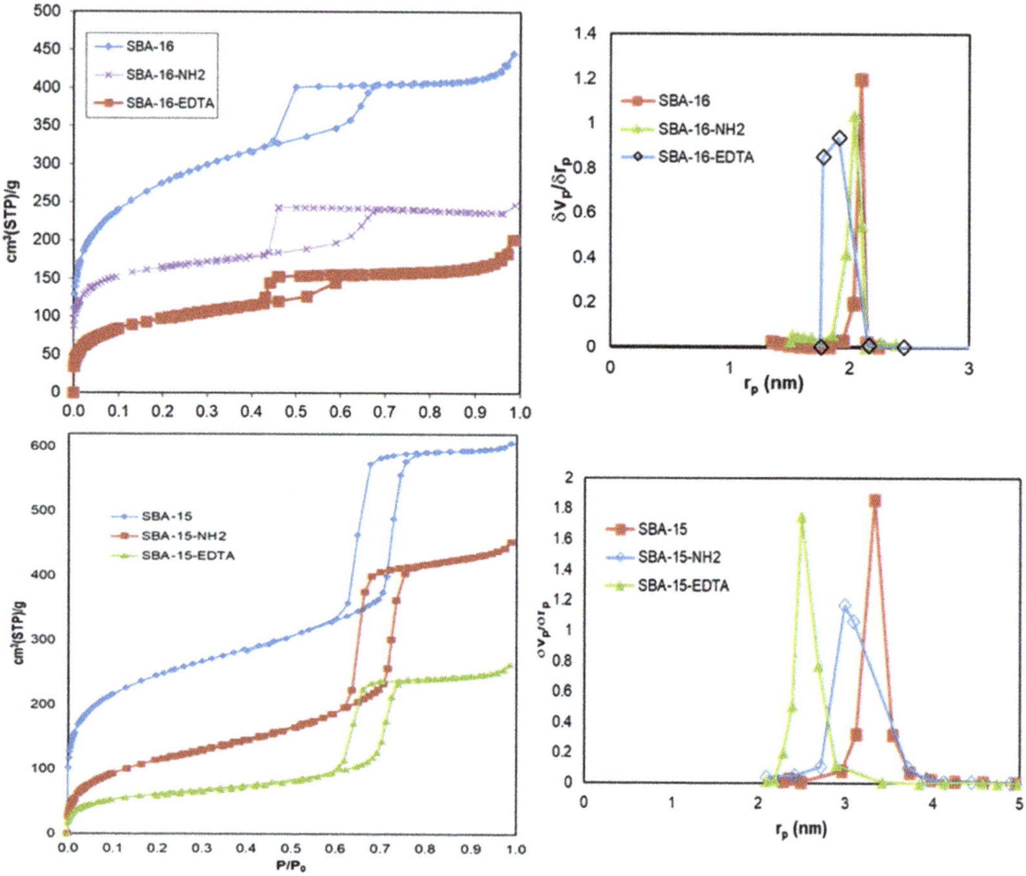

Figure 5. N_2 adsorption–desorption isotherms and pore size distribution of SBA-16 and SBA-15 before and after modification.

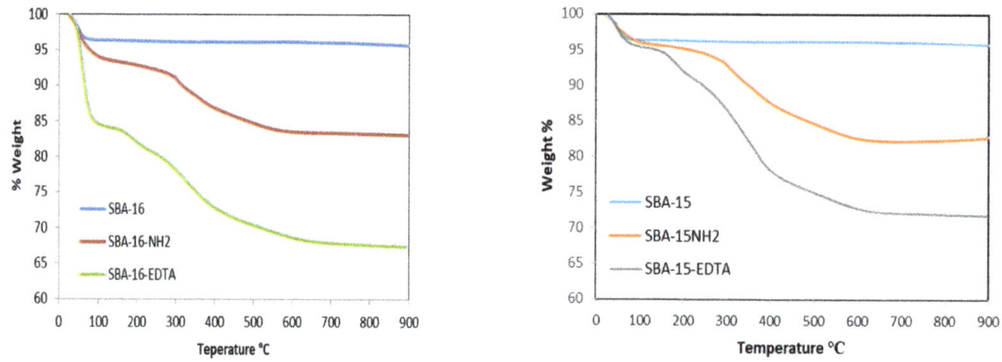

Figure 6. Thermogravimetric curves for SBA-16 and SBA-15 before and after modification.

The infrared spectrum of calcined SBA-15 and SBA-16 (Figure 7) shows typical bands of silanol groups at 3500–3750 cm^{-1} [40]. After modification with aminopropyl groups, the intensity of these bands decreased, while the bands characteristic of aminopropyl groups appeared. These new bands are attributed to the symmetric and asymmetric stretching of CH_2 groups (v_{as} (CH_2) = 2933 cm^{-1}, v_s (CH_2) = 2876 cm^{-1}), as well as NH_2 vibration (v_{as} = 3372 cm^{-1}, v_s = 3300 cm^{-1}) [41]. The band at 1594 cm^{-1} corresponds to NH_2 bending. The anchoring of EDTA on amino groups resulted in the disappearance of NH_2 stretching vibration bands at 3372 and 3300 cm^{-1}. Moreover, the C-O asymmetrical carboxylate stretching vibration was observed at 1675 cm^{-1}. The band at 1744 cm^{-1} was attributed to the stretching vibration of the carboxylic group [42].

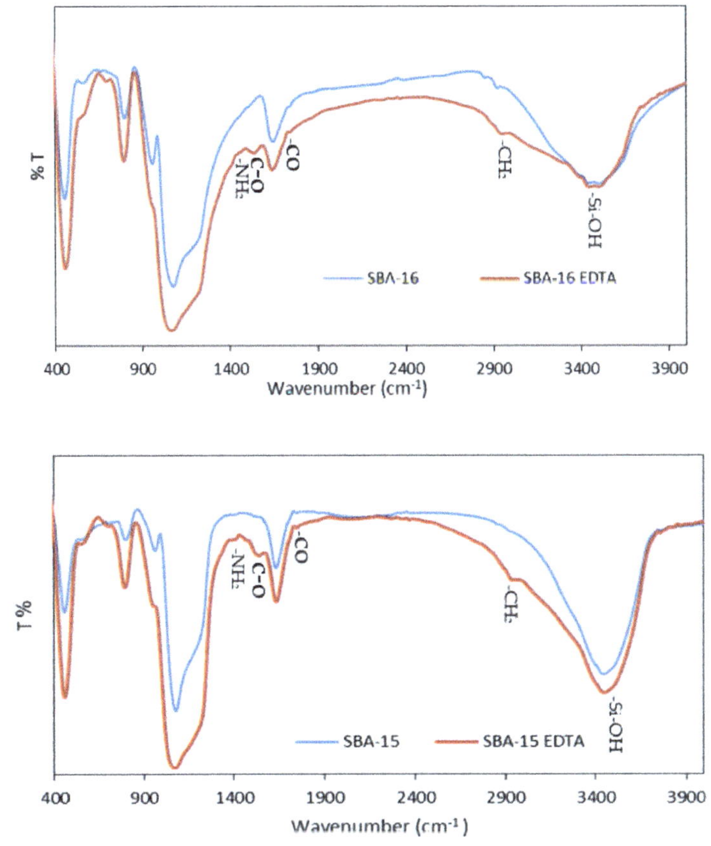

Figure 7. FTIR spectra for SBA-16 and SBA-15 before and after modification.

Figure 8 represents the FTIR spectrum of zinc-loaded SBA-16-EDTA. After zinc adsorption, the band at 1594 cm^{-1} that corresponds to -NH_2 bending decreased, indicating the adsorption of zinc ions on -NH_2 groups that did not react with EDTA. Moreover, the bands at 1675 cm^{-1} and 1744 cm^{-1}, which correspond to the asymmetrical stretching vibration of the carboxylate C-O and at -CO stretching vibration of carboxylic group, respectively, decreased after zinc adsorption owing to the chelation of zinc ions by EDTA. These results prove the adsorption of zinc ions on SBA-16-EDTA and further prove the heterogeneity of the modified mesoporous silica surface.

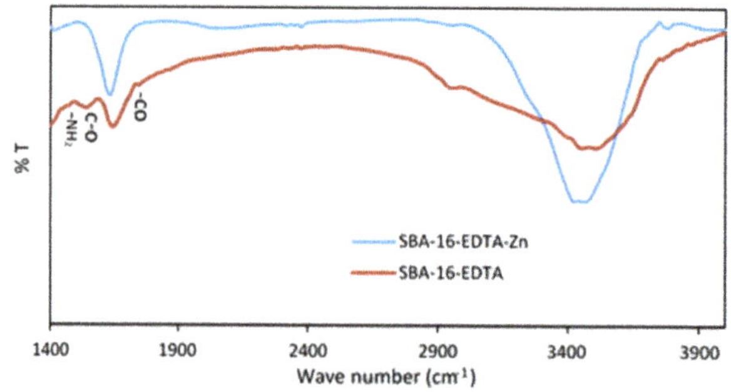

Figure 8. FTIR spectra for SBA-16-EDTA after zinc adsorption.

3.2. Zinc Adsorption Experiments

3.2.1. Effect of Contact Time and pH

For both SBA-16 and SBA-15, equilibrium was reached quickly (within the first 30 min) and the amount of Zn^{2+} adsorbed was much higher for SBA-16 (Figure 9). The obtained results may be due to the difference in structure between SBA-15 and SBA-16. The latter cage-like structure favors the diffusion of zinc ions.

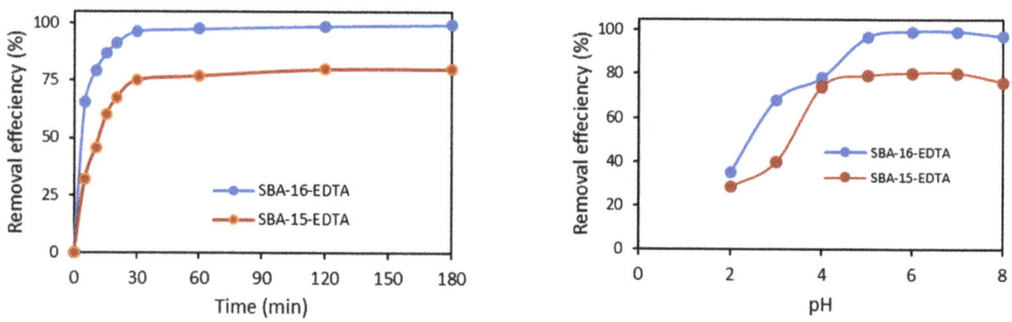

Figure 9. Effect of contact time and pH on Zn^{2+} on EDTA-modified SBA-16 and SBA-15 (at RT and $[Zn^{2+}]_i$ = 30 ppm).

The pH of the solution directly affects zinc ion adsorption because it controls its speciation as well as the adsorbent surface charge. The removal of Zn^{2+} ions on both adsorbents increased as pH increased from 2 to 7 (Figure 9). As the pH increased from 2 to 4, the adsorption efficiency increased from 35% to 78%. At pH 5, the adsorption further increased to 96.6% and reached 99% at pH 6 and 7. Above pH 7, the adsorption capacity slightly decreased (97%). For pH values higher than 3, EDTA molecules have a carboxylate form, thus increasing zinc complexation. Above pH 7, zinc starts to precipitate and form complexes with OH^- ($Zn(OH)_2$).

3.2.2. Adsorption Kinetics

The two kinetic models, pseudo first-order and pseudo-second order, used to calculate the kinetic parameters are expressed in Equations (3) and (4), respectively [43]:

$$\ln (q_e - q_t) = \ln q_e - k_1 t \tag{3}$$

$$\frac{t}{q_t} = \frac{1}{k_2 q_e^2} + \frac{t}{q_e} \tag{4}$$

where q_t and q_e are the quantity of metal ions adsorbed (mg g^{-1}) at time t (min) and at equilibrium, respectively. k_1 (min^{-1}) and k_2 (g mg^{-1} min^{-1}) are the pseudo-first- and pseudo-second-order rate constants.

The theoretical qe values obtained from the pseudo-second-order kinetic model were very close to the experimental ones (Table 2). The obtained results indicated that zinc ion adsorption on both adsorbents followed the pseudo-second-order kinetic model (Figure 10). This model is based on sorption equilibrium capacity and presumes that the sorption capacity is proportional to the number of active sites occupied on the sorbent [44]. This suggests that the adsorption rate mainly depends on the active adsorption site content on the adsorbent surface, and the rate-limiting step is chemisorption involving valence forces through sharing or exchange of electrons between specific sites on adsorbent and metal ions [45]. The chemical interaction between EDTA and metal ions is correlated and in accordance with the kinetic results obtained.

Table 2. Comparison of the first- and the second-order kinetic models for Zn^{2+} adsorption.

		First-Order Kinetic Model			Second-Order Kinetic Model		
	q_e exp (mg g^{-1})	k_1 (min^{-1})	q_e cal (mg g^{-1})	R^2	q_e cal (mg g^{-1})	k_2 (g mg min^{-1})	R^2
SBA-16-EDTA	29.9	0.031	5.7	0.826	30.3	0.014	0.999
SBA-15-EDTA	24	0.102	15.83	0.992	24.7	0.008	0.998

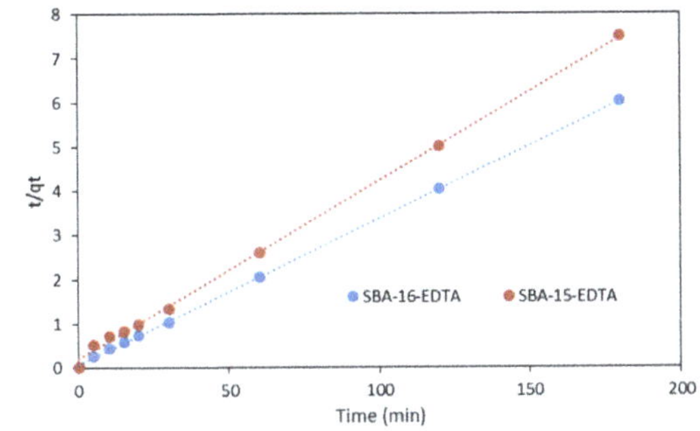

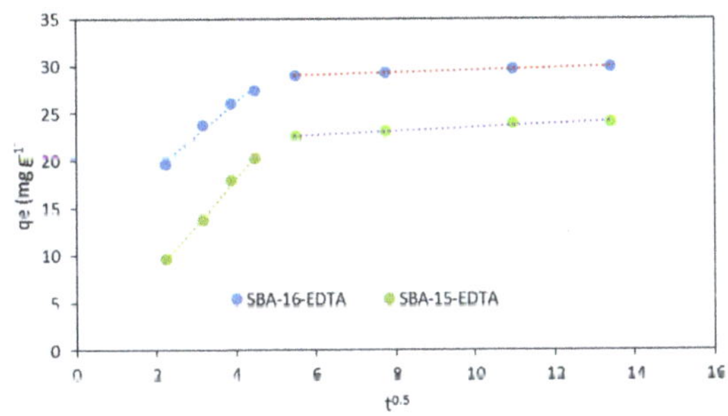

Figure 10. Pseudo-second-order model kinetic model (**above**) and intra-particle diffusion model (**below**) for Zn^{2+} adsorption (pH = 6 at RT).

To determine if the intra-particle diffusion is a rate-limiting step in the zinc adsorption on both adsorbents (SBA-15 and SBA-16), the intra-particle diffusion model proposed by

Weber and Morris [46] was used to analyze the kinetic results. This model is expressed as follows:

$$q_t = K_{id}t^{1/2} + C \tag{5}$$

where k_{id} is the rate constant of intra-particle diffusion (mg g^{-1} $min^{-1/2}$) and C is the intercept (mg g^{-1}). High C values propose that external diffusion has a greater role than the rate-limiting step, because the C value is related to the boundary layer thickness [47]. A plot of the zinc amount adsorbed (q_t) versus $t^{0.5}$ should be linear, and if the line passes through the origin, then intra-particle diffusion is the only rate-controlling step [48]. The obtained results are illustrated in Figure 9 and the parameters are displayed in Table 3. The plots present two linear parts indicating that two steps have occurred. The first sharp part corresponds to the external surface adsorption, while the second part represents the gradual adsorption step, such that the intra-particle diffusion is rate-limiting [49]. As the plot q_t versus $t^{0.5}$ was not a straight line passing through the origin, the process of adsorption is not controlled only by the intra-particle diffusion where film diffusion might have an effect on the kinetics.

Table 3. Parameters of the intra-particle diffusion model.

	k_{id1} (mg $g^{-1}min^{-1/2}$)	C_1 (mg g^{-1})	R^2	k_{id2} (mg $g^{-1}min^{-1/2}$)	C_2 (mg g^{-1})	R^2
SBA-16-EDTA	3.5	12.2	0.981	0.114	28.4	0.991
SBA-15-EDTA	4.8	-	0.994	0.196	21.5	0.925

3.2.3. Adsorption Isotherms

The adsorption behavior for both adsorbents was analyzed by Langmuir and Freundlich isotherm models in order to model the amount of solute adsorbed per unit of adsorbent, q_e, as a function of equilibrium concentration in the bulk solution, C_e, at a constant temperature. The Langmuir and Freundlich, models, are expressed in Equations (6) and (7), respectively:

$$\frac{C_e}{q_e} = \frac{1}{K_L q_{max}} + \frac{C_e}{q_{max}} \tag{6}$$

$$\log q_e = \log K_f + \frac{1}{n}\log C_e \tag{7}$$

where C_e and qmax denote the metal concentration (mg L^{-1}) at the equilibrium state and the adsorption capacity (mg g^{-1}), respectively. The value of n is the inverse of the heterogeneity factor of the adsorption process. Meanwhile, K_L and K_f are the Langmuir (L mg^{-1}) and Freundlich (mg g^{-1}) constants related to the mean free energy of adsorption, respectively [50,51].

The adsorption isotherms of the experimental data are shown in Figure 11 and the parameters of these two models are shown in Table 4. From the linear regression correlation coefficient R^2, it can be deduced that the equilibrium data could be well described by the Freundlish isotherm, so the adsorption is reversible in a heterogeneous system that is not limited to the formation of monolayers [52]. The values of n were all between 1 and 10, indicating that the adsorption performance of zinc ions on both adsorbents was favorable under the studied conditions [53], so both adsorbents can be considered efficient for zinc metal ion removal, with the preference for SBA-16. Moreover, the Freundlich expression is an exponential equation and, therefore, assumes that, as the metal ions' concentration increases, their concentration on the adsorbent surface also increases, indicating a non-ideal adsorption, not limited to monolayer formation.

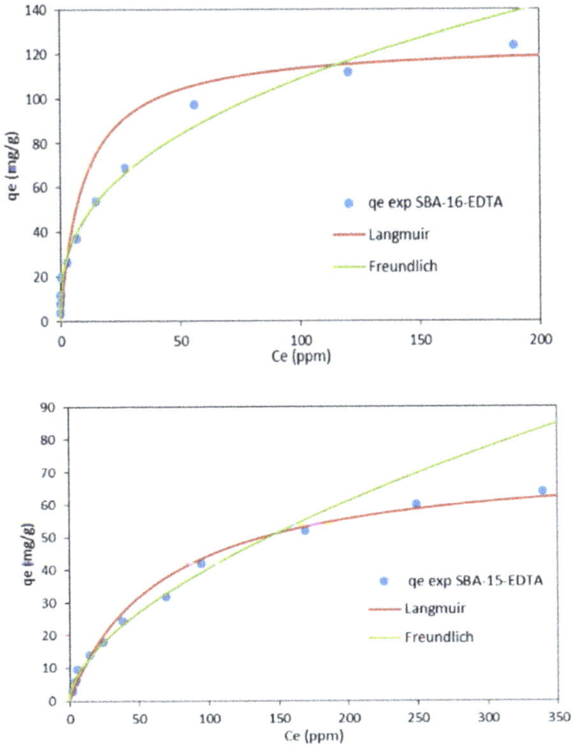

Figure 11. Langmuir and Freundlich adsorption isotherms for Zn^{2+} adsorption on EDTA-modified SBA-16 and SBA-15.

Table 4. Langmuir and Freundlich models for Zn^{2+} adsorption on SBA-16 and SBA-15.

	Langmuir Model			Freundlich Model		
	$q^{exp}{}_{max}$ (mg g^{-1})	K_L (L mg^{-1})	R^2	n	K_f (mg g^{-1})	R^2
SBA-16	184.1	0.1	0.964	3.32	25.12	0.974
SBA-15	107	0.03	0.682	1.28	4	0.985

Contrary to the Langmuir isotherm model, which is commonly used for monolayer adsorption, where most of the adsorption sites have equal affinities toward the asorbate, the Freundlich isotherm model is used to describe a heterogeneous chemisorption process in which the surface is not energetically uniform [54]. Isotherms of this form have been observed for a wide range of heterogeneous surfaces, including activated carbon, silica, clays, metals, and polymers. In the case of SBA-16-EDTA and SBA-15-EDTA, the obtained results showed that the data fitted Freundlich, as previously mentioned, and this is mainly because of the heterogeneous surface of both adsorbents, taking into consideration the presence of some unfunctionalized silanol groups, amino groups along the EDTA fixed on the majority of sites. Such heterogeneity will directly govern the adsorption process on both adsorbents and, as a result, the obtained isotherm as well. It can be concluded that Freundlich fits well over the entire concentration range; however, there is an obvious deviation at higher concentrations. So, in general, the Freundlich isotherm will be a more accurate approximation at lower concentrations [55]. Moreover, the Freundlich expression is an exponential equation and, therefore, assumes that, as the metal ions' concentration increases, their concentration on the adsorbent surface also increases, indicating a non-ideal adsorption, not limited to monolayer formation [56]. Concerning the chemisorption process, the presence of functional groups can be used as evidence in proposing the adsorption mechanism. The Zn-chemisorption mechanism onto EDTA-modified SBA-16 and SBA-15 can be proposed as follows: Zn^{2+} ions bind to carboxylate groups of EDTA via ionic forces with carboxylic oxygen atoms. These oxygen atoms exhibit a negative charge in their structure as a result of the dissociation of carboxylic groups. The negatively charged oxygen atom in carboxylate anions will coordinate with zinc cations, resulting in the formation of metal–carboxylate complexes (COO–Zn) on the adsorbents' surface [57,58].

4. Adsorbents' Regeneration

Batch desorption experiments and the desorption efficiencies were compared for both adsorbents. In two beakers, each containing 40 mL of 1 M HCl and 20 mg of SBA-15-EDTA or SBA-16-EDTA at RT, the mixtures were stirred at 300 rpm for 2 h. Then, the mixtures were filtered and dried in order to be used again. Five regeneration cycles were performed. Figure 12 shows that, after the fifth regeneration cycle, both adsorbents conserved about 85% of their removal efficiency.

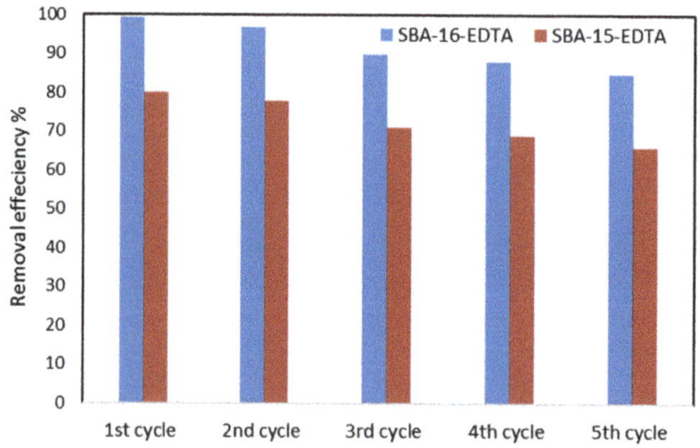

Figure 12. Regeneration and reuse of SBA-16-EDTA and SBA-15-EDTA.

5. Conclusions

In this study, two adsorbents were prepared by EDTA immobilization on SBA-15 and SBA-16 mesoporous silica. The ordered mesostructures of the obtained hybrid organic/inorganic materials were well preserved after modification. The modified adsorbents were characterized by scanning electron microscopy (SEM) and transmission electron microscopy (TEM), X-ray diffraction (XRD), nitrogen (N_2) adsorption–desorption analysis, Fourier transform infrared spectroscopy (FT-IR), and thermogravimetric analysis. The effects of contact time and pH on zinc adsorption on both materials were studied. The adsorption process was fast, indicating the high affinity of adsorbents to chelating zinc ions. The modified SBA-16 showed higher efficiency for eliminating Zn^{2+} compared with SBA-15 owing to its favorable structure characteristics (pore structure). The kinetic data well fitted the pseudo-second-order model, where the rate adsorption process depends on the exchange kinetics between the ligand and zinc ions. The effect of intra-particle diffusion was also investigated. Equilibrium data were also fitted by Freundlich isotherm. Both adsorbents can be regenerated using HCl solution and reused for up to five cycles.

Author Contributions: Z.E. performed the experiments; Z.E., I.B.-G. and Y.P. analyzed the data; Z.E. wrote the manuscript; Y.P. managed the study and the unit research in Poitiers. All authors have read and agreed to the published version of the manuscript.

Funding: This research was funded by Poitiers University grant number 1.

Institutional Review Board Statement: Not applicable.

Informed Consent Statement: Not applicable.

Data Availability Statement: Authors ensure that this manuscript is ethically sound and meet industry-recognized standards that are reflected in MDPI policies.

Acknowledgments: The authors are thankful to the (IC2MP) 'Institut de Chimie des Milieux et Matériaux de Poitiers-UMR 7285', Poitiers University, PRES for the financial support of this work.

Conflicts of Interest: The authors declare no conflict of interest.

References

1. Asberry, H.B.; Kuo, C.; Gung, C.; Conte, E.D.; Suen, S. Characterization of water bamboo husk biosorbents and their application in heavy metals ion trapping. *Microchem. J.* **2014**, *113*, 59–63. [CrossRef]
2. Zou, Y.; Wang, X.; Khan, A.; Wang, P.; Liu, Y.; Alsaedi, A.; Hayat, T.; Wang, X. Environmental remediation and application of nanoscale zerovalent iron and its composites for the removal of heavy metal ions: A review. *Environ. Sci. Technol.* **2016**, *50*, 7290–7304. [CrossRef] [PubMed]
3. Verlicchi, P.; Grillini, V. Surface water and groundwater quality in South Africa and mozambique—Analysis of the Most critical pollutants for drinking purposes and challenges in water treatment selection. *Water* **2020**, *12*, 305. [CrossRef]
4. Taseidifar, M.; Makavipour, F.; Pashley, R.M.; Rahman, A.F.M.M. Removal of heavy metal ions from water using ion flotation. *Environ. Technol. Innov.* **2017**, *8*, 182–190. [CrossRef]
5. García-Niño, W.R.; Pedraza-Chaverrí, J. Protective effect of curcumin against heavy metals-induced liver damage. *Food Chem. Toxicol.* **2014**, *69*, 182–201. [CrossRef]
6. Wang, W.N.; Liang, H.; Wang, A. Effect of pH and Zn^{2+} on subcultured muscle cells from Macrobrachium nipponense. *Methods Cell Sci.* **2000**, *22*, 277–284. [CrossRef]
7. Wadige, C.P.M.M.; Taylor, A.M.; Maher, W.A.; Krikowa, F. Bioavailability and toxicity of zinc from contaminated freshwater sediments: Linking exposure-dose–response relationships of the freshwater bivalve Hyridella australis to zinc-spiked sediments. *Aquat. Toxicol.* **2017**, *156*, 179–190. [CrossRef]
8. Yang, X.; Wan, Y.; Zheng, Y.; He, F.; Yu, Z.; Huang, J.; Wang, H.; Ok, Y.S.; Jiang, Y.; Gao, B. Surface functional groups of carbon-based adsorbents and their roles in the removal of heavy metals from aqueous solutions: A critical review. *Chem. Eng. J.* **2019**, *366*, 608–621. [CrossRef]
9. Bazrafshan, E.; Mohammadi, L.; Moghaddam, A.A.; Mahvi, A.H. Heavy metals removal from aqueous environments by electrocoagulation process—A systematic review. *J. Environ. Health Sci. Eng.* **2015**, *13*, 74. [CrossRef]
10. Zhou, G.; Luo, J.; Liu, C.; Chu, L.; Crittenden, J. Efficient heavy metal removal from industrial melting effluent using fixed-bed process based on porous hydrogel adsorbents. *Water Res.* **2018**, *131*, 246–254. [CrossRef]
11. Qasem, N.A.; Mohammed, R.H.; Lawal, D.U. Removal of heavy metal ions from wastewater: A comprehensive and critical review. *Npj Clean Water* **2021**, *4*, 36. [CrossRef]
12. Aguado, J.; Arsuaga, J.; Arencibia, A.; Lindo, M. Gas_cn, V Aqueous heavy metals removal by adsorption on amine-functionalized mesoporous silica. *J. Hazard. Mater.* **2009**, *163*, 213–221. [CrossRef]
13. Babel, S.; Kurniawan, T.A. Low-cost adsorbents for heavy metals uptake from contaminated water: A review. *J. Hazard. Mater.* **2003**, *97*, 219–243. [CrossRef] [PubMed]
14. Repo, E.; Kurniawan, T.A.; Warchol, J.K.; Sillanpää, M.E. Removal of Co(II) and Ni(II) ions from contaminated water using silica gel functionalized with EDTA and/or DTPA as chelating agents. *J. Hazard. Mater.* **2009**, *171*, 1071–1080. [CrossRef] [PubMed]
15. Onundi, Y.B.; Mamum, A.A.; Al-Khatib, M.F.; Ahmed, Y.M. Adsorption of copper, nickel and lead ions from synthetic semiconductor industrial wastewater by palm shell activated carbon. *Int. J. Environ. Sci. Technol.* **2010**, *7*, 751–758. [CrossRef]
16. Swarnalatha, K.; Ayoob, S. Adsorption Studies on Coir Pith for Heavy Metal Removal. *Int. J. Sustain. Eng.* **2016**, *9*, 259–265. [CrossRef]
17. Demiral, İ.; Samdan, C.; Demiral, H. Enrichment of the surface functional groups of activated carbon by modification. *Method Surf. Interfaces* **2021**, *22*, 100873.
18. Marciniak, M.; Goscianska, J.; Frankowski, M.; Pietrzak, R. Optimal synthesis of oxidized mesoporous carbons for the adsorption of heavy metal ions. *J. Mol. Liq.* **2019**, *276*, 630–637. [CrossRef]
19. Upadhyay, U.; Sreedhar, I.; Singh, S.A.; Patel, C.M.; Anitha, K.L. Recent advances in heavy metal removal by chitosan based adsorbents. *Carbohydr. Polym.* **2021**, *251*, 117000. [CrossRef]
20. Zhang, T.; Wang, W.; Zhao, Y.; Bai, H.; Wen, T.; Kang, S.; Song, G.; Song, S.; Komarneni, S. Removal of heavy metals and dyes by clay-based adsorbents: From natural clays to 1D and 2D nano-composites. *Chem. Eng.* **2020**, *420*, 127574. [CrossRef]
21. Zhou, W.; Deng, J.; Qin, Z.; Huang, R.; Wang, Y.; Tong, S. Construction of MoS_2 nanoarrays and MoO_3 nanobelts: Two efficient adsorbents for removal of Pb(II), Au(III) and Methylene Blue. *J. Environ. Sci.* **2022**, *111*, 38–50. [CrossRef] [PubMed]

22. Barik, B.; Kumar, A.; Nayak, P.S.; Achary, L.S.K.; Rout, L.; Dash, P. Ionic liquid assisted mesoporous silica-graphene oxide nanocomposite synthesis and its application for removal of heavy metal ions from water. *Mater. Chem. Phys.* **2020**, *239*, 122028. [CrossRef]
23. Qiu, B.; Xie, L.; Zeng, J.; Liu, T.; Yan, M.; Zhou, S.; Liang, Q.; Tang, J.; Liang, K.; Kong, B. Interfacially super-assembled asymmetric and H2O2 sensitive multilayer-sandwich magnetic mesoporous silica nanomotors for detecting and removing heavy metal ions. *Adv. Funct. Mater.* **2021**, *31*, 2010694. [CrossRef]
24. Mannaa, M.A.; Altass, H.M.; Salama, R.S. MCM-41 grafted with citric acid: The role of carboxylic groups in enhancing the synthesis of xanthenes and removal of heavy metal ions. *Environ. Nanotechnol. Monit. Manag.* **2021**, *15*, 100410. [CrossRef]
25. Betiha, M.A.; Moustafa, Y.M.; El-Shahat, M.F.; Rafik, E. Polyvinylpyrrolidone-Aminopropyl-SBA-15 schiff Base hybrid for efficient removal of divalent heavy metal cations from wastewater. *J. Hazard. Mater.* **2020**, *397*, 122675. [CrossRef] [PubMed]
26. Gupta, R.; Gupta, S.K.; Pathak, D.D. Selective adsorption of toxic heavy metal ions using guanine-functionalized mesoporous silica [SBA-16-g] from aqueous solution. *Microporous Mesoporous Mater.* **2019**, *288*, 109577. [CrossRef]
27. Kaewprachum, W.; Wongsakulphasatch, S.; Kiatkittipong, W.; Striolo, A.; Cheng, C.K.; Assabumrungrat, S. SDS modified mesoporous silica MCM-41 for the adsorption of Cu^{2+}, Cd^{2+}, Zn^{2+} from aqueous systems. *J. Environ. Chem. Eng.* **2020**, *8*, 102920. [CrossRef]
28. Hussain, A.A.; Nazir, S.; Irshad, R.; Tahir, K.; Raza, M.; Khan, Z.U.H.; Khan, A.U. Synthesis of functionalized mesoporous Ni-SBA-16 decorated with MgO nanoparticles for Cr (VI) adsorption and an effective catalyst for hydrodechlorination of chlorobenzene. *Mater. Res. Bull.* **2021**, *133*, 111059. [CrossRef]
29. Zolfaghari, G.; Zanganeh Asadabadi, A. Synthesis of nanoporous silica SBA-16 and its modification with NH_2 ligands and metal for removal of chromium in industry and environment. *J. Environ. Sci. Technol.* **2021**. [CrossRef]
30. An, F.; Gao, B.; Dai, X.; Wang, M.; Wang, X. Efficient removal of heavy metal ions from aqueous solution using salicylic acid type chelate adsorbent. *J. Hazard. Mater.* **2011**, *192*, 956–962. [CrossRef]
31. Northcott, K.; Kokusen, H.; Komatsu, Y.; Stevens, G. Synthesis and Surface Modification of Mesoporous Silicate SBA-15 for the Adsorption of Metal Ions. *Sep. Sci. Technol.* **2006**, *41*, 1829–1840. [CrossRef]
32. Zhao, D.; Huo, Q.; Feng, J.; Chmelka, B.; Stucky, G. Nonionic triblock and star diblock copolymer and oligomeric surfactant syntheses of highly ordered, hydrothermally stable, mesoporous silica structures. *J. Am. Chem. Soc.* **1998**, *120*, 6024–6036. [CrossRef]
33. Grudzien, R.M.; Grabicka, B.E.; Jaroniec, M. Adsorption studies of thermal stability of SBA-16 mesoporous silicas. *Appl. Surf. Sci.* **2007**, *253*, 5660–5665. [CrossRef]
34. Ezzeddine, Z.; Batonneau-Gener, I.; Pouilloux, Y.; Hamad, H.; Saad, Z.; Kazpard, V. Divalent Heavy Metals Adsorption onto Different Types of EDTA-Modified Mesoporous Materials: Effectiveness and Complexation Rate. *Microporous Mesoporous Mater.* **2015**, *212*, 125–136. [CrossRef]
35. Ghodbane, I.; Nouri, L.; Hamdaoui, O.; Chiha, M. Kinetic and equilibrium study for the sorption of cadmium(II) ions from aqueous phase by eucalyptus bark. *J. Hazard. Mater.* **2008**, *152*, 148–158. [CrossRef] [PubMed]
36. Nikiforova, T.E.; Kozlov, V.A.; Telegin, F.Y. Chemisorption of copper ions in aqueous acidic solutions by modified chitosan. *Mater. Sci. Eng. B* **2021**, *263*, 114778. [CrossRef]
37. Gupta, R.; Layek, S.; Pathak, D.D. Synthesis and characterization of guanine-functionalized mesoporous silica [SBA-16-G]: A metal-free and recyclable solid base catalyst for synthesis of pyran-annulated heterocyclic compounds. *Res. Chem. Intermed.* **2018**, *45*, 1619–1637. [CrossRef]
38. Zhao, D.; Feng, J.; Huo, Q.; Melosh, N.; Fredrickson, G.H.; Chmelka, B.F.; Stucky, G.D. Triblock Copolymer Syntheses of Mesoporous Silica with Periodic 50 to 300 Angstrom Pores. *Science* **1998**, *279*, 548–555. [CrossRef]
39. Thommes, M.; Kaneko, K.; Neimark, A.V.; Olivier, J.P.; Rodriguez-Reinoso, F.; Rouquerol, J.; Sing, K.S. Physisorption of gases, with special reference to the evaluation of surface area and pore size distribution (IUPAC Technical Report). *Pure Appl. Chem.* **2015**, *87*, 1051–1069. [CrossRef]
40. Gallas, J.P.; Goupil, J.M.; Vimont, A.; Lavalley, J.C.; Gil, B.; Gilson, J.P.; Miserque, O. Quantification of water and silanol species on various silicas by coupling IR spectroscopy and in-situ thermogravimetry. *Langmuir* **2009**, *25*, 5825–5834. [CrossRef]
41. Maria, A.; Zhao, X.; Phys, J. Functionalization of SBA-15 with APTES and Characterization of Functionalized Materials. *J. Phys. Chem. B.* **2003**, *107*, 12650–12657. [CrossRef]
42. Yu, J.; Tong, M.; Sun, X.; Li, B. Enhanced and selective adsorption of Pb^{2+} and Cu^{2+} by EDTAD-modified biomass of baker's yeast. *Bioresour. Technol.* **2008**, *99*, 2588–2593. [CrossRef] [PubMed]
43. Febrianto, J.; Kosasih, A.N.; Sunarso, J.; Ju, Y.H.; Indraswati, N.; Ismadji, S. Equilibrium and kinetic studies in adsorption of heavy metals using biosorbent: A summary of recent studies. *J. Hazard. Mater.* **2009**, *162*, 616–645. [CrossRef] [PubMed]
44. Ho, Y.S.; McKay, G. Pseudo-second order model for sorption processes. *Process Biochem.* **1999**, *34*, 451–465. [CrossRef]
45. Ezzeddine, Z.; Batonneau-Gener, I.; Pouilloux, Y.; Hamad, H.; Saad, Z. The applicability of as synthesized mesoporous carbon CMK-3 as a heavy metals adsorbent: Application to real water samples. *Energy Sources Part A Recovery Util. Environ. Eff.* **2021**, *43*, 1675–1690. [CrossRef]
46. Weber, J.; Morris, J. Kinetics of Adsorption on Carbon from Solution. *J. Sanit. Eng. Div.* **1963**, *89*, 31–60. [CrossRef]
47. Lin, J.; Zhan, Y.; Zhu, Z. Adsorption characteristics of copper (II) ions from aqueous solution onto humic acid immobilized surfactant-modified zeolite. *Colloids Surf. A Physicochem. Eng. Asp.* **2011**, *384*, 9–16. [CrossRef]

48. Ji, F.; Li, C.; Tang, B.; Xu, J.; Lu, G.; Liu, P. Preparationof cellulose acetate/zeolite composite fiber and its adsorption behavior for heavymetal ions in aqueous solution. *Chem. Eng. J.* **2012**, *209*, 325–333. [CrossRef]
49. Annadurai, G.; Krishnan, M. Batch equilibrium adsorption of reactive dye on to natural biopolymer. *Iran. Polym. J.* **1997**, *16*, 169–172.
50. Freundlich, H. Uber die adsorption in losungen. *Z. Phys. Chem.* **1906**, *57*, 387–470. [CrossRef]
51. Langmuir, I. The adsorption of gases on plane surface of glass, mica and platinum. *J. Am. Chem. Soc.* **1918**, *40*, 1361–1368. [CrossRef]
52. Vimonses, V.; Lei, S.; Jin, B.; Chow, C.W.K.; Saint, C. Kinetic study and equilibrium isotherm analysis of Congo Red adsorption by clay materials. *Chem. Eng. J.* **2009**, *148*, 354–364. [CrossRef]
53. Hameed, B.H. Equilibrium and kinetic studies of methyl violet sorption by agricultural waste. *J. Hazard. Mater.* **2008**, *154*, 204–212. [CrossRef] [PubMed]
54. Agboola, O.D.; Benson, N.U. Physisorption and chemisorption mechanisms influencing micro (nano) plastics-organic chemical contaminants interactions: A review. *Front. Environ. Sci.* **2021**, *9*, 167. [CrossRef]
55. Umpleby II, R.J.; Baxter, S.C.; Bode, M.; Berch, J.K., Jr.; Shah, R.N.; Shimizu, K.D. Application of the Freundlich adsorption isotherm in the characterization of molecularly imprinted polymers. *Anal. Chim. Acta* **2001**, *435*, 35–42. [CrossRef]
56. Ezzeddine, Z.; Abdallah, L.; Hamad, H. Heavy Metals Adsorption on Cyanex-272 Modified Activated Carbon: Efficacy and Selectivity. *Univers. J. CarbonRes.* **2023**, *1*, 22–37.
57. Zhu, S.; Khan, M.A.; Kameda, T.; Xu, H.; Wang, F.; Xia, M.; Yoshioka, T. New insights into the capture performance and mechanism of hazardous metals Cr3+ and Cd2+ onto an effective layered double hydroxide based material. *J. Hazard. Mater.* **2022**, *426*, 128062. [CrossRef]
58. Zhu, S.; Chen, Y.; Khan, M.A.; Xu, H.; Wang, F.; Xia, M. In-depth study of heavy metal removal by an etidronic acid-functionalized layered double hydroxide. *ACS Appl. Mater. Interfaces* **2022**, *14*, 7450–7463. [CrossRef]

Disclaimer/Publisher's Note: The statements, opinions and data contained in all publications are solely those of the individual author(s) and contributor(s) and not of MDPI and/or the editor(s). MDPI and/or the editor(s) disclaim responsibility for any injury to people or property resulting from any ideas, methods, instructions or products referred to in the content.

toxics

MDPI

Article

Comparison of ZnO Nanoparticles Prepared by Precipitation and Combustion for UV and Sunlight-Driven Photocatalytic Degradation of Methylene Blue

Sucheewan Krobthong, Tipawan Rungsawang and Sutthipoj Wongrerkdee *

Department of Physical and Material Sciences, Faculty of Liberal Arts and Science, Kasetsart University Kamphaeng Saen Campus, Kamphaeng Saen, Nakhon Pathom 73140, Thailand
* Correspondence: sutthipoj.s@ku.ac.th

Abstract: ZnO nanoparticles (NPs) were comparatively synthesized via precipitation and combustion techniques. The ZnO NPs synthesized via precipitation and combustion exhibited similar polycrystalline hexagonal wurtzite structures. The large crystal sizes of ZnO NPs were obtained from the ZnO precipitation in comparison with those from the ZnO combustion, while the particle sizes were in the same range. The functional analysis implied that the ZnO structures had surface defects. Moreover, absorbance measurement showed the same absorbance range in ultraviolet light. In the photocatalytic degradation of methylene blue, ZnO precipitation exhibited higher degradation performance than ZnO combustion. This was attributed to the larger crystal sizes of ZnO NPs, which provided an enduring carrier movement at semiconductor surfaces and reduced electron-hole recombination. Thus, the crystallinity of ZnO NPs can be considered an important factor in photocatalytic activity. Furthermore, precipitation is an interesting synthesizing method for preparing ZnO NPs with large crystal sizes.

Keywords: ZnO; precipitation; combustion; photocatalyst; methylene blue

1. Introduction

Synthetic dyes are a concern in wastewater because they can have negative impacts on the environment and human health [1]. Some dyes contain harmful chemicals that can be harmful to fish and other aquatic organisms when they are present at high concentrations in the water. Many dyes are resistant to traditional wastewater treatment processes [2], which means that they can pass through treatment plants and appear in rivers, lakes, and other bodies of water. Thus, wastewater containing synthetic dyes is a consideration for degrading and reducing toxic chemicals. The photocatalytic degradation of contaminating dyes is considered an interesting process for wastewater treatment because it can break dye molecules into smaller and less toxic compounds [3]. Photocatalytic degradation is a simple method of wastewater treatment that can be scaled up, such as in its applications in agricultural farms, industrial estates, and ranches [4]. The photocatalytic degradation of dyes requires semiconductor materials and processing in the presence of light [5]. There are several materials that can be employed as photocatalysts, including titanium dioxide (TiO_2), tin oxide (SnO_2), and zinc oxide (ZnO) [6]. These popular materials are often studied for their photocatalytic properties due to their wide bandgap energy (E_g) and high surface area resulting in efficient light absorption and enhanced catalytic activity, respectively. Among these, ZnO is a well-known multifunctional-semiconductor material that has been widely investigated for its usage as a photocatalyst [7] because of its good photocatalytic activity, low-cost material, and non-toxicity [8–10]. Importantly, suitable synthesis methods have been studied, such as microwave synthesis, laser-ablation, hydrothermal, sol-gel, precipitation, and combustion [11–15].

Precipitation and combustion are interesting methods because they are low-cost and rapid processes. ZnO was synthesized for precipitation using a variety of zinc sources, including $Zn(CH_3COO)_2 \cdot 2H_2O$ and $Zn(NO_3)_2 \cdot 6H_2O$ [16–18]. These ZnOs had hexagonal wurtzite structures with a variety of morphologies. Absorbance spectra cover both Ultraviolet (UV) and visible regions, suggesting that they can be activated by UV or visible light. The ZnO nanostructures were demonstrated to be interesting photocatalysts, and they showed acceptable photocatalytic degradation of methylene blue (MB). In studies of combustion methods, synthesized ZnO NPs have been presented for use as photocatalysts [19–21]. Although the ZnO NPs were prepared at different temperatures, there were similar hexagonal wurtzites. However, surface impurities or surface defects were obtained more frequently than those of ZnO prepared by precipitation. For the photocatalytic test, it presented a potential for degrading dyes and several toxic molecules.

The mentioned literature has proven that ZnO is an interesting photocatalyst to degrade dyes and pesticides under UV or sunlight exposure. However, ZnO NPs synthesized by precipitation and combustion have not been directly compared in terms of photocatalyst application. Thus, the present work aims to directly compare the structural and optical properties and photocatalytic activity of ZnO NPs prepared by precipitation and combustion. These methods were selected to investigate the effect of the preparation method on the characteristics of ZnO NPs and their photocatalytic activity. It is well-known that small particle sizes play an important role in offering large surfaces for photocatalytic activity [22]. The combustion method is aimed at producing small ZnO NPs to observe their effect on photocatalytic activity. On the other hand, the high crystallinity of ZnO NPs synthesized by the precipitation method is considered to offer long durations of carrier transport at the ZnO surface, which potentially increases activity duration at the surface. Therefore, the methods were comparatively investigated.

2. Materials and Methods

ZnO NPs were synthesized via the precipitation process by mixing a 0.2 M aqueous solution of $Zn(NO_3)_2 \cdot 6H_2O$ with a 0.4 M aqueous solution of NH_4HCO_3 in DI water. The starting aqueous solutions were separately prepared, stirred, and heated at 70 °C for 1 h. Then, the NH_4HCO_3 solution was added dropwise into the $Zn(NO_3)_2 \cdot 6H_2O$ solution at a dropping rate of 2 mL/min [23]. The mixed solution was heated and stirred for an additional hour and then cooled down. Afterward, the obtained white product was filtered, heated at 70 °C for 3 h, and ground. The obtained product was then calcined at 550 °C for 6 h and ground again to obtain "ZnO precipitation". For the combustion process, a 1 M aqueous solution of $Zn(NO_3)_2 \cdot 6H_2O$ was prepared by dissolving the compound in 20 mL of ethylene glycol (EG) and stirring for 1 h at room temperature [11]. The resulting aqueous solution was then heated to 350 °C for 20 min on a hotplate. Meanwhile, cleaned glass slides were heated to the same temperature on another hotplate to serve as substrates for the combustion process. The $Zn(NO_3)_2 \cdot 6H_2O$ solution was then dropped onto the heated substrates and allowed to undergo combustion for 20 min, followed by cooling down to room temperature. The products of the combustion were scratched from the substrates. The resulting material was identified as "ZnO combustion" after being ground. The precipitation and combustion methods are briefly illustrated in Figure 1.

The obtained ZnO NPs, ZnO precipitation, and ZnO combustion, were characterized by a Scanning Electron Microscope (SEM, JSM-6610 LV, JOEL, Bangkok, Thailand) for morphological observation. Microstructural analysis was conducted using Transmission Electron Microscopy (TEM, TECNAI G^2 20 S-TWIN, FEI, Nakhon Ratchasima, Thailand). Particle size distribution in DI water was analyzed using Dynamic Light Scattering (DLS, Beckman Coulter, Delsa Nano C, Bangkok, Thailand). XRD (D8 Advance, Bruker, Khon Kaen, Thailand), which was conducted to analyze the crystalline structures of the ZnO NPs. Selected Area Electron Diffraction (SAED) was performed to analyze the crystalline structure. Raman spectrometer (Thermo Scientific, DXR SmartRaman, Bangkok, Thailand) was used to investigate the vibrational nature of the ZnO NPs. Fourier Transform Infrared

Spectrophotometry (FTIR, Spectrum Two, PerkinElmer, Nakhon Pathom, Thailand) was employed to examine the surface functional groups present on the NPs. A UV–VIS–NIR spectrophotometer (Hitachi, UH1450, Bangkok, Thailand) was utilized to monitor the absorbance spectra, and a spectrofluorometer was used to observe the fluorescence spectra.

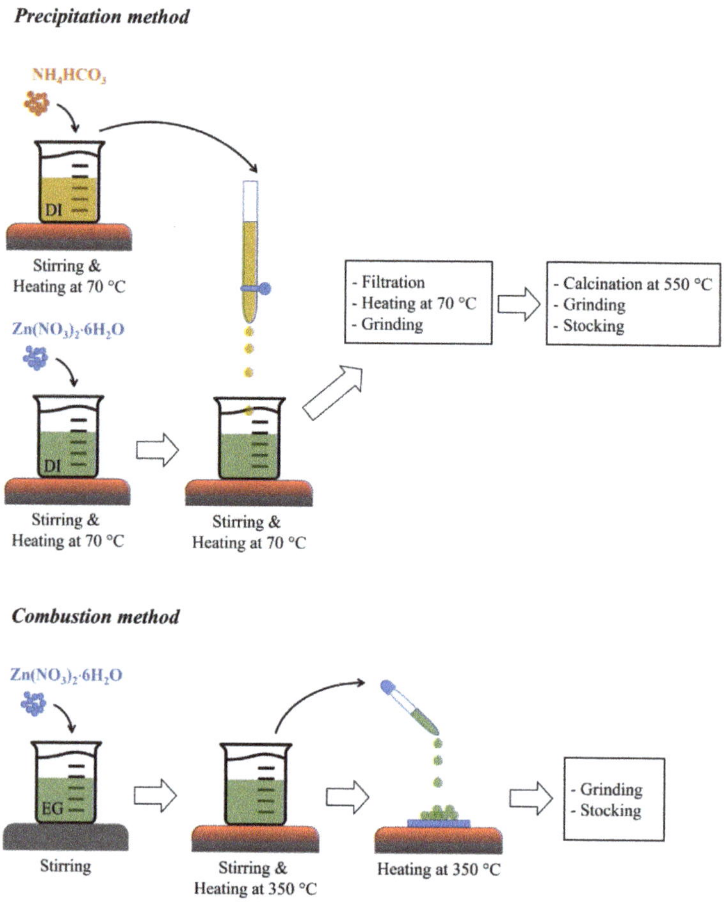

Figure 1. The illustration of ZnO preparation via precipitation and combustion methods.

In the photocatalytic activity test, the produced ZnO NPs via precipitation and combustion were utilized as photocatalysts for MB degradation. The starting MB solution was prepared by dissolving 5 mg of MB in 1 L of DI water (pH ~ 6.93). It was then stirred in the dark for 1 h. The solution was divided into two beakers, each containing 100 mL. Each beaker was loaded with a portion of 0.1 g of the ZnO NPs and stirred for 10 min in the dark. The mixtures were then placed in the photocatalytic activity station and left for 30 min in the dark to reach adsorption-desorption equilibrium. Afterward, the MB solution containing ZnO NPs was irradiated with UV light (Philips TUV 30W, 3 lamps, Nakhon Pathom, Thailand) to activate the photocatalytic activity. The solutions were sampled to measure absorbance using a UV–VIS spectrophotometer (BioMate 160, Nakhon Pathom, Thailand) and evaluate the photocatalytic activity. Furthermore, the photocatalytic degradation of MB was tested in natural sunlight to demonstrate its effectiveness and usability.

3. Results and Discussion

The morphology and microstructure of prepared ZnO NPs are presented in Figures 2 and 3, respectively. The large NP sizes appeared for ZnO precipitation, while the aggregations of numerous NPs were observed for the ZnO combustion. This refers to the fact that ZnO precipitation has a lower NP boundary than ZnO combustion. The large NP sizes of ZnO precipitation can provide long paths for carrier (electron-hole pairs) movement at the ZnO surface [4]. Moreover, the large NP sizes also reduce recombination due to carrier-boundary collision. The different structures of ZnO NPs prepared by precipitation and combustion in this study might be due to differences in reaction process, temperature, and reaction time. DLS measurements, shown in Figure 4, were used to investigate the particle sizes of ZnO NPs prepared by precipitation and combustion in an indirect manner. The particle sizes of the ZnO precipitation were found in terms of hydrodynamic diameters to be 700–1400 nm, while the diameters of the ZnO combustion were 700–3000 nm. The average sizes of ZnO precipitation and ZnO combustion were 1022 and 1462 nm, respectively. It is noted that the DLS results present larger particle sizes of ZnO NPs than SEM and TEM images. This could be an agglomeration of NPs in DI water, which results in a larger hydrodynamic size detection. The large distribution for ZnO combustion could be attributed to NP impurities, which cause low stability in DI water and result in stronger agglomeration. Although the ranges are different, there is a similar nature to the aggregates of ZnO NPs. The aggregation effect might reduce the surface area of ZnO NPs, which would slow down the photocatalytic activity.

Figure 2. SEM morphological images of ZnO prepared by (**a**) precipitation and (**b**) combustion.

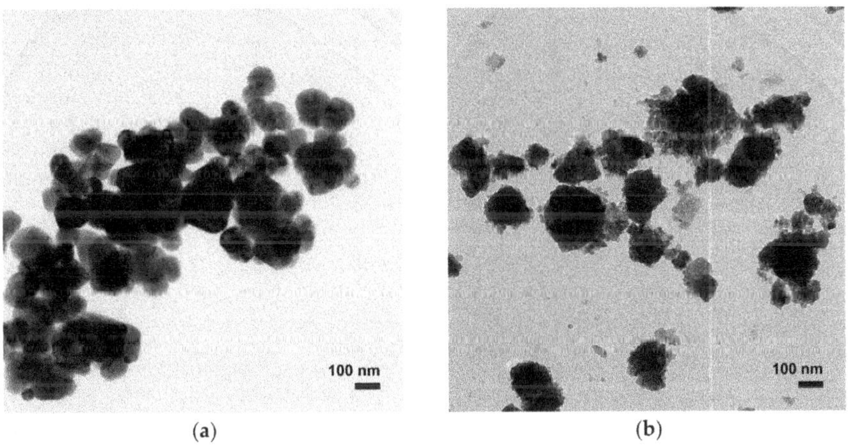

Figure 3. TEM microstructural images of ZnO prepared by (**a**) precipitation and (**b**) combustion.

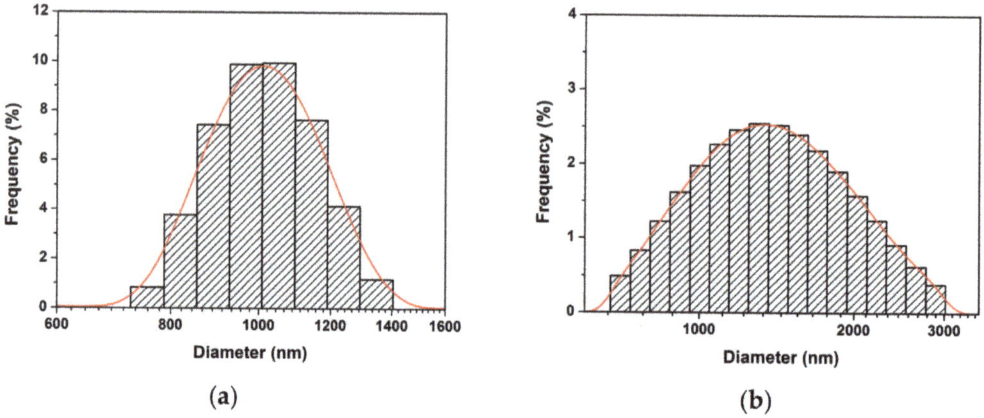

Figure 4. Hydrodynamic diameter of aggregates of ZnO NPs prepared by (**a**) precipitation and (**b**) combustion.

The crystal structure determined by XRD patterns is presented in Figure 5. The XRD patterns are well-matched with standard JCPDS no. 36-1451 [24,25], which indicates the hexagonal wurtzite structure of ZnO. The appeared additional peak at 2θ~23.5° for ZnO combustion can refer to the orthorhombic phase of $Zn(OH)_2$ (JCPDS no. 01-071-2215) caused by the incomplete reaction [11]. The XRD peaks for ZnO precipitation were narrower than those for ZnO combustion, indicating that the crystal size of ZnO precipitation is larger than that of the ZnO combustion. The crystal size (D) was calculated using the Scherrer equation (Equation (1)) [26–28]:

$$D = k\lambda/\beta\cos\theta \quad (1)$$

where k and λ are the shape factor of XRD measurement (0.89) and the wavelength of the X-ray source (1.5406 Å), respectively. The β and θ are the full width at half maximum (FWHM) and the diffraction angle for each plane, respectively. The crystal sizes calculated from the three major peaks of (100), (002), and (101) planes were 45.5 ± 10.0 and 10.4 ± 2.6 nm for ZnO precipitation and ZnO combustion, respectively, in agreement with the SEM results. The larger crystal size for ZnO precipitation reflects better crystallization in comparison with ZnO combustion because the precipitation process provides a longer chemical reaction time for crystallization. On the other hand, the reaction can be limited by the short duration of the combustion process. The SAED patterns (Figure 6) exhibited similar ring-like patterns for ZnO precipitation and ZnO combustion, indicating polycrystalline structures due to random phase growth [23].

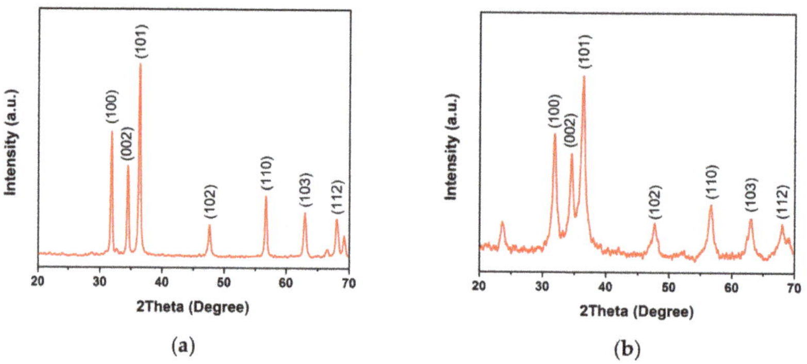

Figure 5. XRD patterns of ZnO NPs prepared by (**a**) precipitation and (**b**) combustion.

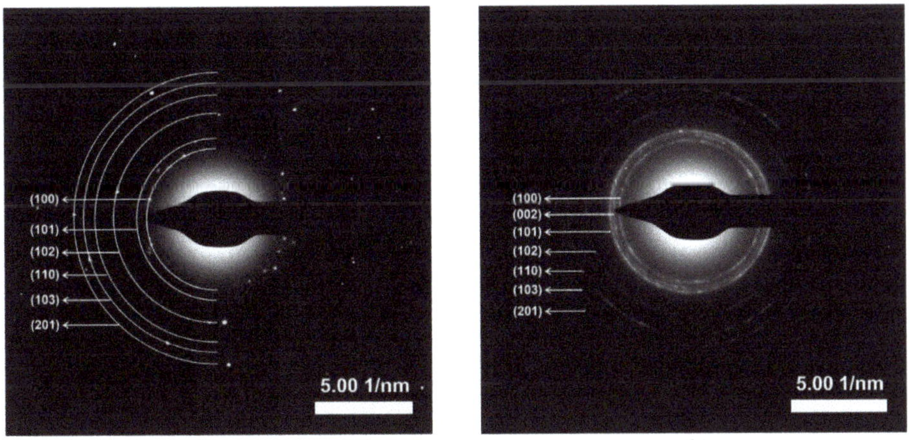

(a) (b)

Figure 6. SAED ring-like patterns of ZnO NPs prepared by (**a**) precipitation and (**b**) combustion.

Figure 7a depicts the Raman shift for vibrational analysis of ZnO NPs. The ZnO NPs revealed dominant peaks at 100 and 440 cm^{-1}, which assigned the E_2 (low) and E_2 (high) modes, respectively [29,30]. The E_2 (low) vibration represented the zinc crystal lattice in ZnO, while the E_2 (high) vibration represented the oxygen vibration. The peaks at 334, 384, 542, and 585 cm^{-1} were assigned to the E_2 (high)–E_2 (low), A_1 (TO), A_1 (LO), and E_1 (LO) modes, respectively [31]. The peak at 223 could be assigned to 2TA or $2E_2$ (low) modes. [30,32]. These peaks indicate the disordered structures for both ZnO NPs prepared by precipitation and combustion. Moreover, the E_2(low) and E_2(high) modes show low intensity compared with the ZnO precipitation. These behaviors suggest the presence of oxygen defects in the ZnO combustion. In Figure 7b, FTIR spectra in the range of 4000–450 cm^{-1} were analyzed to investigate functional groups of the ZnO NPs. The peaks at 490 cm^{-1} contributed to the ZnO structure, due to the Zn-O vibration [33]. The peaks at 819 and 1647 cm^{-1} corresponded to C=O stretching [11]. The peaks at 1320 and 1435 cm^{-1} corresponded to C-O stretching and C-H blending, respectively [11]. The peaks at 3391 and 3497 cm^{-1} were assigned to the O-H stretching. The appearance of these peaks implies the presence of residuals.

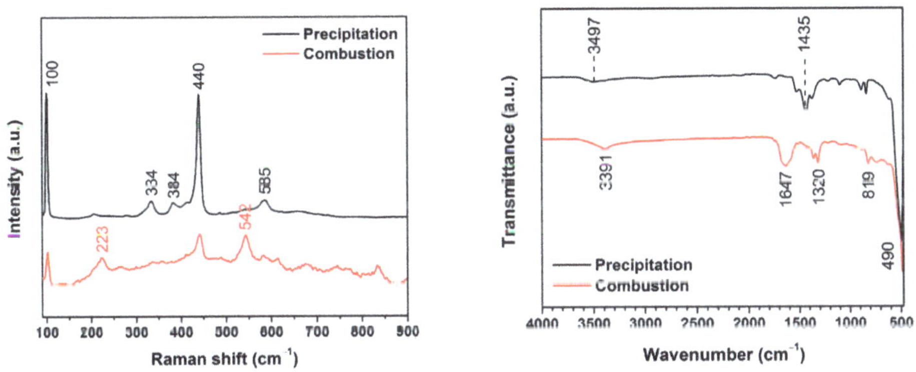

(a) (b)

Figure 7. Vibrational analysis of ZnO NPs prepared by precipitation and combustion: (**a**) Raman shift and (**b**) FTIR spectra.

The optical absorbance was measured (Figure 8a) to survey the energy absorption of the ZnO NPs. The band gap energy (E_g) of ZnO was estimated using Equation (2) [11,17]:

$$(\alpha h\nu)^2 = A(h\nu - E_g) \tag{2}$$

where α, h, ν, and A are the absorption coefficient, Planck's constant, frequency, and constant, respectively. The $(\alpha h\nu)^2$ vs. hν curve, Figure 8b, was used to calculate the E_g value. E_g values of the ZnO precipitation and the ZnO combustion are 3.18 and 3.30 eV, respectively. The difference in E_g for ZnO precipitation and ZnO combustion might be due to structural disorders, chemical defects, and impurities. The E_g values correspond to the range of the UV region, which indicates that these ZnO NPs can be utilized as photocatalysts under UV irradiation. Note that the lower E_g for the ZnO precipitation refers to the fact that it can be excited with lower incident energy in comparison with the ZnO combustion. The fluorescence spectra of ZnO NPs excited at 325 nm are shown in Figure 8c. The ZnO precipitation and the ZnO combustion present a similar emission peak in the UV region at 366 nm. The behavior might be electron-hole recombination, referring to the nature of ZnO. However, surface defects are detected according to the minor broad peaks at around 387–407 nm.

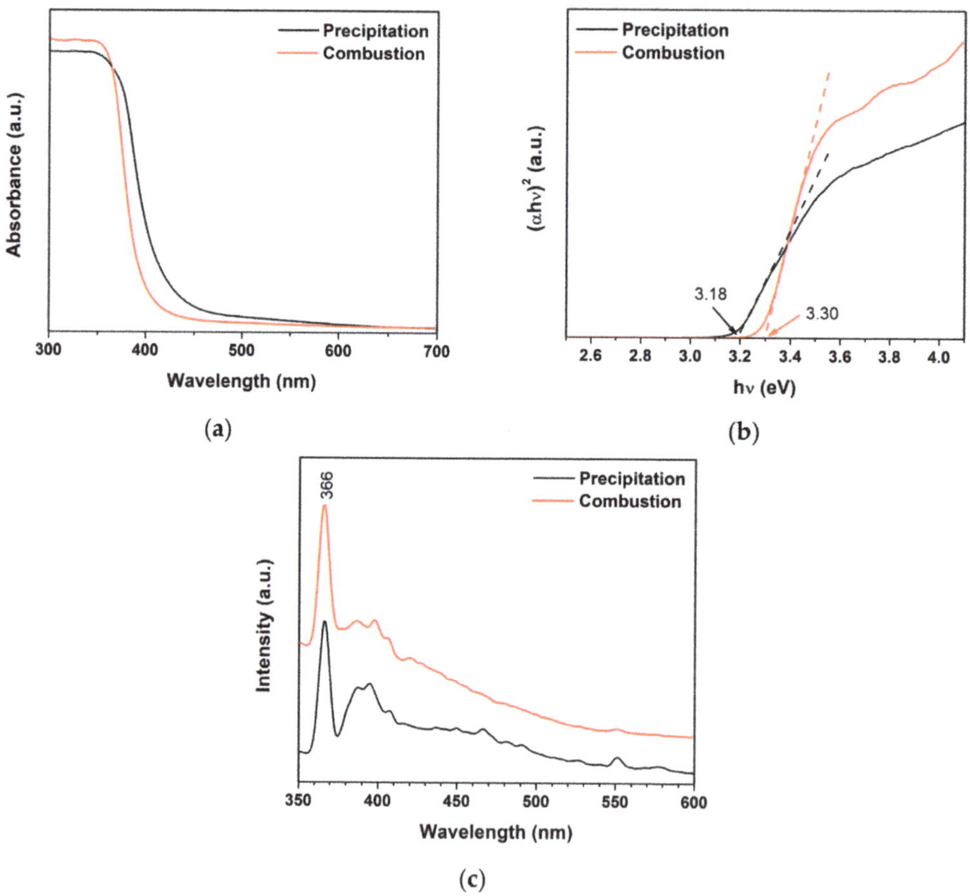

Figure 8. (**a**) absorbance, (**b**) band gap energy analysis, and (**c**) fluorescence spectra of ZnO NPs.

The ZnO precipitation and the ZnO combustion were preliminarily employed to degrade MB present in water under UV irradiation for investigating the photocatalytic degradation performance. The measured absorbance of the MB solution is shown in Figure 9. The absorbance decreased as the UV irradiation time increased, which indicates more degraded MB molecules for the longer irradiation periods. It is seen that the absorbance of the MB solution containing the ZnO precipitation decreased more in comparison with the solution containing the ZnO combustion. As a note on the degradation of MB under UV light without photocatalyst (blank), the absorbance of MB showed a non-significant change, indicating low degradation activity in the absence of a photocatalyst.

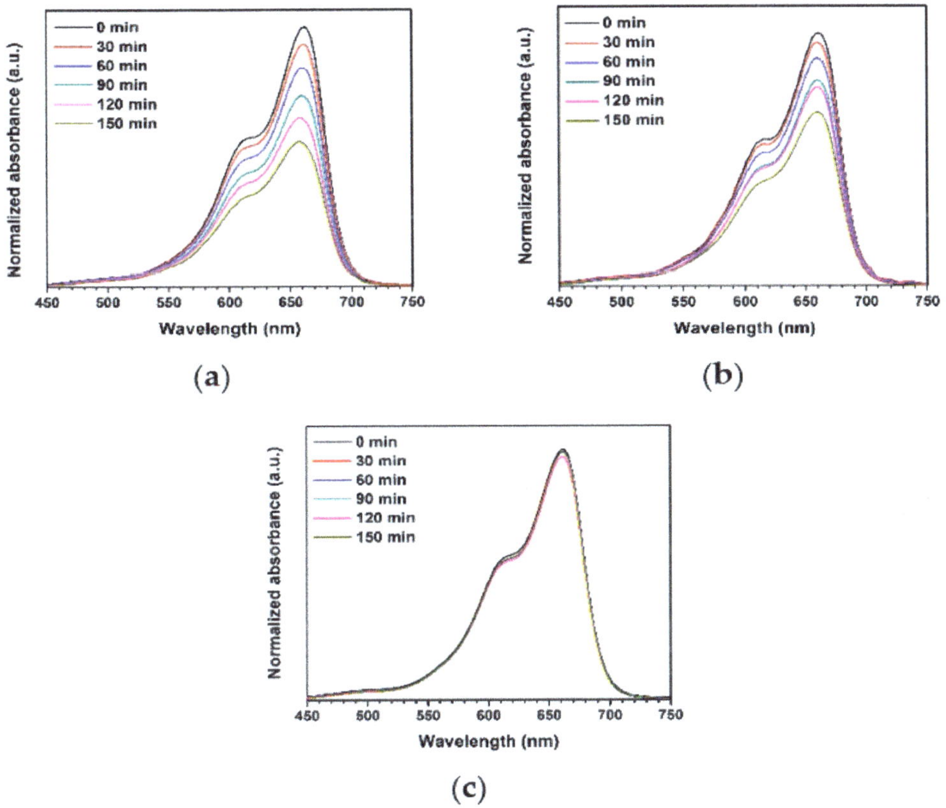

Figure 9. Normalized absorbance of MB solution under UV light using different photocatalysts: (a) ZnO precipitation, (b) ZnO combustion, and (c) blank.

As shown in Figure 10a, the interval to the initial absorbance (A/A_0) ratio of MB was plotted to observe the change in MB after photocatalytic activity. When compared to ZnO combustion, the A/A_0 of MB had a significantly lower value for ZnO precipitation. For a better comparison, the apparent degradation rate constant (k_{app}) was analyzed using Equation (3) [9,28]:

$$\ln (A_0/A) = k_{app}t \quad (3)$$

where t is the irradiation time. The k_{app} was estimated from the slope of Figure 10b and listed in Table 1. The ZnO precipitation showed higher k_{app} values for the photocatalytic degradation of MB than for that of the ZnO combustion around 1.6 times. The result corresponds with the larger crystal sizes of the ZnO precipitation compared to those of the ZnO combustion.

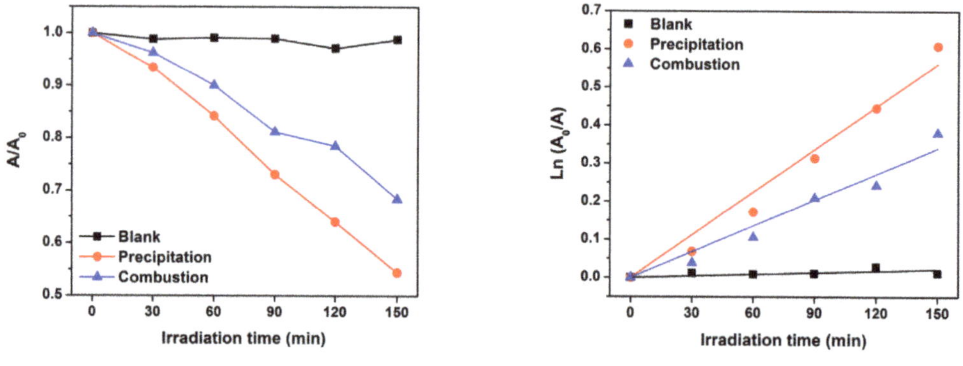

Figure 10. Photocatalytic performance of MB under UV light: (**a**) plot of A/A_0 analysis, and (**b**) apparent degradation rate constant analysis.

Table 1. Apparent degradation rate constant for photocatalytic degradation of MB under UV and natural sunlight irradiation using ZnO NPs.

Photocatalyst	k_{app} (10^{-3} min^{-1})	
	UV	Sunlight
Blank	0.14	3.13
ZnO precipitation	3.74	35.31
ZnO combustion	2.27	8.56

The application of ZnO photocatalysts was further investigated to degrade MB under natural sunlight irradiation (average sunlight intensity of 933.3 ± 70.8 W/m^2) in order to demonstrate their effectiveness and usability. The absorbance of MB was monitored and analyzed as shown in Figure 11. The A/A_0 ratio rapidly decreased for the ZnO precipitation, suggesting that the ZnO precipitation and the ZnO combustion have a positive effect on the photocatalytic degradation of MB under sunlight activation. It should be noted that in the blank condition, MB decreased with increasing irradiation time. This indicates stronger photolysis of MB under sunlight compared with UV light. Moreover, the k_{app} value reveals that the ZnO precipitation had the highest k_{app} value (Table 1). The result exhibits a similar trend to those under UV light. Thus, it can be confirmed that the ZnO precipitation is more effective in degrading MB compared to the ZnO combustion. It can be interpreted that the ZnO prepared by precipitation has a greater photocatalytic performance in comparison with the ZnO prepared by combustion. This effect can be due to the large crystal sizes of the ZnO NPs. It is well-known that large crystal sizes can provide a long lifetime for carrier movement at semiconductor surfaces and reduce electron-hole recombination [4]. On the other hand, short lifetime and recombination can occur for the small crystal size because carriers may hit many boundaries during the movement leading to low carriers for photocatalytic activity. Thus, it can be concluded that the crystallinity of the ZnO NPs prepared by precipitation has an impact on the photocatalytic activity for degrading MB molecules in water. Furthermore, precipitation is an interesting synthesizing method for preparing ZnO NPs with large crystal sizes in comparison with the combustion method. This study demonstrates the acceptable performance of ZnO photocatalysts for wastewater treatment [34–36]. On the other hand, the photocatalytic activity of ZnO NPs prepared by combustion exhibited lower performance in comparison with precipitation. This behavior might be caused by the presence of defects such as structural disorders, chemical defects, and impurities in the ZnO structure. This leads to the inhibition of carrier movement at ZnO surfaces, which reduces the photocatalytic activity.

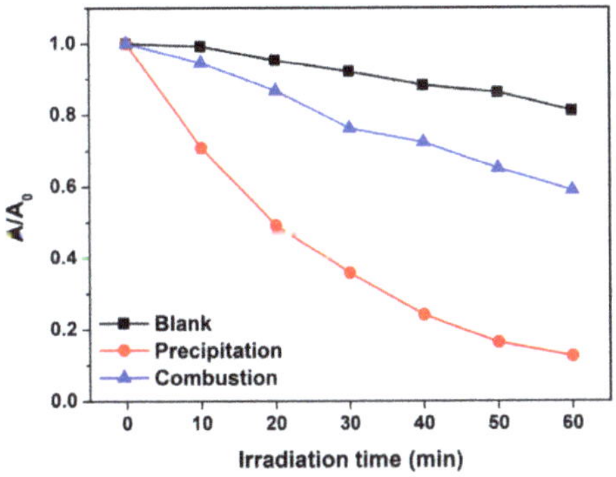

(a)

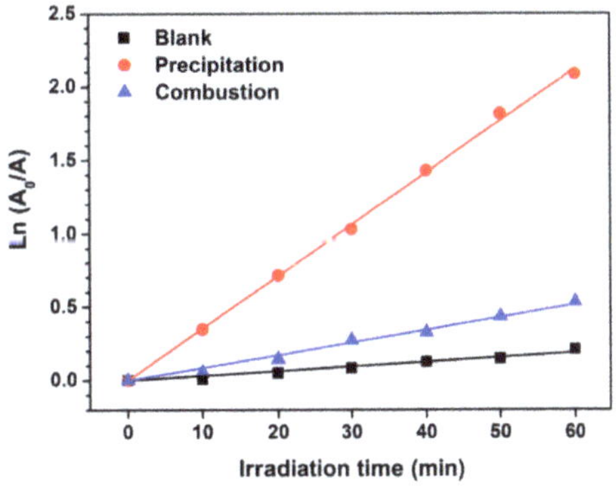

(b)

Figure 11. Photocatalytic performance of MB degradation under natural sunlight: (**a**) plot of A/A_0 analysis and (**b**) apparent degradation rate constant analysis.

4. Conclusions

ZnO NPs were comparatively synthesized via precipitation and combustion methods. The ZnO NPs prepared by precipitation and combustion exhibited similar polycrystalline hexagonal wurtzite structures. Hydrodynamic sizes of ZnO precipitation and ZnO combustion were 1022 and 1462 nm, respectively. The large crystal sizes of ZnO NPs were obtained for the ZnO precipitation in comparison with those of the ZnO combustion. However, the functional analysis implied the presence of ZnO structures with surface defects. Moreover, the absorbance measurement showed the same absorbance range in the UV region. For photocatalytic degradation of MB, the ZnO precipitation showed higher degradation efficiencies and apparent degradation rates constant than the ZnO combustion. This was due to the large crystal sizes of the ZnO NPs, which provided an enduring carrier movement at semiconductor surfaces and reduced electron-hole recombination. Thus, the crystallinity of the ZnO NPs can be considered an important factor in photocatalytic activity. Furthermore, precipitation is an interesting synthesizing method for preparing ZnO NPs with large crystal sizes.

Author Contributions: Conceptualization, S.K., T.R. and S.W.; methodology, T.R. and S.W.; validation, T.R. and S.W.; formal analysis, S.K., T.R. and S.W.; investigation, S.K. and S.W.; resources, T.R. and S.W.; data curation, S.K. and S.W.; writing—original draft preparation, S.K. and S.W.; writing—review and editing, S.W.; supervision, S.W.; project administration, S.W.; funding acquisition, S.W. All authors have read and agreed to the published version of the manuscript.

Funding: This research was partially funded by the Kasetsart University Research and Development Institute (KURDI), grant no. FF(S-KU)39.66.

Institutional Review Board Statement: Not applicable.

Informed Consent Statement: Not applicable.

Data Availability Statement: Not applicable.

Acknowledgments: The authors would like to acknowledge the Research Promotion and Technology Transfer Center (RPTTC), Faculty of Liberal Arts and Science, Kasetsart University Kamphaeng Saen Campus for FTIR measurement.

Conflicts of Interest: The authors declare no conflict of interest.

References

1. Naz, T.; Rasheed, A.; Ajmal, S.; Sarwar, N.; Rasheed, T.; Baig, M.M.; Zafar, M.S.; Kang, D.J.; Dastgeer, G. A facile approach to synthesize ZnO-decorated titanium carbide nanoarchitectures to boost up the photodegradation performance. *Ceram. Int.* **2021**, *47*, 33454–33462. [CrossRef]
2. Sanakousar, F.M.; Vidyasagar, C.C.; Jimenez-Perez, V.M.; Prakash, K. Recent progress on visible-light-driven metal and non-metal doped ZnO nanostructures for photocatalytic degradation of organic pollutants. *Mater. Sci. Semicond.* **2022**, *140*, 106390. [CrossRef]
3. Hosseini, A.; Karimi, H.; Foroughi, J.; Sabzehmeidani, M.M.; Ghaedi, M. Heterogeneous photoelectro-Fenton using ZnO and TiO_2 thin film as photocatalyst for photocatalytic degradation Malachite Green, Heterogeneous photoelectro-Fenton using ZnO and TiO_2 thin film as photocatalyst for photocatalytic degradation Malachite Green. *Appl. Surf. Sci.* **2021**, *6*, 100126. [CrossRef]
4. Phophayu, S.; Pimpang, P.; Wongrerkdee, S.; Sujinnapram, S.; Wongrerkdee, S. Modified graphene quantum dots-zinc oxide nanocomposites for photocatalytic degradation of organic dyes and commercial herbicide. *J. Reinf. Plast. Compos.* **2020**, *39*, 81–94. [CrossRef]
5. Biswal, H.J.; Srivastava, T.; Vundavilli, P.R.; Gupta, A. Facile fabrication of hydrophobic ZnO nanostructured nickel microtubes through pulse electrodeposition as promising photocatalyst for wastewater remediation. *J. Manuf. Process.* **2022**, *75*, 538–551. [CrossRef]
6. Ishchenko, O.; Rogé, V.; Lamblin, G.; Lenoble, D.; Fechete, I. TiO_2, ZnO, and SnO_2-based metal oxides for photocatalytic applications: Principles and development. *C. R. Chim.* **2021**, *24*, 103–124. [CrossRef]
7. Noman, M.T.; Amor, N.; Petru, M. Synthesis and applications of ZnO nanostructures (ZONSs): A review. *Crit. Rev. Solid State Mater. Sci.* **2022**, *47*, 99–141. [CrossRef]
8. Manojkumar, P.; Pranav, S.; Premchand, C.; Lokeshkumar, E.; Rameshbabu, N. Development of surface-modified galvanised steel as an immobilised photocatalyst for textile wastewater treatment. *Inorg. Chem. Commun.* **2022**, *145*, 110056. [CrossRef]
9. Lincho, J.; Zaleska-Medynska, A.; Martins, R.C.; Gomes, J. Nanostructured photocatalysts for the abatement of contaminants by photocatalysis and photocatalytic ozonation: An overview. *Sci. Total Environ.* **2022**, *837*, 155776. [CrossRef]
10. Le Pivert, M.; Martin, N.; Leprince-Wang, Y. Hydrothermally Grown ZnO Nanostructures for Water Purification via Photocatalysis. *Crystals* **2022**, *12*, 308. [CrossRef]
11. Wongrerkdee, S.; Wongrerkdee, S.; Boonruang, C.; Sujinnapram, S. Enhanced Photocatalytic Degradation of Methylene Blue Using Ti-Doped ZnO Nanoparticles Synthesized by Rapid Combustion. *Toxics* **2023**, *11*, 33. [CrossRef]
12. El-Gendy, A.O.; Nawaf, K.T.; Ahmed, E.; Samir, A.; Hamblin, M.R.; Hassan, M.; Mohamed, T. Preparation of zinc oxide nanoparticles using laser-ablation technique: Retinal epithelial cell (ARPE-19) biocompatibility and antimicrobial activity when activated with femtosecond laser. *J. Photochem. Photobiol. B Biol.* **2022**, *234*, 112540. [CrossRef]
13. Khudiar, S.S.; Nayef, U.M.; Mutlak, F.A.H. Preparation and characterization of ZnO nanoparticles via laser ablation for sensing NO_2 gas. *Optik* **2021**, *246*, 167762. [CrossRef]
14. Wojnarowicz, J.; Chudoba, T.; Lojkowski, W. A Review of Microwave Synthesis of Zinc Oxide Nanomaterials: Reactants, Process Parameters and Morphologies. *Nanomaterials* **2020**, *10*, 1086. [CrossRef]
15. Arya, S.; Mahajan, P.; Mahajan, S.; Khosla, A.; Datt, R.; Gupta, V.; Young, S.J.; Oruganti, S.K. Review—Influence of Processing Parameters to Control Morphology and Optical Properties of Sol-Gel Synthesized ZnO Nanoparticles. *ECS J. Solid State Sci. Technol.* **2021**, *10*, 023002. [CrossRef]
16. Stanley, R.; Alphas Jebasingh, J.; Manisha Vidyavathy, S. Cost-effective and sunlight-driven degradation of anionic and cationic dyes with pure ZnO nanoparticles. *Int. J. Environ. Sci. Technol.* **2022**, *19*, 11249–11262. [CrossRef]
17. Sivakumar, S.; Robinson, Y.; Mala, N.A. Studies on photocatalytic performance and supercapacitor applications of undoped and Cu-doped ZnO nanoparticles. *Appl. Surf. Sci. Adv.* **2022**, *12*, 100344. [CrossRef]
18. Balcha, A.; Yadav, O.P.; Dey, T. Photocatalytic degradation of methylene blue dye by zinc oxide nanoparticles obtained from precipitation and sol-gel methods. *Environ. Sci. Pollut. Res.* **2016**, *23*, 5485–25493. [CrossRef]
19. AlAbdulaal, T.H.; Ganesh, V.; AlShadidi, M.; Hussien, M.S.A.; Bouzidi, A.; Algarni, H.; Zahran, H.Y.; Abdel-wahab, M.S.; Yahia, I.S.; Nasr, S. The Auto-Combustion Method Synthesized Eu_2O_3-ZnO Nanostructured Composites for Electronic and Photocatalytic Applications. *Materials* **2022**, *15*, 3257. [CrossRef]
20. Manimozhi, R.; Rajkumar, K.; Sabarish, K.; Talwar, M.N.; Gnana Prakash, A.P. Solution combustion synthesized $ZnO/Na_2Ti_6O_{13}$ composite for degradation of 4-nitrophenol under solar irradiation. *Carbon Lett.* **2022**, *32*, 1355–1363. [CrossRef]
21. Yadav, S.; Jindal, J.; Mittal, A.; Sharma, S.; Kumari, K.; Kumar, N. Facile solution combustion synthesized, Li doped ZnO nanostructures for removal of abiotic contaminants. *J. Phys. Chem. Solids* **2021**, *157*, 110217. [CrossRef]
22. Kusiak-Nejman, E.; Wojnarowicz, J.; Morawski, A.W.; Narkiewicz, U.; Sobczak, K.; Gierlotka, S.; Lojkowski, L. Size-dependent effects of ZnO nanoparticles on the photocatalytic degradation of phenol in a water solution. *Appl. Surf. Sci.* **2021**, *541*, 148416. [CrossRef]
23. Sujinnapram, S.; Wongrerkdee, S. Synergistic effects of structural, crystalline, and chemical defects on the photocatalytic performance of Y-doped ZnO for carbaryl degradation. *J. Environ. Sci.* **2023**, *124*, 667–677. [CrossRef] [PubMed]
24. Wongrerkdee, S.; Krobthong, S. Synthesis, Characterization, and Photocatalytic Property of Ba-Doped ZnO Nanoparticles Synthesized Using Facile Precipitation. *Integr. Ferroelectr.* **2022**, *224*, 192–204. [CrossRef]

25. Wang, Z.; Li, J. Facile Liquid-Phase Synthesis of a High-Performance Cd-Doped ZnO-Quantum-Dot-Based Photocatalyst. *ECS J. Solid State Sci. Technol.* **2021**, *10*, 124003. [CrossRef]
26. Holzwarth, U.; Gibson, N. The Scherrer equation versus the 'Debye-Scherrer equation'. *Nat. Nanotechnol.* **2011**, *6*, 534. [CrossRef]
27. Krobthong, S.; Wongrerkdee, S.; Pimpang, P.; Moungsrijun, S.; Sujinnapram, S.; Nilphai, S.; Rungsawang, T.; Wongrerkdee, S. ZnO Nanoparticles Coprecipitation with Aluminum and Copper Ions for Efficient Photocatalytic Degradation of Commercial Glyphosate. *Integr. Ferroelectr.* **2022**, *222*, 69–83. [CrossRef]
28. Wang, M.; Li, C.; Liu, B.; Qin, W.; Xie, Y. Influence of Calcination Temperature on Photocatalyst Performances of Floral Bi_2O_3/TiO_2 Composite. *Catalysts* **2022**, *12*, 1635. [CrossRef]
29. Ganesh, V.; Hussien, M.S.A.; Shaik, U.P.; Ade, R.; Mohammed, M.I.; AlAbdulaal, T.H.; Zahran, H.Y.; Yahia, I.S.; Abdel-wahab, M.S. Impact of Mo-Doping on the Structural, Optical, and Electrocatalytic Degradation of ZnO Nanoparticles: Novel Approach. *Crystals* **2022**, *12*, 1239. [CrossRef]
30. Sapkota, R.; Duan, P.; Kumar, T.; Venkataraman, A.; Papadopoulos, C. Thin Film Gas Sensors Based on Planetary Ball-Milled Zinc Oxide Nanoinks: Effect of Milling Parameters on Sensing Performance. *Appl. Sci.* **2021**, *11*, 9676. [CrossRef]
31. Kisan, B.; Kumar, J.; Alagarsamy, P. Experimental and first-principles study of defect-induced electronic and magnetic properties of ZnO nanocrystals. *J. Phys. Chem. Solids* **2020**, *146*, 109580. [CrossRef]
32. Phan, T.-L.; Zhang, Y.D.; Yang, D.S.; Nghia, N.X.; Thanh, T.D.; Yu, S.C. Defect-induced ferromagnetism in ZnO nanoparticles prepared by mechanical milling. *Appl. Phys. Lett.* **2013**, *102*, 072408. [CrossRef]
33. Moungsrijun, S.; Wongrerkdee, S. Investigation of structural, optical, and electrical properties of ZnO thin films for electrooptical devices. *Suranaree J. Sci. Technol.* **2022**, *29*, 030085.
34. Sujinnapram, S.; Nilphai, S.; Moungsrijun, S.; Krobthong, S.; Wongrerkdee, S. Clustered ZnO nanoparticles synthesized via precipitation for photocatalytic degradation of methyl orange and glyphosate. *Dig. J. Nanomater. Biostruct.* **2021**, *16*, 317–329. [CrossRef]
35. Rungsawang, T.; Sujinnapram, S.; Nilphai, S.; Wongrerkdee, S. Influence of yttrium doping on ZnO nanoparticles for enhanced photocatalytic degradation of methylene blue. *Dig. J. Nanomater. Biostruct.* **2021**, *16*, 1209–1217. [CrossRef]
36. Zhang, S.; Malik, S.; Ali, N.; Khan, A.; Bilal, M.; Rasool, K. Covalent and Non-covalent Functionalized Nanomaterials for Environmental Restoration. *Top. Curr. Chem.* **2022**, *380*, 44. [CrossRef]

Disclaimer/Publisher's Note: The statements, opinions and data contained in all publications are solely those of the individual author(s) and contributor(s) and not of MDPI and/or the editor(s). MDPI and/or the editor(s) disclaim responsibility for any injury to people or property resulting from any ideas, methods, instructions or products referred to in the content.

toxics

MDPI

Article

Facilitated Adsorption of Mercury(II) and Chromium(VI) Ions over Functionalized Carbon Nanotubes

Gururaj M. Neelgund [1,*], Erica A. Jimenez [1], Ram L. Ray [2] and Mahaveer D. Kurkuri [3]

1 Department of Chemistry, Prairie View A&M University, Prairie View, TX 77446, USA
2 College of Agriculture and Human Sciences, Prairie View A&M University, Prairie View, TX 77446, USA
3 Centre for Research in Functional Materials (CRFM), JAIN (Deemed-to-be University), Jain Global Campus, Bengaluru 562 112, Karnataka, India
* Correspondence: gmneelgund@pvamu.edu

Abstract: By considering the importance of water and its purity, herein, a powerful adsorbent has been developed for the adsorption of two toxic contaminants that commonly exist in water, viz., divalent mercury and hexavalent chromium. The efficient adsorbent, CNTs–PLA–Pd, was prepared by covalent grafting polylactic acid to carbon nanotubes and subsequent deposition of palladium nanoparticles. The CNTs–PLA–Pd could adsorb Hg(II), and Cr(VI) entirely exists in water. The adsorption rate for Hg(II) and Cr(VI) was rapid at initial stage, followed by gradual decrease, and attained the equilibrium. The Hg(II) and Cr(VI) adsorption was perceived within 50 min and 80 min, respectively with CNTs–PLA–Pd,. Further, experimental data for Hg(II) and Cr(VI) adsorption was analyzed, and kinetic parameters were estimated using pseudo–first and second–order models. The adsorption process of Hg(II) and Cr(VI) followed the pseudo–second–order kinetics, and the rate–limiting step in the adsorption was chemisorption. The Weber−Morris intraparticle pore diffusion model revealed that the Hg(II) and Cr(VI) adsorption over CNTs–PLA–Pd occurs through multiple phases. The experimental equilibrium parameters for the Hg(II) and Cr(VI) adsorption were estimated by Langmuir, Freundlich, and Temkin isotherms models. All three models were well suited and demonstrated that Hg(II) and Cr(VI) adsorption over CNTs–PLA–Pd transpires through monolayer molecular covering and chemisorption.

Keywords: carbon nanotubes; polylactic acid; palladium; mercury; chromium; adsorption

1. Introduction

Water is a fundamental need of humans for survival and mandatory for other processes such as domestic, agriculture, and industry. Therefore, pure water is vital for a healthy life. However, excessive industrialization and urbanization are causing water pollution by releasing heavy metals containing effluents. Heavy metals are toxic, bio–accumulative, and naturally non–degradable. Heavy metals' ingestion at trace levels can also lead to irreparable detrimental impacts such as neurological system disorder, cancer, gastrointestinal disorder, muscular weakness, brain damage, shortness of breath, asthma, hypotension, liver and thyroid damage, allergies, and bone defects [1,2]. The accumulation of heavy metals in the body can cause carcinogenic, teratogenic, and mutagenic effects [3]. Therefore, exposure to heavy metals and their content effluents is not safe. Among different heavy metals, chromium, and mercury are top–priority hazardous because of their acute and chronic detrimental outcomes, which are substantially harmful [1,2]. Chromium is recognized as teratogenic, mutagenic, and carcinogenic [4,5]. It is likely to produce genotoxic DNA effects in the nucleus of cells and can cause cancer, allergies, gastritis, and organ damage [6,7]. Generally, chromium exists with two stable oxidation states, tri and hexavalent, in water [8]. These species differ in physicochemical properties and chemical and biochemical reactivity [8]. Out of both species, trivalent Cr(III) is a biologically essential nutrient that plays a significant role in the metabolism process of glucose, protein, and lipids by enhancing insulin activity [9]. However, hexavalent Cr(VI) is highly poisonous to humans and other living organisms [10]. Generally, Cr(VI) is introduced to the environment by effluents produced by industries like leather tanning, electroplating, metal finishing, textile finishing, steel fabrication, wood preservation, pulp processing, paint, dyes, paper, fertilizers, and photography [11]. Likewise, releasing Cr(VI) content effluents is perilous.

Another heavy metal that has a higher toxicity than chromium is mercury, which is highly volatile and environmentally persistent [12]. In water, mercury exists in the divalent oxidation state as Hg(II) and is significantly hazardous, non–degradable, and bio–accumulative [13]. Hg(II) tends to accumulate in the body tissues, muscles, lipids, liver, and other organs [14]. The accumulation of Hg(II) at the minute level in the body would induce substantial damage to the central nervous, immune, and reproductive systems [13]. It can also cause severe impairment to kidney, renal, lung, and liver function [15]. Hg(II) ions have a high affinity to thiol moieties of enzymes and proteins and could break normal human cells to cause fatal consequences that include organ dysfunction, chronic diseases, metabolic disruption, breakdown of the chromosome, immune and nervous system damage, and may cause death [14,16–20]. Moreover, Hg(II) ions have a high tendency to react with organic and inorganic matter to generate a variety of harmful compounds [21]. The reaction of mercury with organic matter can form the lethal methyl mercury (CH_3Hg). The CH_3Hg is exceptionally venomous, has a high binding affinity to proteins, and can accumulate in body tissues to lead to identical consequences like Hg(II) [21]. Hg(II) can enter the environment through myriad anthropogenic sources such as volcanic explosions, weathering of rock, erosion of natural deposits, geothermal processes, wild forest fires, and vaporization from soil and water [14,22]. In addition, it can be introduced by industrial operations like gold refineries, petroleum refineries, electroplating, leather tanning, metal finishing, pharmaceutical industries, and runoff from landfills and croplands [14,23]. Nevertheless, releasing Hg(II) is not innocuous.

Several functional materials have been developed to remove heavy metals from water to prevent their consequences and promote water purification [14,21,24–31]. These materials promoted the removal of heavy metals from water by various techniques like adsorption [32,33], ion exchange [34,35], membrane separation [36], coagulation [37], chemical precipitation [38], extraction [39], dialysis [40], electrochemical separation [41] and more. Among these, adsorption is the most effective, versatile, economically feasible, environmentally sustainable, and technologically promising [1–3,7–10]. The additional advantage of adsorption is its high selectivity and enrichment factor [42]. Because of its importance, several adsorbents such as activated carbon [1], biochar [1], hydroxyapatite/Fe_3O_4/polydopamine [2], polyacrylonitrile–based porous carbon [3], vesicular basalt rock [6], activated carbon [7], ferrihydrite [8], exopolysaccharides [9], chitosan/bentonite composite [10], amine–incorporated UIO–66–NH_2 [20], silica nanoparticles [24], graphene oxide–$MnFe_2O_4$ Magnetic Nanohybrids [25], zinc oxide [26], ZnS [31], and carbon nanotubes [43–45] were developed and verified their efficiency in removal of several heavy metals. Compared to these, carbon nanotubes (CNTs) are fascinating due to their unique physical, chemical, and mechanical properties [46,47]. CNTs possess the interesting nanometric one–dimensional structure, high surface–to–volume ratio, high accessible surface area, high aspect ratio, and higher physical and chemical stability [46,47]. These unique properties made CNTs a prospective candidate for adsorption. In addition, the distinct open tubular architecture of CNTs having quantized π electron clouds facilitates the adsorption process [48,49]. Beyond their excellent properties, the application of CNTs in adsorption is hindered due to their hydrophobic nature and the tendency to entangle. To come up with these limitations, functionalization is a suitable approach to transform the hydrophobic CNTs into hydrophilic ones and prevent them from entanglement. It could be attained by grafting the CNTs with a suitable polymer like polylactic acid (PLA). PLA is a vital biopolymer that has crucial properties like non–toxicity, biocompatibility, biodegradability, thermoplasticity, and exceptional mechanical properties required for adsorption [50]. PLA is produced by condensation polymerization of lactic acid, which is derived by the fermentation of sugars from carbohydrate sources such as corn, sugarcane, and tapioca [51]. By considering the importance of CNTs and PLA, herein, CNTs have been functionalized by the covalent grafting of PLA chains. Further, the adsorption capability of PLA–grafted CNTs was improved through the deposition of palladium nanoparticles. Palladium is classically known for its catalytic properties; instead, it has been explored as a potential adsorbent. The efficiency of produced CNTs–PLA–Pd was estimated by the adsorption rate of two toxic pollutants, Hg(II) and Cr(VI), which commonly persist in the water. The adsorption capability, its dynamics, and control mechanisms for the Hg(II) and Cr(VI) adsorption over CNTs–PLA–Pd were evaluated using the pseudo–first and second–order kinetic models. The Weber–Morris intraparticle pore diffusion model was used to find the reaction pathways and the rate–controlling step of the adsorption process. Further, the adsorption equilibrium was determined by fitting the experimental results with Langmuir, Freundlich, and Temkin isotherm models.

2. Experimental Analysis

2.1. Materials

The CNTs and chemicals used were purchased from Millipore Sigma. Ultrapure water obtained by the Milli–Q Plus system (Millipore) was used to prepare aqueous solutions.

2.2. Preparation of CNTs–PLA–Pd

The CNTs–PLA (10 mg) was prepared according to the reported procedure [52] and dispersed in 7 mL DI water by sonication for 10 min. Then an aqueous solution of $PdCl_2$ (5 mL, 0.01 mol L^{-1}) was added, and the mixture was stirred at room temperature for 2 h. A freshly prepared 5 mL aqueous solution of $NaBH_4$ (0.05 mol L^{-1}) was slowly added and stirred at ambient conditions for 4 h. The resulting CNTs–PLA–Pd was centrifuged and purified by washing with DI water. It was dried in a vacuum for 12 h at 40 °C.

2.3. Adsorption Experiments

The stock solution of Hg(II) and Cr(VI) with 1 g L^{-1} was prepared in DI water using mercury(II) chloride ($HgCl_2$) and potassium dichromate ($K_2Cr_2O_7$), respectively. Further, the stock solution was diluted to desired concentrations using DI water. The kinetic adsorption experiments of Hg(II) and Cr(VI) were carried out to find the contact time needed to reach equilibrium. In the typical experiment, 100 mg of CNTs–PLA–Pd was dispersed into 500 mL of Hg(II) and Cr(VI) solution with a concentration of 0.5 mg L^{-1} having an initial pH of 5.7 and 5.1, respectively. It was allowed to stir at room temperature, and an adequate quantity of samples was collected after the required contact time. The adsorbent, CNTs–PLA–Pd was separated by centrifugation, and the concentration of residual Hg(II) and Cr(VI) in the solution was estimated using the atomic absorption spectrometer. The efficiency in Hg(II) and Cr(VI) adsorption as a function of time was monitored for 120 min, and the adsorbed amount of Hg(II) and Cr(VI) by CNTs–PLA–Pd was calculated using the following equation.

$$q_t = \frac{(C_0 - C_t)\ V}{M} \tag{1}$$

where q_t is the amount of pollutant adsorbed (mg/g) at time t; C_0 is the initial concentration of pollutant in solution (mg L^{-1}), and C_t is the concentration of pollutant in solution (mg L^{-1}) at time t; V is the volume of the solution (L), and M is the amount of adsorbent (g).

Further, the efficiency of CNTs–PLA–Pd towards Hg(II) and Cr(VI) adsorption was determined by:

$$\text{Removal efficiency (\%)} = \frac{(C_0 - C_t)\ V}{C_0} \times 100 \tag{2}$$

For adsorption isotherms experiments, 10 mg of CNTs–PLA–Pd was added to 50 mL Hg(II) and Cr(VI) solution and allowed to stir at room temperature for 24 h to attain equilibrium. The concentration of Hg(II) and Cr(VI) solution was varied from 2 to 10 mg L^{-1}

to obtain the adsorption isotherms. After reaching the equilibrium, CNTs–PLA–Pd was separated by centrifugation, and the concentration of Hg(II) and Cr(VI) in the solution was estimated by the atomic absorption spectrometer. The Hg(II) and Cr(VI) adsorption rate at equilibrium, q_e (mg g^{-1}), was evaluated by:

$$q_e = \frac{(C_0 - C_e)\,V}{M} \quad (3)$$

where q_e is the amount of Hg(II) and Cr(VI) adsorbed (mg/g) at equilibrium.

2.4. Characterization

The XRD of samples was obtained using the Scintag X–ray diffractometer, model PAD X, equipped with a Cu Kα photon source (45 kV, 40 mA) at a scanning rate of 3°/min. The ATR–FTIR spectra were perceived by Smith's ChemID diamond attenuated total reflection (DATR) spectrometer. The SEM images of samples were acquired with the JEOL JXA–8900 microscope, and the X-ray photoelectron spectra (XPS) were acquired by the Perkin Elmer PHI 5600 ci X-ray photoelectron spectrometer. The Hg(II) and Cr(VI) concentration was measured using a Varian SpectrAA 220FS atomic absorption spectrometer.

3. Results and Discussion

The ATR–FTIR spectrum of CNTs–COOH (Figure 1a) exhibited the bands at 612 and 1634 cm^{-1} corresponding to the A2u and E1u phonon modes of CNTs [53]. It showed the band due to the O–H bonds at 3360 cm^{-1} and the band of C=O bonds of –COOH groups existing over the surface of CNTs at 1698 cm^{-1} [54]. The band of the C=C was found at 1559 cm^{-1} [54]. Compared to the spectra of CNTs–COOH and PLA (Figure 1b), CNTs–PLA (Figure 1c) displayed important characteristic bands. In particular, the bands that appeared at 734 and 864 cm^{-1} were due to C=O stretching, and the band at 1040 cm^{-1} was due to C–CH_3 stretching [52,55]. The significant intensity band that appeared at 1080 cm^{-1} was due to C–O–C vibrations, and the absorption at 1126 cm^{-1} was owing to CH_3 vibrations [55]. The band of C–C–O with high frequency was established at 1179 cm^{-1}, and the band at 1262 cm^{-1} corresponds to C–H and C–O–C [52,55]. The doublet exists at 1355 and 1379 cm^{-1}, and the singlet at 1449 cm^{-1} was assigned to CH_3 vibrations. The band for C=O stretching was positioned at 1744 cm^{-1} with high intensity. The CH and CH_3 vibrations band was displayed as a doublet at 2935 and 2988 cm^{-1}, respectively. The absorption of –OH stretching was situated at 3500 cm^{-1}. The spectrum of CNTs–PLA–Pd (Figure 1d) revealed the distinctive bands found in the spectrum of CNTs–PLA with diminished intensity and minute shifting in their position. The powder XRD of CNTs–PLA–Pd, presented in Figure 2, demonstrated the well–resolved reflection peaks at 26.4, 39.5, 45.8, 46.1, and 67.2°. The peaks located at 26.4 and 45.8° were assigned to (0 0 2) and (1 0 0) planes of CNTs, respectively [56]. The peaks situated at 39.5, 46.1, and 67.2° were due to the (1 1 1), (2 0 0), and (2 2 0) planes of Pd (JCPDS No. 89–4897). The FESEM images of CNTs–PLA–Pd, illustrated in Figure 3, show the effective exfoliation of CNTs. It was approached by the successful grafting of PLA chains to the surface of CNTs. The repulsion between the polymeric chains of PLA caused the efficient exfoliation of CNTs. The presence of individual CNT is observable in Figure 3a. However, it is perceptibly visible in Figure 3b. The CNTs have a width of about 50–75 nm and a length of several μm. Some CNTs have a distracting surface, which could be created by the harsh treatment of pristine CNTs with a mixture of concentrated sulfuric and nitric acids applied before the grafting of PLA. Furthermore, the existence of densely populated Pd nanoparticles over the surface of PLA–grafted CNTs is evident in Figure 3. Nonetheless, Pd nanoparticles have assembled and formed some clusters that led to the seeding of twinned or multiple–twinned nanoparticles, which commonly occur in palladium [52,57–59]. The grafting of PLA to CNTs has provided a suitable environment for depositing the Pd nanoparticles and directed their strong adherence. The resilient anchoring of Pd nanoparticles to the surface of PLA–grafted CNTs prevented them from detachment.

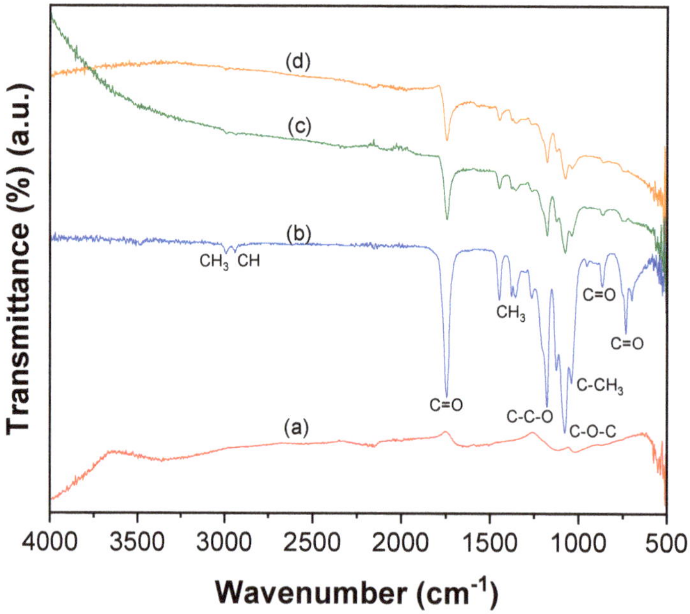

Figure 1. ATR–FTIR spectra of (**a**) CNTs–COOH, (**b**) PLA, (**c**) CNTs–PLA, and (**d**) CNTs–PLA–Pd.

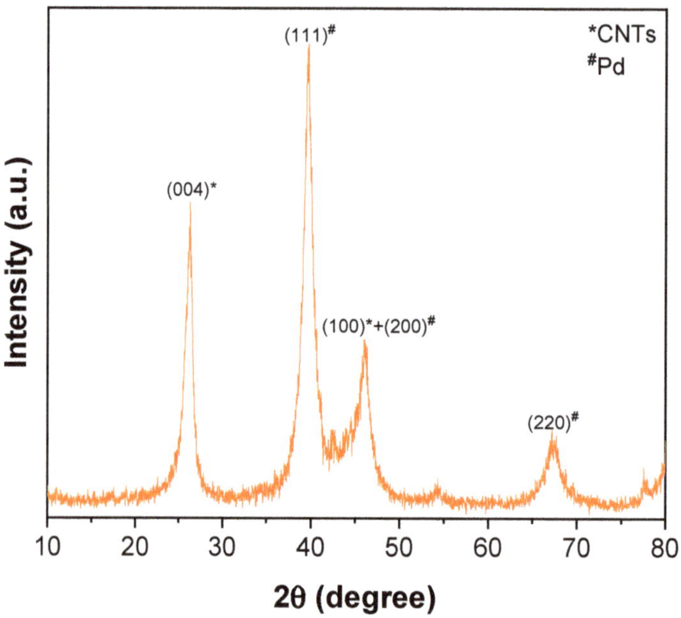

Figure 2. XRD of CNTs–PLA–Pd.

The XPS survey spectrum of CNTs–PLA–Pd shown in Figure 4a revealed the presence of C, O, and Pd. The high-resolution spectrum of C1s (Figure 4b) displayed a peak at 284.3 eV, divulging into four distinct peaks by Gaussian fitting. The peak at 284.3 eV was assigned to C–C bonds of sp^2 hybridized carbon atoms of CNTs [60,61]. The peak at 286.7 eV was due to C–C bonds occurring in structurally defective sp^3 hybridized carbon atoms and C–O bonds of CNTs [60,61]. While the low–intensity peaks that appeared at 288.3 and 290.1 eV were owing to carbonyl, C=O, and carboxyl carbon O=C–O, respectively [60,61]. The spectrum of Pd 3d (Figure 4c) demonstrated the spin–orbit split doublet with a high–energy band of Pd $3d_{3/2}$ at 340.6 eV and a low–energy band of Pd $3d_{5/2}$ at 335.3 eV [62,63]. The difference in the binding energy between Pd $3d_{3/2}$ and Pd $3d_{5/2}$ peaks was 5.3 eV, which validates the occurrence of Pd in the zero valent of the metallic state [60]. The deconvoluted O 1 s peaks (Figure 4d) show several types of oxygen species. The peak of lattice oxygen (O_2^{2-}) was situated at 530.1 eV. The peaks for C=O and C–O functional groups were located at 531.7 eV and 533.6 eV, respectively [64]. The peak at 535.5 eV is likely due to the chemisorbed oxygen or adsorbed molecular H_2O [64].

Figure 3. FESEM images of (**a**,**b**) CNTs–PLA–Pd.

The CNTs–PLA–Pd as adsorbent, its efficiency was revealed by evaluating the adsorption rate of Hg(II), and Cr(VI) present in water. It is known that the adsorption rate depends on the contact time between the adsorbent and adsorbate. So, the adsorption rate of Hg(II) was measured as a function of contact time and compared with CNTs and PLA (Figure 5). The CNTs–PLA–Pd was able to completely adsorb the Hg(II) within 50 min. However, CNTs and PLA were able to adsorb 74 and 58% of Hg(II), respectively. Therefore, the adsorption ability of CNTs and PLA was significantly improved after their blending and deposition of Pd nanoparticles in CNTs–PLA–Pd. Also, the efficiency of CNTs–PLA–Pd was estimated by the adsorption of Cr(VI) (Figure 6). The plot perceived for q_t versus t is shown in Figure S1. It represented that the adsorption rate of Hg(II) and Cr(VI) enhanced with contact time and finally red the equilibrium. Initially, the adsorption rate was high, followed by gradual reduction and final attainment of equilibrium. Within the initial 30 min, about 96 and 78% of Hg(II) and Cr(VI), respectively, were adsorbed. The complete adsorption of Hg(II) was accomplished in 50 min and Cr(VI) within 80 min. The rapid adsorption that occurred initially for 30 min could be due to the abundant availability of the active sites that exist over the surface of CNTs–PLA–Pd. With the progress of time, the active sites are being saturated by adsorption of the high quantity of Hg(II) and Cr(VI) ions and the repulsive forces transpire between solute molecules in the solid and bulk phases [65]. The results were analyzed by Pseudo–first and second–order models. The pseudo–first–order equation applied can be expressed as:

$$n(q_e - q_t) = \ln q_e - k_1 t \quad (4)$$

where q_e and q_t are the quantity of Hg(II) and Cr(VI) (mg g^{-1}) at equilibrium and particular time t (min), respectively. k_1 (min^{-1}) is the pseudo–first–order rate constant. The slope of the plot, $\ln(q_e - q_t)$ versus t, provides the value of k_1, and the intercept gives the theoretical value of q_e.

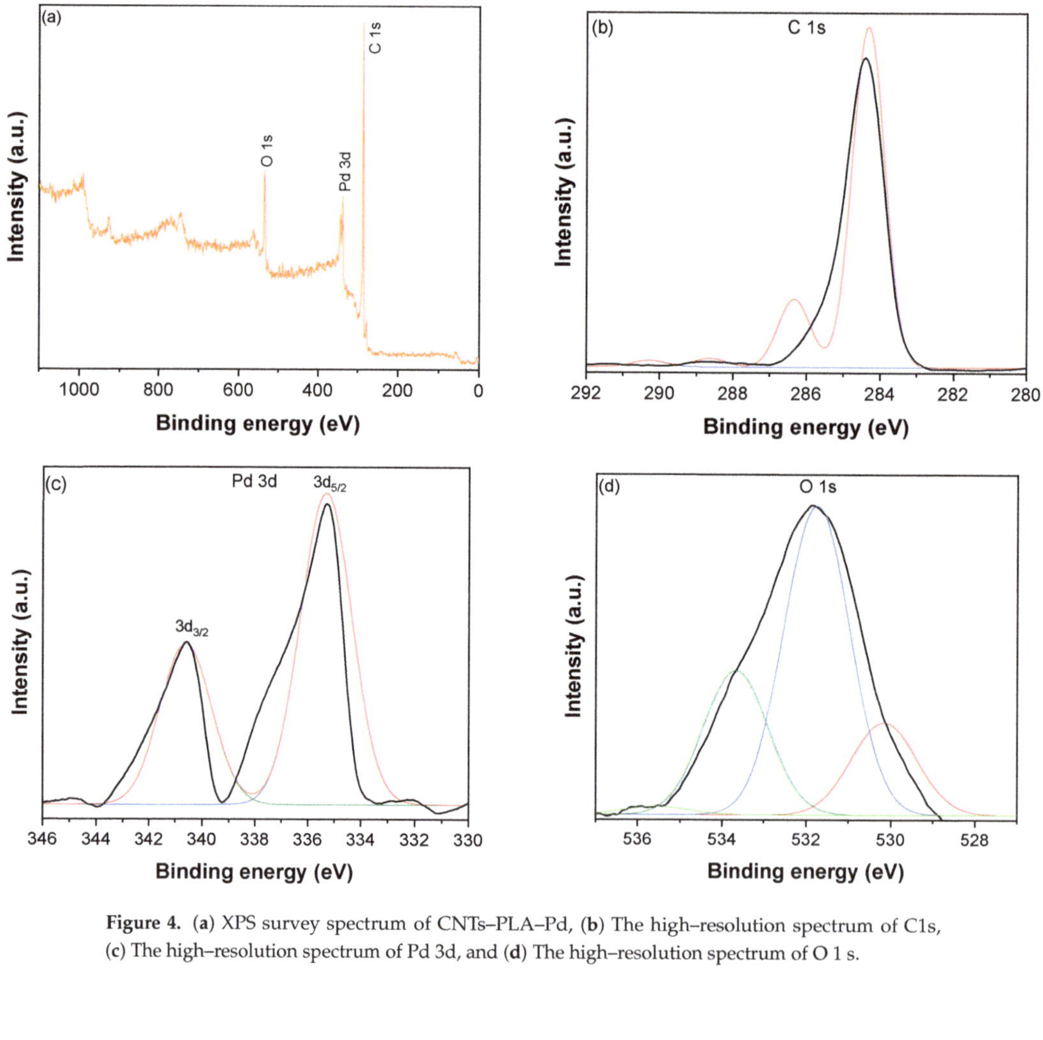

Figure 4. (**a**) XPS survey spectrum of CNTs–PLA–Pd, (**b**) The high–resolution spectrum of C1s, (**c**) The high–resolution spectrum of Pd 3d, and (**d**) The high–resolution spectrum of O 1 s.

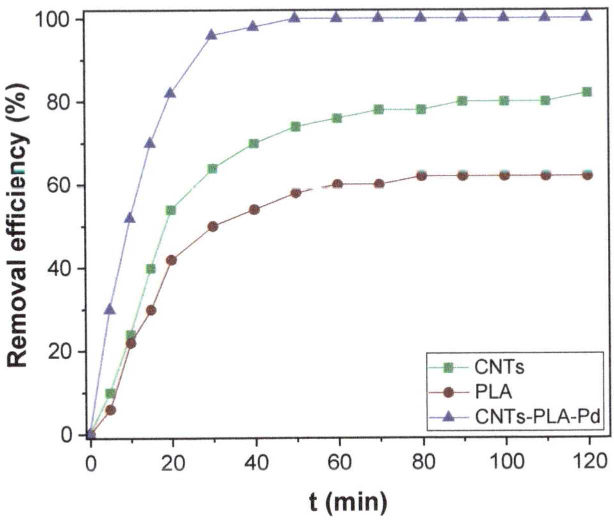

Figure 5. The adsorption kinetics of Hg(II) over CNTs, PLA, and CNTs–PLA–Pd.

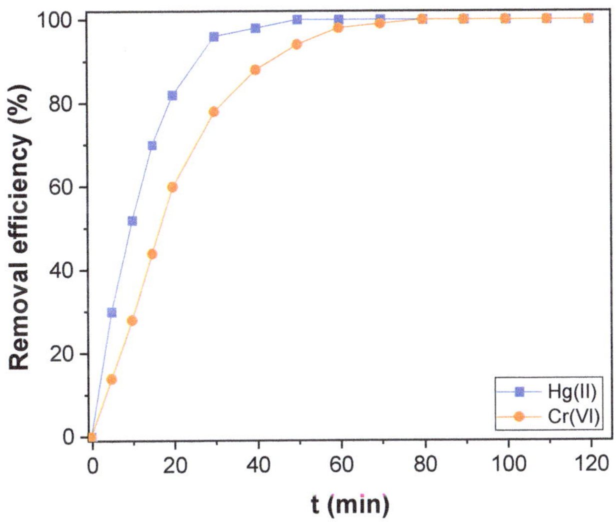

Figure 6. The adsorption kinetics of Hg(II) and Cr(VI) over CNTs–PLA–Pd.

The pseudo–second–order equation used can be represented by:

$$\frac{t}{q_t} = \frac{1}{k_2\, q_e^2} + \frac{t}{q_e} \tag{5}$$

where k_2 [g (mg min)$^{-1}$] is the second–order rate constant, which can be perceived by the plot of t/q_t against t. The slope is $1/q_e$, and the intercept is $1/k_2\, q_e{}^2$.

The pseudo–first and second–order plots obtained for Hg(II) adsorption and Cr(VI) are shown in Figures 7 and 8, respectively and the kinetic parameters determined are presented in Table 1. The correlation coefficient (R^2) perceived for pseudo–first–order kinetics was 0.9850 and 0.9841 for Hg(II) and Cr(VI), respectively. For pseudo–second–order kinetics, it was 0.9991 and 0.9930 for Hg(II) and Cr(VI), respectively. The R^2 value of pseudo–second–order kinetics was higher than first–order kinetics. Besides, the value of q_e (exp) agreed with the value of q_e (cal) estimated from pseudo–second–order kinetics rather than the value determined by pseudo–first–order kinetics for both Hg(II) and Cr(VI) (Table 1). Therefore, the Hg(II) and Cr(VI) adsorption could be better explained by pseudo–second–order kinetics rather than first–order kinetics. Hence chemisorption is the rate–limiting step in the Hg(II) and Cr(VI) adsorption over the surface of CNTs–PLA–Pd [66].

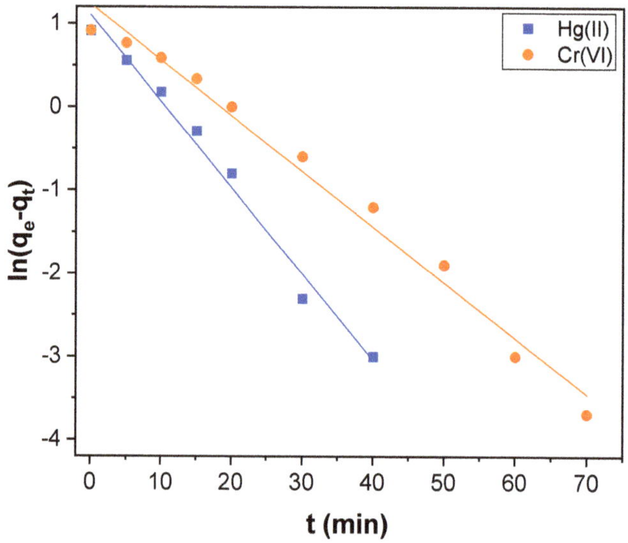

Figure 7. The pseudo–first–order kinetics for Hg(II) and Cr(VI) adsorption over CNTs–PLA–Pd.

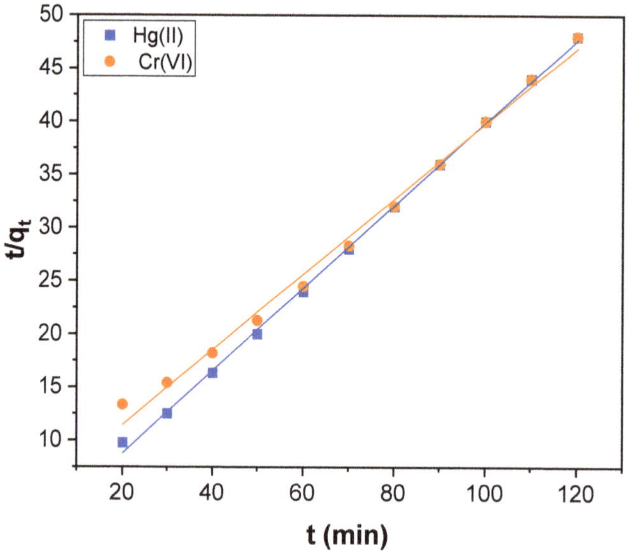

Figure 8. The pseudo–second–order kinetics for Hg(II) and Cr(VI) adsorption over CNTs–PLA–Pd.

Table 1. Parameters calculated for Hg(II) and Cr(VI) adsorption from adsorption kinetic models.

Adsorbate	q_e (exp) mg g^{-1}	Pseudo–First Order Kinetic Model			Pseudo–Second–Order Kinetic Model		
		q_e (cal) mg g^{-1}	k_1 (min^{-1})	R^2	q_e (cal) mg g^{-1}	k_2 (g mg^{-1} min^{-1})	R^2
Hg(II)	2.500	1.0040	0.1036	0.9850	2.5686	0.1542	0.9991
Cr(VI)	2.500	1.2270	0.0668	0.9841	2.8190	0.0290	0.9930

Further, reaction pathways and the rate–controlling step in the Hg(II) and Cr(VI) adsorption were evaluated by the application of the Weber−Morris intraparticle pore diffusion model using the following equation [67].

$$q_t = k_{id}\, t^{0.5} + c \tag{6}$$

where k_{id} (mg $g^{-1}min^{-1}$) is the intraparticle diffusion rate constant, which can be calculated from the linear plot of q_t versus $t^{0.5}$ c (mg g^{-1}) is the intraparticle diffusion constant, estimated from the intercept, and is directly proportional to the boundary layer thickness. It is assumed that the higher the intercept value, the more significant the contribution of the surface adsorption in the rate–controlling step. If the regression of q_t versus $t^{0.5}$ plot is linear and passes through the origin, intraparticle diffusion plays a significant role in controlling the kinetics of the adsorption process. If it does not pass through the origin, in that case, intraparticle diffusion is not the only rate–limiting step. Instead, it is contributed by the boundary layer effect [68].

The plot of q_t versus $t^{0.5}$, shown in Figure 9, does not pass through the origin of the coordinates (Figure S2). So, intraparticle diffusion is not the sole rate–limiting step in the adsorption process of Hg(II) and Cr(VI). The intercept value obtained from Figure 9 was low (Table S1), and the intraparticle diffusion plot (Figure S2) revealed two straight lines. It shows that the intraparticle diffusion was not only a rate–controlling step in the Hg(II) and Cr(VI) adsorption; it was contributed by the boundary layer effect also [68,69]. The multilinearity of Figure 9 exhibits that the adsorption process of Hg(II) and Cr(VI) transpires through multiple phases instead of a single [66]. Among these multilinear steps, the initial phase occurred through the rapid adsorption of Hg(II) and Cr(VI) ions. The second phase could be the diffusion of Hg(II) and Cr(VI) ions into the pores of CNTs–PLA–Pd, and the third phase could be due to the equilibrium of adsorption that causes chemical reaction/bonding [66].

To reveal the Hg(II) and Cr(VI) adsorption in detail, the experimental equilibrium parameters were determined by three isotherm models viz., the Langmuir, the Freundlich, and the Temkin models. The Langmuir model assumes that the adsorbed molecules form a monolayer and that the adsorption can occur at a fixed number of adsorption sites and all equivalent in adsorption abilities. Each molecule also has a constant enthalpy and adsorption activation energy. In other words, all molecules have an equal affinity to entire adsorption sites [70]. With the help of the Langmuir isotherm model, it is possible to find the maximum adsorption capacity of the adsorbent using the linear form of the Langmuir isotherm model represented by the following equation:

$$\frac{C_e}{q_e} = \frac{C_e}{q_m} + \frac{1}{K_L\, q_m} \tag{7}$$

where q_e (mg g^{-1}) is the amount of adsorbed Hg(II) and Cr(VI) per unit mass of CNTs–PLA–Pd; C_e (mg L^{-1}) is the concentration of Hg(II) and Cr(VI) at equilibrium; q_m is the maximum amount of the Hg(II) and Cr(VI) adsorbed per unit mass of CNTs–PLA–Pd to form a complete monolayer on the surface–bound at high C_e. K_L is the Langmuir adsorption constant related to the free energy of adsorption. The linear fitting for the Langmuir plot of specific adsorption (C_e/q_e) versus the equilibrium concentration (C_e) obtained for Hg(II) and Cr(VI) adsorption is shown in Figure 10. The parameters calculated by the Langmuir isotherm model are tabulated in Table 2. The maximum adsorption capacity (q_m) calculated for the Hg(II) and Cr(VI) adsorption was 263.2 and 196.1 mg g^{-1}, respectively. The value of K_L for Hg(II) and Cr(VI) was 0.6041 and 0.2567 L mg^{-1}, respectively.

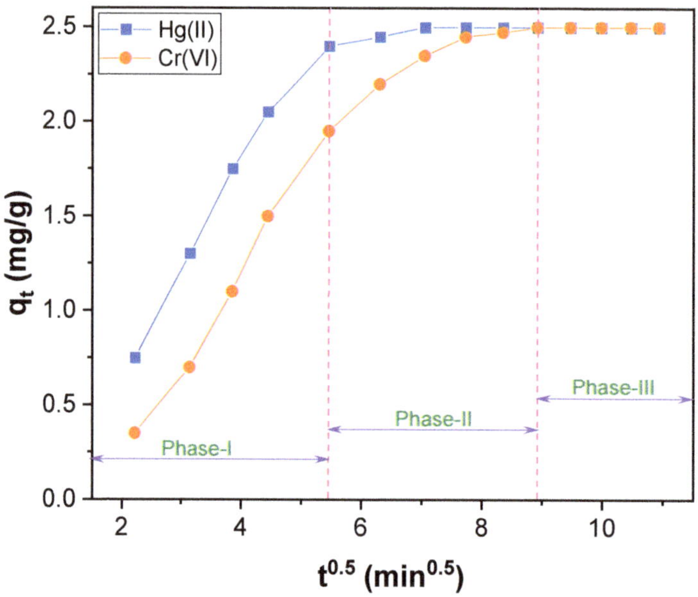

Figure 9. The Weber–Morris intraparticle diffusion plot for Hg(II) and Cr(VI) adsorption over CNTs–PLA–Pd.

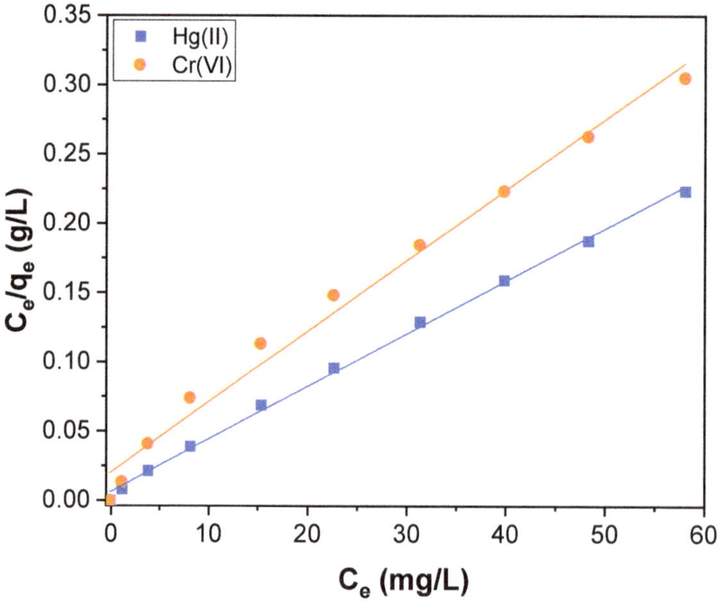

Figure 10. The Langmuir isotherm plot for Hg(II) and Cr(VI) adsorption over CNTs–PLA–Pd.

Table 2. Parameters calculated for Hg(II) and Cr(VI) adsorption from Langmuir, Freundlich, and Temkin adsorption isotherms models.

Adsorbate	Langmuir Isotherm			Freundlich Isotherm			Temkin Isotherm		
	q_m (mg g^{-1})	K_L (L mg^{-1})	R^2_{Lan}	K_F (mg g^{-1})	n	R^2_{Fre}	A	B	R^2_{Tem}
Hg(II)	263.2	0.6041	0.9980	145.7	6.685	0.9822	109.1	30.00	0.9959
Cr(VI)	196.1	0.2567	0.9880	130.5	7.966	0.9970	260.5	22.41	0.9970

The Freundlich isotherm is derived from the assumption that the adsorption sites are distributed exponentially concerning the heat of adsorption [71]. It provides the relationship between the equilibrium of liquid and solid phase capacity based on the multilayer adsorption properties consisting of the heterogeneous surface of the adsorbent. The Freundlich isotherm, which supports multilayer adsorption, agrees with the Langmuir model over moderate ranges of concentrations but differs at low and high concentrations. The linear form of the Freundlich isotherm used is denoted by the below equation:

$$\ln q_e = \ln K_F + \frac{\ln C_e}{n} \quad (8)$$

where q_e (mg g^{-1}) is the amount of Hg(II) and Cr(VI) adsorbed at equilibrium. K_F and n are Freundlich constants. K_F symbolizes the affinity of the adsorbent, and n signifies the adsorption intensity. C_e (mg L^{-1}) is the concentration of Hg(II) and Cr(VI) at equilibrium. The value of K_F and n were evaluated by the slope and intercept of the Freundlich isotherm plot shown in Figure 11, and the assessed values are presented in Table 2. The magnitude of the exponent, 1/n, indicates the favorability of adsorption. If the value of n ranging from 1 to 10 indicates the favorable conditions for the process of adsorption [72]. For the Hg(II) and Cr(VI) adsorption, the calculated value of n was 6.685 and 7.966, respectively. It reveals that Hg(II) and Cr(VI) adsorption over CNTs–PLA–Pd occurs in favorable conditions.

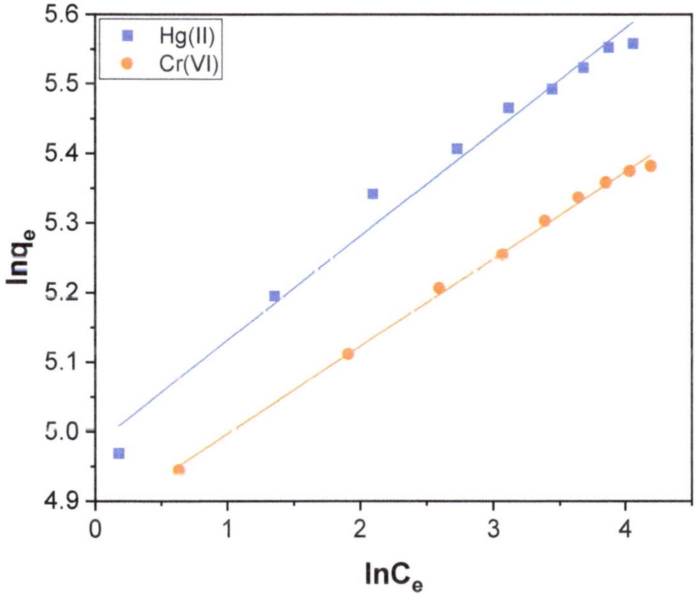

Figure 11. The Freundlich isotherm plot for Hg(II) and Cr(VI) adsorption over CNTs–PLA–Pd.

The Temkin isotherm model assumes that during adsorption, the heat of all molecules decreases linearly with the increase in coverage of the adsorbent surface and that adsorption is characterized by a uniform distribution of binding energies up to maximum binding energy [73]. The Temkin isotherm model used is described by following equation:

$$q_e = B \ln A + B \ \ln \ C_e \quad (9)$$

where $B = RT/K_T$, K_T is the Temkin constant related to the heat of adsorption (J mol^{-1}); A is the Temkin isotherm constant (L g^{-1}), R is the gas constant (8.314 J/mol K), and T the absolute temperature (K). The Temkin isotherm fitting plot of q_e versus ln C_e is shown in Figure 12, and the evaluated parameters are listed in Table 2.

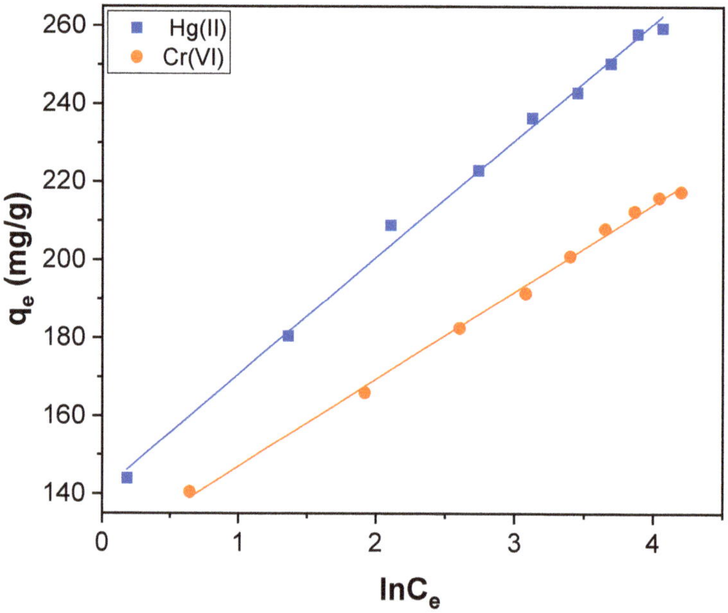

Figure 12. The Temkin isotherm plot for Hg(II) and Cr(VI) adsorption over CNTs–PLA–Pd.

All the tested Langmuir, Freundlich, and Temkin isotherm models have a better fit for Hg(II) and Cr(VI) adsorption with a high value of R^2. In particular, for adsorption of Hg(II), R^2_{Lan}, $R^2_{Fre,}$ and R^2_{Tem} were 0.9980, 0.9822, and 0.9959, respectively. Similarly, for adsorption of Cr(VI), R^2_{Lan}, $R^2_{Fre,}$ and R^2_{Tem} were 0.9880, 0.9970, and 0.9970, respectively. Therefore, all three models are well-suited and validated with Hg(II) and Cr(VI) adsorption. It indicates that the adsorption mechanism of Hg(II) and Cr(VI) was associated with the basic assumption of all three isotherm models. In general, the Hg(II) and Cr(VI) adsorption over CNTs–PLA–Pd occurred through monolayer molecular covering and chemisorption together [74]. The maximum adsorption capacity (q_m) of CNTs–PLA–Pd estimated for Hg(II) and Cr(VI) adsorption was compared to the reported values for different adsorbents and presented in Tables 3 and 4, respectively [75–83]. The stability of CNTs–PLA–Pd was estimated by recovering and applying it in three subsequent cycles of adsorption of Hg(II). The adsorption capability of recycled CNTs–PLA–Pd was not significantly reduced in all four measured cycles. Therefore, CNTs–PLA–Pd is a sturdy adsorbent suitable for repeated use.

Table 3. Comparison of maximum adsorption capacity (q_m) of CNTs–PLA–Pd estimated for Hg(II) adsorption with reported value for different adsorbents.

Adsorbent	Maximum Adsorption Capacity q_m (mg/g)	pH	Ref
RM–nZVI	94.58	5.0	[14]
UIO–66–NCS	250.0	5.0	[20]
UIO–66–IT	580.0	5.5	[20]
PAMC	76.3	6.0	[75]
CHAP–SH	282.7	4.5	[76]
M–PAL	203.4	4.0	[77]
ACM-5	191.9	5.8	[78]
MCS–4N	140.3	4.5	[79]
MCS–5N	109.7	4.5	[79]
CNTs–PLA–Pd	263.2	5.7	This work

Table 4. Comparison of maximum adsorption capacity (q_m) of CNTs–PLA–Pd estimated for Cr(VI) adsorption with reported value for different adsorbents.

Adsorbent	Maximum Adsorption Capacity q_m (mg/g)	pH	Ref
BWAC	59.23	2.0	[7]
Chitosan/bentonite composite	16.38	3.0	[10]
NH_2–ASNs	34.0	2.0	[24]
NH_2–MSNs	42.2	2.0	[24]
Fe_3O_4 NPs	56.61	5.0	[80]
Oil palm bagasse	115.45	2.0	[81]
Yam peels	50.12	2.0	[81]
Al–GNSC	13.458	4.0	[82]
Cu (I)–MOF	96.0	6.0	[83]
CNTs–PLA–Pd	263.2	5.7	This work

4. Conclusions

In conclusion, an efficient adsorbent, CNTs–PLA–Pd, was successfully produced by covalent grafting of PLA to CNTs and subsequent deposition of Pd nanoparticles. Thus, developed CNTs–PLA–Pd could remove Hg(II) and Cr(VI) entirely from the water within 50 min and 80 min, respectively. The initial rate for Hg(II) and Cr(VI) adsorption over CNTs–PLA–Pd was rapid, gradually reducing, and finally attained equilibrium. The analysis of experimental results revealed that the Hg(II) and Cr(VI) adsorption occurs through pseudo–second–order kinetics. It is identified that the chemisorption was the rate–limiting step in the adsorption process. The Weber–Morris intraparticle pore diffusion model exhibited that the adsorption of Hg(II) and Cr(VI) over CNTs–PLA–Pd occurs through multiple phases. The q_m value for Hg(II) and Cr(VI) adsorption was 263.2 and 196.1 mg g^{-1}, respectively, and the K_L value was 0.6041 and 0.2567 L mg^{-1} for Hg(II) and Cr(VI) adsorption, respectively. The K_F value estimated for Hg(II) and Cr(VI) adsorption was 145.7 and 130.5 mg g^{-1}, respectively, while the value of n was 6.685 and 7.966 for Hg(II) and Cr(VI) adsorption, respectively. The Langmuir, Freundlich, and Temkin models showed that the Hg(II) and Cr(VI) adsorption transpires with monolayer molecular covering and chemisorption.

Supplementary Materials: The following supporting information can be downloaded at: https://www.mdpi.com/article/10.3390/toxics11060545/s1, Table S1: Parameters calculated from the intra-particle diffusion plot provided in Figure S2; Figure S1: The plot perceived q_t as a function of time for Hg(II) and Cr(VI) adsorption over CNTs–PLA–Pd; Figure S2: The Intraparticle diffusion model for Hg(II) and Cr(VI) adsorption over CNTs–PLA–Pd.

Author Contributions: G.M.N.: data curation, writing original draft, review, and editing funding acquisition. E.A.J.: investigation, data curation. R.L.R.: writing, review, and editing, M.D.K.: writing, review, and editing. All authors have read and agreed to the published version of the manuscript.

Funding: This research was funded by the U.S.-Egypt Science and Technology Joint Fund from the National Academy of Sciences and the departmental grant (L-0002-20181021) from the Welch Foundation, Texas, United States.

Institutional Review Board Statement: Not applicable.

Informed Consent Statement: Not applicable.

Data Availability Statement: Not applicable.

Acknowledgments: The author (GMN) acknowledges the financial support from the National Academy of Sciences and the Welch Foundation, Texas, United States.

Conflicts of Interest: The authors declare no conflict of interest.

References

1. Khurshid, H.; Mustafa, M.R.U.; Isa, M.H. Adsorption of chromium, copper, lead and mercury ions from aqueous solution using bio and nano adsorbents: A review of recent trends in the application of AC, BC, nZVI and MXene. *Environ. Res.* **2022**, *212*, 113138. [CrossRef] [PubMed]
2. Foroutan, R.; Peighambardoust, S.J.; Ahmadi, A.; Akbari, A.; Farjadfard, S.; Ramavandi, B. Adsorption mercury, cobalt, and nickel with a reclaimable and magnetic composite of hydroxyapatite/Fe_3O_4/polydopamine. *J. Environ. Chem. Eng.* **2021**, *9*, 105709. [CrossRef]
3. Feng, B.; Shen, W.; Shi, L.; Qu, S. Adsorption of hexavalent chromium by polyacrylonitrile–based porous carbon from aqueous solution. *R. Soc. Open Sci.* **2018**, *5*, 171662. [CrossRef]
4. Raghunathan, V.K.; Tettey, J.N.A.; Ellis, E.M.; Grant, M.H. Comparative chronic in vitro toxicity of hexavalent chromium to osteoblasts and monocytes. *J. Biomed. Mater. Res. A* **2009**, *88 Pt A*, 543–550. [CrossRef]
5. Cohen, M.D.; Kargacin, B.; Klein, C.B.; Costa, M. Mechanisms of chromium carcinogenicity and toxicity. *Crit. Rev. Toxicol.* **1993**, *23*, 255–281. [CrossRef] [PubMed]
6. Alemu, A.; Lemma, B.; Gabbiye, N. Adsorption of chromium (III) from aqueous solution using vesicular basalt rock. *Cogent Environ. Sci.* **2019**, *5*, 1650416. [CrossRef]
7. Dula, T.; Siraj, K.; Kitte, S.A. Adsorption of hexavalent chromium from aqueous solution using chemically activated carbon prepared from locally available waste of bamboo (*Oxytenanthera abyssinica*). *SRN Environ. Chem.* **2014**, *2014*, 438245. [CrossRef]
8. Dzieniszewska, A.; Kyziol-Komosinska, J.; Pająk, M. Adsorption and bonding strength of chromium species by ferrihydrite from acidic aqueous solutions. *PeerJ* **2020**, *8*, e9324. [CrossRef]
9. Kailasam, S.; Arumugam, S.; Balaji, K.; Kanth, S.V. Adsorption of chromium by exopolysaccharides extracted from lignolytic phosphate solubilizing bacteria. *Int. J. Biol. Macromol.* **2022**, *206*, 788–798. [CrossRef]
10. Yang, J.; Huang, B.; Lin, M. Adsorption of hexavalent chromium from aqueous solution by a chitosan/bentonite composite: Isotherm, kinetics, and thermodynamics studies. *J. Chem. Eng. Data* **2020**, *65*, 2751–2763. [CrossRef]
11. Ayoub, G.M.; Damaj, A.; El-Rassy, H.; Al-Hindi, M.; Zayyat, R.M. Equilibrium and kinetic studies on adsorption of chromium (VI) onto pine-needle-generated activated carbon. *SN Appl. Sci.* **2019**, *1*, 1562. [CrossRef]
12. Zhu, Y.; Wu, J.; Wang, H.; Wang, J.; Shen, H.; Ying, Z. Interference effect of experimental parameters on the mercury removal mechanism of biomass char under an oxy–fuel atmosphere. *ACS Omega* **2021**, *6*, 35124–35133. [CrossRef] [PubMed]
13. Hu, X.; Yan, L.; Wang, Y.; Xu, M. Ion–imprinted sponge produced by ice template–assisted freeze drying of salecan and graphene oxide nanosheets for highly selective adsorption of mercury (II) ion. *Carbohydr. Polym.* **2021**, *258*, 117622. [CrossRef] [PubMed]
14. Sahu, M.K.; Patel, R.K.; Kurwadkar, S. Mechanistic insight into the adsorption of mercury (II) on the surface of red mud supported nanoscale zero–valent iron composite. *J. Contam. Hydrol.* **2022**, *246*, 103959. [CrossRef]
15. Zhang, X.; Hao, Y.; Wang, X.; Chen, Z.; Li, C. Competitive adsorption of cadmium(II) and mercury(II) Ions from aqueous solutions by activated carbon from *xanthoceras sorbifolia* bunge hull. *J. Chem.* **2016**, *2016*, 4326351. [CrossRef]
16. Chen, X.; Fang, J.; Liao, S.; Mia, R.; Li, W.; Gao, C.; Tian, D.; Li, W. A smart chitosan nonwoven fabric coated with coumarin–based fluorophore for selective detection and efficient adsorption of mercury (II) in water. *Sens. Actuators B* **2021**, *342*, 130064. [CrossRef]
17. Tchounwou, P.B.; Yedjou, C.G.; Patlolla, A.K.; Sutton, D.J. Heavy metal toxicity and the environment. *Mol. Clin. Environ. Toxicol.* **2012**, *101*, 133–164. Available online: https://link.springer.com/chapter/10.1007%2F978-3-7643-8340-4_6 (accessed on 31 May 2022).
18. Zhang, Z.; Wu, D.; Guo, X.; Qian, X.; Lu, Z.; Xu, Q.; Yang, Y.; Duan, L.; He, Y.; Feng, Z. Visible study of mercuric ion and its conjugate in living cells of mammals and plants. *Chem. Res. Toxicol.* **2005**, *18*, 1814–1820. Available online: http://refhub.elsevier.com/S0925-4005(18)31621-6/sbref0035 (accessed on 2 June 2022). [CrossRef]
19. Park, J.-D.; Zheng, W. Human exposure and health effects of inorganic and elemental mercury. *J. Prev. Med. Public Health* **2012**, *45*, 344–352. [CrossRef]

20. Awad, F.S.; Bakry, A.M.; Ibrahim, A.A.; Lin, A.; El-Shall, M.S. Thiol– and amine–incorporated UIO–66–NH_2 as an efficient adsorbent for the removal of mercury(II) and phosphate ions from aqueous solutions. *Ind. Eng. Chem. Res.* **2021**, *60*, 12675–12688. [CrossRef]
21. Shen, J.; Zhang, S.; Zeng, Z.; Huang, J.; Shen, Y.; Guo, Y. Synthesis of magnetic short–channel mesoporous silica SBA–15 modified with a polypyrrole/polyaniline copolymer for the removal of mercury Ions from aqueous solution. *ACS Omega* **2021**, *6*, 25791–25806. [CrossRef] [PubMed]
22. Schroeder, W.H.; Munthe, J. Atmospheric mercury—An overview. *Atmos. Environ.* **1998**, *32*, 809–822. [CrossRef]
23. Ma, L.; Han, L.; Chen, S.; Hu, J.; Chang, L.; Bao, W.; Wang, J. Rapid synthesis of magnetic zeolite materials from fly ash and iron–containing wastes using supercritical water for elemental mercury removal from flue gas. *Fuel Process. Technol.* **2019**, *189*, 39–48. [CrossRef]
24. Jang, E.-H.; Pack, S.P.; Kim, I.; Chung, S. A systematic study of hexavalent chromium adsorption and removal from aqueous environments using chemically functionalized amorphous and mesoporous silica nanoparticles. *Sci. Rep.* **2020**, *10*, 5558. [CrossRef]
25. Kumar, S.; Nair, R.R.; Pillai, P.B.; Gupta, S.N.; Iyengar, M.A.R.; Sood, A.K. Graphene oxide−$MnFe_2O_4$ Magnetic Nanohybrids for efficient removal of lead and arsenic from water. *ACS Appl. Mater. Interfaces* **2014**, *6*, 17426–17436. [CrossRef]
26. Ramos-Hernández, L.E.; Pérez-Aguilar, N.V.; Ovando-Medina, V.M.; Oyervides-Muñoz, E.; Arcibar-Orozco, J.A. Arcibar–Orozco, Photoinduced adsorption of Cr(VI) ions in nano–zinc oxide and nano–zinc oxide/polypyrrole composite. *J. Appl. Polym. Sci.* **2022**, *139*, e52225. [CrossRef]
27. Liu, H.; Zhang, Z.; Shen, F.; Zhou, Y.; Liu, J.; Yang, H. Two–dimensional WS_2 as a new mercury removal material: Mercury conversion pathway and effect of defect. *Fuel* **2022**, *307*, 121864. [CrossRef]
28. Vicente-Martínez, Y.; Caravaca, M.; Soto-Meca, A. Simultaneous adsorption of mercury species from aquatic environments using magnetic nanoparticles coated with nanomeric silver functionalized with L–Cysteine. *Chemosphere* **2021**, *282*, 131128. [CrossRef]
29. Bayuo, J.; Rwiza, M.J.; Sillanpa, M.; Mteia, K.M. Removal of heavy metals from binary and multicomponent adsorption systems using various adsorbents—A systematic review. *RSC Adv.* **2023**, *13*, 13052–13093. [CrossRef]
30. Wan, K.; Wang, G.; Xue, S.; Xiao, Y.; Fan, J.; Li, L.; Miao, Z. Preparation of humic acid/L-cysteine–codecorated magnetic Fe_3O_4 nanoparticles for selective and highly efficient adsorption of mercury. *ACS Omega* **2021**, *6*, 7941–7950. [CrossRef]
31. Yang, Y.; Huang, R.; Xu, W.; Zhang, J.; Li, C.; Song, J.; Zhu, T. Different crystal forms of ZnS nanomaterials for the adsorption of elemental mercury. *Environ. Sci. Technol.* **2021**, *55*, 6965–6974. [CrossRef]
32. Selvi, K.; Pattabhi, S.; Kadirvelu, K. Removal of Cr(VI) from aqueous solution by adsorption onto activated carbon. *Bioresour. Technol.* **2001**, *80*, 87–89. [CrossRef]
33. Yang, T.; Han, C.; Tang, J.; Luo, Y. Removal performance and mechanisms of Cr(VI) by an in-situ self-improvement of mesoporous biochar derived from chicken bone. *Environ. Sci. Pollut. Res.* **2020**, *27*, 5018–5029. [CrossRef]
34. Li, L.-L.; Feng, X.-Q.; Han, R.-P.; Zang, S.-Q.; Yang, G. Cr(VI) removal via anion exchange on a silver–triazolate MOF. *J. Hazard. Mater.* **2017**, *321*, 622–628. [CrossRef]
35. Galán, B.; Castañeda, D.; Ortiz, I. Removal and recovery of Cr(VI) from polluted ground waters: A comparative study of ion exchange technologies. *Water Res.* **2005**, *39*, 4317–4324. [CrossRef]
36. Kozlowski, C.A.; Walkowiak, W. Removal of chromium(VI) from aqueous solutions by polymer inclusion membranes. *Water Res.* **2002**, *36*, 4870–4876. [CrossRef]
37. Adhoum, N.; Monser, L.; Bellakhal, N.; Belgaied, J.-E. Treatment of electroplating wastewater containing Cu^{2+}, Zn^{2+} and Cr(VI) by electrocoagulation. *J. Hazard. Mater.* **2004**, *112*, 207–213. [CrossRef] [PubMed]
38. Golbaz, S.; Jafari, A.J.; Rafiee, M.; Kalantary, R.R. Separate and simultaneous removal of phenol, chromium, and cyanide from aqueous solution by coagulation/precipitation: Mechanisms and theory. *Chem. Eng. J.* **2014**, *253*, 251–257. [CrossRef]
39. Dupont, L.; Guillon, E. Removal of hexavalent chromium with a lignocellulosic substrate extracted from wheat bran. *Environ. Sci. Technol.* **2003**, *37*, 4235–4241. [CrossRef] [PubMed]
40. Peng, C.; Meng, H.; Song, S.; Lu, S.; Lopez-Valdivieso, A. Elimination of Cr(VI) from electroplating wastewater by electrodialysis following chemical precipitation. *Sep. Sci. Technol.* **2004**, *39*, 1501–1517. [CrossRef]
41. Kongsricharoern, N.; Polprasert, C. Chromium removal by a bipolar electro–chemical precipitation process. *Water Sci. Technol.* **1996**, *34*, 109–116. [CrossRef]
42. Mulani, K.; Daniels, S.; Rajdeo, K.; Tambe, S.; Chavan, N. Adsorption of chromium(VI) from aqueous solutions by coffee polyphenol–formaldehyde/acetaldehyde resins. *J. Polym.* **2013**, *2013*, 798368. [CrossRef]
43. Qu, G.; Zhou, J.; Liang, S.; Li, Y.; Ning, P.; Pan, K.; Ji, W.; Tang, H. Thiol–functionalized multi–walled carbon nanotubes for effective removal of Pb(II) from aqueous solutions. *Mater. Chem. Phys.* **2022**, *278*, 125688. [CrossRef]
44. Egbosiuba, T.C.; Egwunyenga, M.C.; Tijani, J.O.; Mustapha, S.; Abdulkareem, A.S.; Kovo, A.S.; Krikstolaityte, V.; Veksha, A.; Wagner, M.; Lisak, G. Activated multi–walled carbon nanotubes decorated with zero valent nickel nanoparticles for arsenic, cadmium and lead adsorption from wastewater in a batch and continuous flow modes. *J. Hazard. Mater.* **2022**, *423*, 126993. [CrossRef] [PubMed]
45. Wang, Z.; Xu, W.; Jie, F.; Zhao, Z.; Zhou, K.; Liu, H. The selective adsorption performance and mechanism of multiwall magnetic carbon nanotubes for heavy metals in wastewater. *Sci. Rep.* **2021**, *11*, 16878. [CrossRef]

46. Guldi, D.M.; Rahman, G.M.A.; Zerbetto, F.; Prato, M. Carbon nanotubes in electron donor–acceptor nanocomposites. *Accounts Chem. Res.* **2005**, *38*, 871–878. [CrossRef]
47. Lin, Y.; Taylor, S.; Li, H.; Fernando, K.A.S.; Qu, L.; Wang, W.; Gu, L.; Zhou, B.; Sun, Y.-P. Advances toward bioapplications of carbon nanotubes. *J. Mater. Chem.* **2004**, *14*, 527–541. [CrossRef]
48. Pathak, A.; Gupta, B.D. Palladium nanoparticles embedded PPy shell coated CNTs towards a high performance hydrazine detection through optical fiber plasmonic sensor. *Sens. Actuators B* **2021**, *326*, 128717. [CrossRef]
49. Schroeder, V.; Savagatrup, S.; He, M.; Lin, S.; Swager, T.M. Carbon nanotube chemical sensors. *Chem. Rev.* **2019**, *119*, 599–663. [CrossRef]
50. Silva, M.M.; Lopes, P.E.; Li, Y.; Pötschke, P.; Ferreira, F.N.; Paiva, M.C. Polylactic acid/carbon nanoparticle composite filaments for sensing. *Appl. Sci.* **2021**, *11*, 2580. [CrossRef]
51. Nagarajan, V.; Mohanty, A.K.; Misra, M. Perspective on polylactic acid (PLA) based sustainable materials for durable applications: Focus on toughness and heat resistance. *ACS Sustain. Chem. Eng.* **2016**, *4*, 2899–2916. [CrossRef]
52. Neelgund, G.M.; Oki, A. Pd nanoparticles deposited on poly(lactic acid) grafted carbon nanotubes: Synthesis, characterization, and application in Heck C–C coupling reaction. *Appl. Catal. A Gen.* **2011**, *399*, 154–160. [CrossRef] [PubMed]
53. Neelgund, G.M.; Aguilar, S.F.; Kurkuri, M.D.; Rodrigues, D.F.; Ray, R.L. Elevated adsorption of lead and arsenic over silver nanoparticles deposited on poly(amidoamine) grafted carbon nanotubes. *Nanomaterials* **2022**, *12*, 3852. [CrossRef] [PubMed]
54. Neelgund, G.M.; Aguilar, S.F.; Jimenez, E.A.; Ray, R.L. Adsorption efficiency and photocatalytic activity of silver sulfide nanoparticles deposited on carbon nanotubes. *Catalysts* **2023**, *13*, 476. [CrossRef]
55. Yuniarto, K.; Purwanto, Y.A.; Purwanto, S.; Welt, B.A.; Purwadaria, H.K.; Sunarti, T.C. Infrared and Raman studies on polylactide acid and polyethylene glycol-400 blend. *AIP Conf. Proc.* **2016**, *1725*, 020101. [CrossRef]
56. Neelgund, G.M.; Oki, A. Photocatalytic activity of CdS and Ag_2S quantum dots deposited on poly(amidoamine) functionalized carbon nanotubes. *Appl. Catal. B Environ.* **2011**, *110*, 99–107. [CrossRef]
57. Neelgund, G.M.; Oki, A. Contribution of polylactic acid and Pd nanoparticles in the enhanced photothermal effect of carbon nanotubes. *Chemistryselect* **2020**, *5*, 11020–11028. [CrossRef]
58. Xiong, Y.; Xia, Y. Shape–controlled synthesis of metal nanostructures: The case of palladium. *Adv. Mater.* **2007**, *19*, 3385–3391. [CrossRef]
59. Winjobi, O.; Zhang, Z.; Liang, C.; Li, W. Carbon nanzotube supported platinum–Palladium nanoparticles for formic acid oxidation. *Electrochim. Acta* **2010**, *55*, 4217–4221. [CrossRef]
60. Wang, Y.; He, Q.; Ding, K.; Wei, H.; Guo, J.; Wang, Q.; O'Connor, R.; Huang, X.; Luo, Z.; Shen, T.D.; et al. Multiwalled carbon nanotubes composited with palladium nanocatalysts for highly efficient ethanol oxidation. *J. Electrochem. Soc.* **2015**, *162*, F755–F763. [CrossRef]
61. Li, Y.; Zhou, W.; Wang, H.; Xie, L.; Liang, Y.; Wei, F.; Idrobo, J.-C.; Pennycook, S.J.; Dai, H. An oxygen reduction electrocatalyst based on carbon nanotube–graphene complexes. *Nat. Nanotechnol.* **2012**, *7*, 394–400. [CrossRef] [PubMed]
62. Mahendiran, C.; Rajesh, D.; Maiyalagan, T.; Prasanna, K. Pd nanoparticles-supported carbon nanotube-encapsulated NiO/MgO composite as an enhanced electrocatalyst for ethanol electrooxidation in alkaline medium. *Chemistryselect* **2017**, *2*, 11438–11444. [CrossRef]
63. She, Y.; Lu, Z.; Fan, W.; Jewell, S.; Leung, M.K.H. Leung, Facile preparation of PdNi/rGO and its electrocatalytic performance towards formic acid oxidation. *J. Mater. Chem. A* **2014**, *2*, 3894–3898. [CrossRef]
64. Tan, H.T.; Chen, Y.; Zhou, C.; Jia, X.; Zhu, J.; Chen, J.; Rui, X.; Yan, Q.; Yang, Y. Palladium nanoparticles supported on manganese oxide–CNT composites for solvent-free aerobic oxidation of alcohols: Tuning the properties of Pd active sites using MnOx. *Appl. Catal. B* **2012**, *119–120*, 166–174. [CrossRef]
65. Joshi, S.; Sharma, M.; Kumari, A.; Shrestha, S.; Shrestha, B. Arsenic removal from water by adsorption onto iron oxide/nano-porous carbon magnetic composite. *Appl. Sci.* **2019**, *9*, 3732. [CrossRef]
66. Zeng, H.; Zhai, L.; Qiao, T.; Yu, Y.; Zhang, J.; Li, D. Efficient removal of As(V) from aqueous media by magnetic nanoparticles prepared with iron-containing water treatment residuals. *Sci. Rep.* **2020**, *10*, 9335. [CrossRef]
67. Weber, W.J., Jr.; Morris, J.C. Kinetics of adsorption on carbon from solution. *J. Sanit. Eng. Div. Proceed. Am. Soc. Civ. Eng.* **1963**, *89*, 31–59. [CrossRef]
68. Hameed, B.; Salman, J.; Ahmad, A. Adsorption isotherm and kinetic modeling of 2,4-D pesticide on activated carbon derived from date stones. *J. Hazard. Mater.* **2009**, *163*, 121–126. [CrossRef]
69. Hamayun, M.; Mahmood, T.; Naeem, A.; Muska, M.; Din, S.; Waseem, M. Equilibrium and kinetics studies of arsenate adsorption by $FePO_4$. *Chemosphere* **2013**, *99*, 207–215. [CrossRef]
70. Langmuir, I. The constitution and fundamental properties of solids and liquids Part I. solids. *J. Am. Chem. Soc.* **1916**, *38*, 2221–2295. [CrossRef]
71. Freundlich, H.M.F. Over the adsorption in solution. *J. Phys. Chem.* **1906**, *57*, 385–471.
72. Ploychompoo, S.; Chen, J.; Luo, H.; Liang, Q. Fast and efficient aqueous arsenic removal by functionalized MIL-100(Fe)/rGO/d-MnO_2 ternary composites: Adsorption performance and mechanism. *J. Environ. Sci.* **2020**, *91*, 22–34. [CrossRef] [PubMed]
73. Temkin, M.J.; Pyzhev, V. Recent modifications to Langmuir isotherms. *Acta Physiochim. USSR* **1940**, *12*, 217–222.

74. Tabuchi, A.; Ogata, F.; Uematsu, Y.; Toda, M.; Otani, M.; Saenjum, C.; Nakamura, T.; Kawasaki, N. Granulation of nickel-aluminum–zirconium complex hydroxide using colloidal silica for adsorption of chromium(VI) ions from the liquid phase. *Molecules* **2022**, *27*, 2392. [CrossRef]
75. Al-Yaari, M.; Saleh, T.A. Mercury Removal from water using a novel composite of polyacrylate-modified carbon. *ACS Omega* **2022**, *7*, 14820–14831. [CrossRef]
76. Wang, Q.; Zhu, S.; Xi, C.; Jiang, B.; Zhang, F. Adsorption and removal of mercury(II) by a crosslinked hyperbranched polymer modified via sulfhydryl. *ACS Omega* **2022**, *7*, 12231–12241. [CrossRef]
77. Kang, C.; Gao, L.; Zhu, H.; Lang, C.; Jiang, J.; Wei, J. Adsorption of Hg(II) in solution by mercaptofunctionalized palygorskite. *Environ. Sci. Pollut. Res.* **2021**, *28*, 66287–66302. [CrossRef]
78. Li, Y.; Li, W.; Liu, Q.; Meng, H.; Lu, Y.; Li, C. Alkynyl carbon materials as novel and efficient sorbents for the adsorption of mercury(II) from wastewater. *J. Environ. Sci.* **2018**, *68*, 169–176. [CrossRef]
79. Shen, W.; Fang, Y.; Azeem, M.; Gao, Y.; Li, X.; Zhao, P.; Ali, A.; Li, M.; Li, R. Chitosan crosslinked with polyamine-co-melamine for adsorption of Hg^{2+}: Application in purification of polluted water. *Int. J. Biol. Macromol.* **2021**, *181*, 778–785. [CrossRef]
80. Geneti, S.T.; Mekonnen, G.A.; Murthy, H.C.A.; Mohammed, E.T.; Ravikumar, C.R.; Gonfa, B.A.; Sabir, F.K. Biogenic synthesis of magnetite nanoparticles using leaf extract of thymus schimperi and their application for monocomponent removal of chromium and mercury ions from aqueous solution. *J. Nanomater.* **2022**, *2022*, 5798824. [CrossRef]
81. Villabona-Ortíz, A.; Tejada-Tovar, C.; González-Delgado, D. Elimination of chromium (VI) and nickel (II) ions in a packed column using oil palm bagasse and yam peels. *Water* **2022**, *14*, 1240. [CrossRef]
82. Vaddi, D.R.; Gurugubelli, T.R.; Koutavarapu, R.; Lee, D.-Y.; Shim, J. Bio-stimulated adsorption of Cr(VI) from aqueous solution by groundnut shell activated carbon@Al embedded material. *Catalysts* **2022**, *12*, 290. [CrossRef]
83. Qi, H.; Niu, X.; Wu, H.; Liu, X.; Chen, Y. Adsorption of chromium (VI) by Cu (I)-MOF in water: Optimization, kinetics, and thermodynamics. *J. Chem.* **2021**, *2021*, 4413095. [CrossRef]

Disclaimer/Publisher's Note: The statements, opinions and data contained in all publications are solely those of the individual author(s) and contributor(s) and not of MDPI and/or the editor(s). MDPI and/or the editor(s) disclaim responsibility for any injury to people or property resulting from any ideas, methods, instructions or products referred to in the content.

toxics

MDPI

Article

Sustainable Development of ZnO Nanostructure Doping with Water Hyacinth-Derived Activated Carbon for Visible-Light Photocatalysis

Sucheewan Krobthong [1], Tipawan Rungsawang [1], Naphatson Khaodara [1], Napat Kaewtrakulchai [2], Kanit Manatura [3], Khewika Sukiam [4], Donchida Wathinputthiporn [5], Sawitree Wongrerkdee [6], Chatdanai Boonruang [7,8] and Sutthipoj Wongrerkdee [1,*]

1 Department of Physical and Material Sciences, Faculty of Liberal Arts and Science, Kasetsart University, Kamphaeng Saen Campus, Nakhon Pathom 73140, Thailand
2 Kasetsart Agricultural and Agro-Industrial Product Improvement Institute, Kasetsart University, Bangkok 10900, Thailand
3 Department of Mechanical Engineering, Faculty of Engineering at Kamphaeng Saen, Kasetsart University, Kamphaeng Saen Campus, Nakhon Pathom 73140, Thailand
4 Department of Tourism and Aviation Business, Faculty of Hospitality Industry, Kasetsart University, Kamphaeng Saen Campus, Nakhon Pathom 73140, Thailand
5 Department of Agricultural Extension and Communication, Faculty of Agriculture at Kamphaeng Saen, Kasetsart University, Kamphaeng Saen Campus, Nakhon Pathom 73140, Thailand
6 Faculty of Engineering, Rajamangala University of Technology Lanna Tak, Tak 63000, Thailand
7 Department of Physics and Materials Science, Faculty of Science, Chiang Mai University, Chiang Mai 50200, Thailand
8 Center of Excellence in Materials Science and Technology, Chiang Mai University, Chiang Mai 50200, Thailand
* Correspondence: sutthipoj.s@ku.ac.th

Abstract: Water hyacinth (Wh) is an aquatic weed considered a nuisance in agricultural and fishing activities. Therefore, this study proposed repurposing this plant into activated carbon (AC). First, the ZnO-AC was precipitated and applied as a photocatalyst for degrading methylene blue. The preliminary photocatalytic test under UV irradiation identified the optimum ZnO-AC photocatalyst to degrade methylene blue (MB). The ZnO-AC photocatalyst recorded the highest degradation rate constant of 11.49×10^{-3} min^{-1}, which was almost two-fold higher than that of ZnO (5.55×10^{-3} min^{-1}). Furthermore, photocatalytic degradation of MB and carbaryl under sunlight irradiation by ZnO-AC demonstrated degradation rate constants of 74.46×10^{-3} min^{-1} and 8.43×10^{-3} min^{-1}, respectively. To investigate the properties of ZnO-AC, several techniques were performed. ZnO-AC and ZnO exhibited similar results in morphology, crystalline structure, and Raman characteristics. However, ZnO-AC presented smaller pore diameters than those of ZnO, which enlarged pore surface area, and the presence of carbon-related groups implied the presence of AC on ZnO-AC surfaces. This can be attributed to the presence of AC on the ZnO surface, increasing the capture of surrounding toxic molecules and elevating the reaction density. This mechanism is attributed to promoting the degradation of toxic molecules. Therefore, using Wh as a carbon source for the transformation of AC can alternatively solve the problems of aquatic weed management and carbon storage strategies, and the application of AC in ZnO-AC photocatalysts can enhance photocatalysis.

Keywords: water hyacinth; activated carbon; ZnO; photocatalysis

1. Introduction

Numerous methodologies have been developed to manage wastewater contaminants. Photocatalysis is one of the most promising approaches due to its efficacy, non-selectivity, and efficiency, which enable repeated usage of the photocatalyst [1–3]. Furthermore, photocatalysis facilitates the removal of various contaminants via mineralization or decomposition of intricate pollutants, such as landfill leachate, into simpler components like water, carbon dioxide, and inorganic ions. Consequently, photocatalysis has been proven to be effective for mineralizing complex contaminants, offering a viable alternative technology capable of degrading and eliminating complex pollutants. Notably, researchers have greatly utilized this method, particularly in developing nano-photocatalysts. Nanostructured photocatalysts are superior to their conventional counterparts as they offer larger surface areas and a higher prevalence of lattice defects [4]. These features provide additional reactive sites and excellent light absorption properties, thus improving photocatalytic efficiency. Several conventional techniques for fabricating nanoscale photocatalysts include hydrothermal, solvothermal, sol-gel, and precipitation methods have been investigated [5–7]. These approaches enable meticulous regulation of the photocatalyst's particle size and morphology.

Metal oxide semiconductors such as TiO_2, SnO_2, and ZnO have been investigated in recent studies. Reports have highlighted ZnO as a crucial semiconducting photocatalyst due to its photosensitivity, stability, low toxicity, availability, excellent electron mobility, cost-effectiveness, and flexibility for use in several synthesis techniques [4,8]. However, the rapid recombination of electron–hole pairs in its structure limits its photocatalytic application [9]. Therefore, reducing recombination and inducing carrier migration in photocatalysts may offer a solution to this issue. These improvements could be achieved by fabricating composite structures, such as fabricating ZnO/Cu-DPA nanocomposites with different levels of Cu-DPA [10]. Photocatalytic activities of nanocomposites are assessed based on the ability to degrade methylene blue (MB) under visible-light irradiation. An earlier study reported that the maximum degradation efficiency and degradation rate of the optimum ZnO/Cu DPA nanocomposite (78.5%, 23×10^{-3} min^{-1}) were higher than those for ZnO (71.8%, 16×10^{-3} min^{-1}). This enhancement was caused by the formation of a p–n heterojunction within the composite system. In the ZnO/Cu-DPA nanocomposite, Cu_2O is responsible for the photocatalytic activity due to its strong absorption of visible light, promoting optimal separation of electron–hole pairs. Moreover, heterostructure formation with ZnO prevented the recombination of photogenerated charge carriers and confined the photo-excited charges at intrinsic defect sites within the ZnO structure. The ZnO/Cu-DPA nanocomposite facilitates the conversion of dissolved oxygen (O_2) into hydroxyl ($^{\bullet}OH$) radicals, suggesting that $^{\bullet}OH$ is the primary active radical involved in the photocatalytic decomposition of MB. The ZnO/CuO/g-C_3N_4 (ZCG) heterostructure nanocomposite was synthesized via the co-crystallization method and employed in a photocatalytic process to eliminate MB dye from wastewater under visible light [11]. The MB degradation record was as high as 97.46% within 50 min, exhibiting superior performance to other single photocatalysts such as ZnO, CuO, and g-C_3N_4. This enhanced performance is attributed to the formation of heterojunction structures, promoting efficient charge transfer and reducing recombination rates [12]. Likewise, the ZnO-graphene oxide (GO) composite photocatalyst for vanillic acid (VA) degradation recorded a degradation efficiency of 99% and 35% under solar light and visible-LED, respectively [13]. This outcome is possibly caused by the adsorption of GO on ZnO. In addition, GO extends light absorption into the visible region, increasing the light harvesting effect and boosting photocatalytic activity.

In this study, carbon-based materials were composited with ZnO to improve photocatalytic activity. Activated carbon (AC) is an alternative to graphene-based materials, and it can be prepared at a low cost from any organic waste, such as agricultural wastes or weeds. Water hyacinth (Wh) is an aquatic weed deemed a nuisance in agricultural and fishing activities. Repurposing this plant as a raw material to produce AC that is composited with ZnO is a promising approach to producing low-cost, carbon-based materials. Furthermore, this strategy allows for the proper waste management of aquatic weeds, positively impacting the environment and farmers. The AC was prepared using a hydrothermal process with activating chemicals. Therefore, this study precipitated ZnO with AC to produce ZnO-AC for photocatalysis.

2. Materials and Methods

The zinc acetate solution was prepared by dissolving 4.39 g zinc acetate 2-hydrate ($Zn(CH_3COO)_2{\cdot}2H_2O$; KEMAUS, AR, MW 219.49) in 100 mL de-ionized (DI) water. Meanwhile, AC mixtures (1, 3, 5, and 10% of the total mass) were added to zinc acetate solution in separate beakers. The AC was prepared via high-temperature carbonization and KOH activation of dried Wh waste, as described in earlier studies [14]. The beakers were heated to 70 °C and stirred continuously for 1 h. Subsequently, the precipitation process was carried out. First, the ammonium solution was prepared by dissolving 3.16 g ammonium bicarbonate (NH_4HCO_3, DAEJUNG, AR, MW 79.06) in 100 mL DI water, followed by heating and stirring under similar conditions as for the zinc acetate preparation. The ammonium solution was added to the zinc acetate or mixture solutions drop-by-drop while heating and stirring to precipitate $Zn(OH)_2$ or $Zn(OH)_2$-AC sediment. After 1 h of precipitation, the $Zn(OH)_2$ or $Zn(OH)_2$-AC sediments were filtered using filter paper for 12 h, dried at 70 °C for 1 h, and ground for 3 h. Finally, the $Zn(OH)_2$ and $Zn(OH)_2$-AC sediments were calcined at 600 °C for 6 h and ground for 1 h to obtain ZnO or ZnO-AC.

The ZnO or ZnO-AC were utilized as photocatalysts to degrade methylene blue (MB) under ultra-violet (UV) light. The MB was dissolved in DI water at an initial concentration of 5 mg/L, stirring under dark conditions at room temperature for 30 min. Subsequently, 100 mg ZnO or ZnO-AC was added into 100 mL MB solution while stirring at room temperature in the dark. The MB solution containing ZnO or ZnO-AC was allowed to undergo adsorption–desorption stabilization for 30 min, followed by irradiation under UV light to activate photocatalysis to degrade MB molecules. The MB solution (3 mL) was sampled at different intervals, and the absorbance was measured via UV–Vis spectroscopy.

To characterize ZnO and ZnO-AC, morphology was observed using a transmission electron microscope (TEM; JEOL JEM-2100 Plus; JEOL Ltd., Tokyo, Japan). The crystalline structure was evaluated using an x-ray diffractometer (XRD; Rigaku, SmartLab, Rigaku, Japan). In addition, a Raman spectrometer (Thermo Scientific, DXR SmartRaman, Waltham, MA, USA) was used to analyze vibrational characteristics. Meanwhile, a surface area and pore-size analyzer (Quantachrome, Autosorb iQ-C-XR-XR-XR, Graz, Austria) was used to determine porosity properties based on the N_2 adsorption–desorption method. The functional group was analyzed via the Fourier transform infrared (FTIR) spectrometer (Spectrum Two, PerkinElmer, Norwalk, CT, USA), while x-ray photoelectron spectroscopy (XPS, Kratos, Axis Ultra DLD, Kratos Analytical, Ltd., Manchester, UK) was used to monitor surface composition and chemical states.

3. Results and Discussion

Both ZnO and ZnO-AC were utilized as UV-activated photocatalysts to degrade MB molecules. Figure 1a–e illustrates the absorbance of MB samples after photocatalysis at different UV-irradiation intervals. There was a decreasing trend in absorbance with irradiation time, indicating that MB molecule degradation in water was catalyzed by the photocatalysts used in this study. The activity of MB molecules under UV irradiation without photocatalysts (blank) is shown in Figure 1f, demonstrating no significant changes in absorbance. This finding implies that MB molecules cannot be degraded effectively without photocatalysts, thus confirming the efficacy of ZnO and ZnO-AC in this study. Additionally, the remaining concentration to initial concentration (C_t/C_0) ratios of MB solutions were plotted comparatively, as shown in Figure 2a, to determine the optimum photocatalyst. The degradation efficiency (DE), which corresponds to the ratio of decomposition to the initial MB molecules, was calculated using Equation (1) [15–17].

$$DE\ (\%) = (1 - (C_t/C_0)) \times 100\% \quad (1)$$

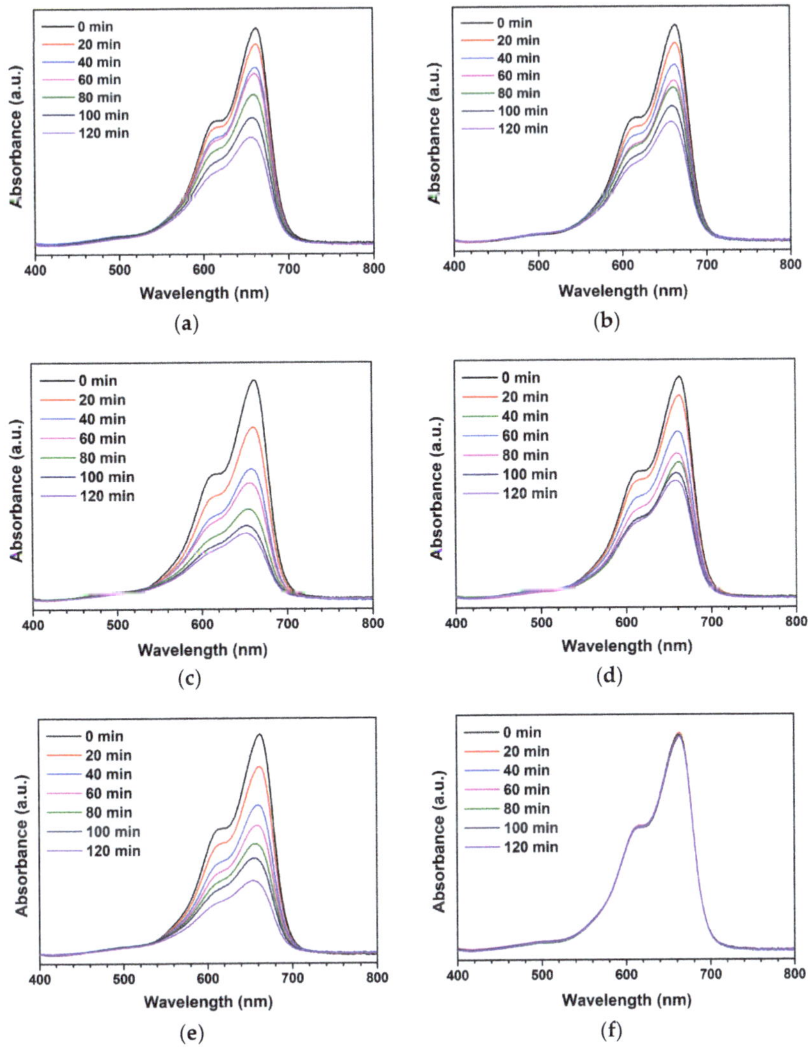

Figure 1. Absorbance of MB after photocatalytic activity at varying intervals of UV-irradiation time using different photocatalysts: (**a**) ZnO, (**b**) ZnO-AC1%, (**c**) ZnO-AC3%, (**d**) ZnO-AC5%, (**e**) ZnO-AC10%, and (**f**) blank.

The decreasing C_t/C_0 trend for the ZnO-AC3% sample was the lowest compared to other conditions, which correlated with higher degradation efficiency (Figure 2b). Further quantitative analysis of photocatalytic performance was performed to determine the degradation rate constant (k_r) (Figure 2c) using Equation (2). In addition, half-life (τ) was calculated to estimate the irradiation time required to reduce MB molecules by half of the initial concentration using Equation (3) [18,19]

$$\ln(C_0/C_t) = k_r t \quad (2)$$

$$\tau = (1/k_r)\ln(2) \quad (3)$$

where t is irradiation time (min). Table 1 presents the degradation rate constant and half-life for each sample.

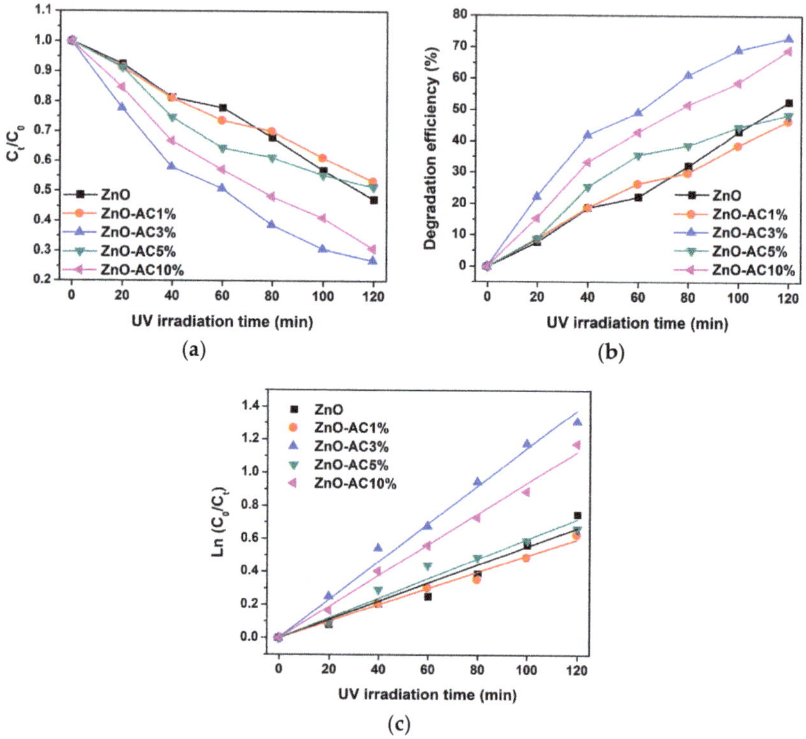

Figure 2. Analysis of photocatalytic degradation of MB under UV irradiation: (**a**) C_t/C_0 ratio, (**b**) degradation efficiency, and (**c**) degradation rate constant.

Table 1. Analytical parameters of MB degradation under UV irradiation.

Sample	k_r (10^{-3} min^{-1})	τ (min)	R^2
ZnO	5.55	125	0.9814
ZnO-AC1%	4.98	139	0.9960
ZnO-AC3%	11.49	60	0.9966
ZnO-AC5%	5.99	116	0.9884
ZnO-AC10%	9.38	74	0.9975

The highest degradation rate constant and lowest half-life were recorded for the ZnO-AC3% sample, suggesting its superiority as a photocatalyst. This result matches a previous study describing a ZnO/AC nanocomposite photocatalyst [20]. In that study, AC was prepared from *Prosopis juliflora*. This could suggest that the optimal content of AC in ZnO-AC structures is 3%. In comparison to other ZnO-based photocatalyst materials (Table 2), ZnO-AC3% can be considered for use in efficient photocatalyst applications.

Table 2. Previously reported analytical degradation rate constant of MB degradation.

Sample	MB Concentration	Photocatalyst Dosage in MB Solution	Light Source	k_r (10^{-3} min^{-1})	Ref.
Ti-ZnO	5 mg/L	0.1 g/100 mL	UV	2.54	[5]
GQDs-ZnO	10 mg/L	0.2 g/100 mL	UV	3.79	[21]
rGO@ZnO	15 mg/L	20 mg/100 mL	Sunlight	5.03	[22]
N-ZnO	10 mg/L	0.1 g/100 mL	Sunlight	19.6	[23]
Cu-ZnO	10 mg/L	25 mg/100 mL	UV	25.4	[24]
ZnO-AC3%	5 mg/L	0.1 g/100 mL	UV	11.49	This work

GQDs: graphene quantum dots; Ti: titanium; rGO: reduced graphene oxide; N: nitrogen; Cu: copper.

Thus, the ZnO-AC3% sample was further analyzed to determine its ability to degrade MB under natural sunlight [19]. Furthermore, the photocatalytic degradation of carbaryl (CBR) insecticide was also investigated by evaluating the potential of photocatalyst application in agricultural chemical degradation, demonstrating a simple way for environmental redemption. The CBR concentration was prepared at an initial concentration of 1 mg/L in DI water. Meanwhile, the average sunlight intensity was 917 W/m^2. The absorbance of MB and CBR after photocatalytic degradation and the degradation rate constants of MB and CBR under natural sunlight with facilitation by photocatalysts are illustrated in Figure 3. The degradation rate constant was 50.61×10^{-3} min^{-1} for ZnO and 74.46×10^{-3} min^{-1} for ZnO-AC3%, indicating the superior performance of the latter in MB degradation. A similar trend was observed for CBR degradation, where ZnO-AC3% recorded a degradation rate constant of 8.43×10^{-3} min^{-1} and this was 5.44×10^{-3} min^{-1} for ZnO. Therefore, ZnO-AC3% has been proven effective as a photocatalyst for MB and CBR degradation, offering a potential application in contaminated areas under natural sunlight.

The ZnO-AC3% was also utilized as a representative ZnO-AC structure for comparison with pristine ZnO for TEM analysis (Figure 4). The images revealed that both photocatalysts had spherical-like nanostructures and comparable sizes, suggesting no discernible impact of AC on the morphological structure of ZnO. Nonetheless, other characteristics of AC structures, such as flake-, sheet-, or plate-like structures, were not identified due to the limited sample availability.

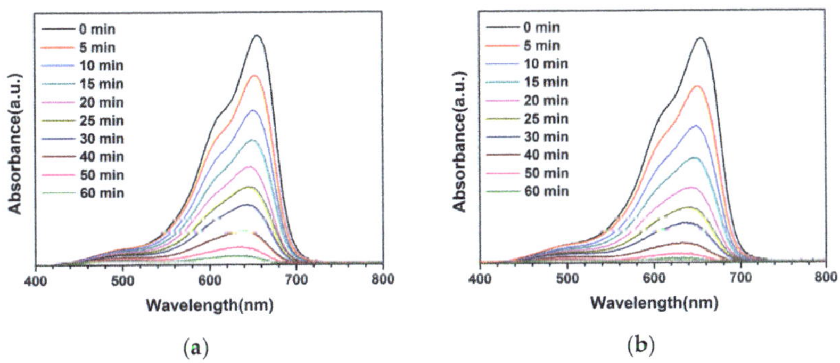

Figure 3. *Cont.*

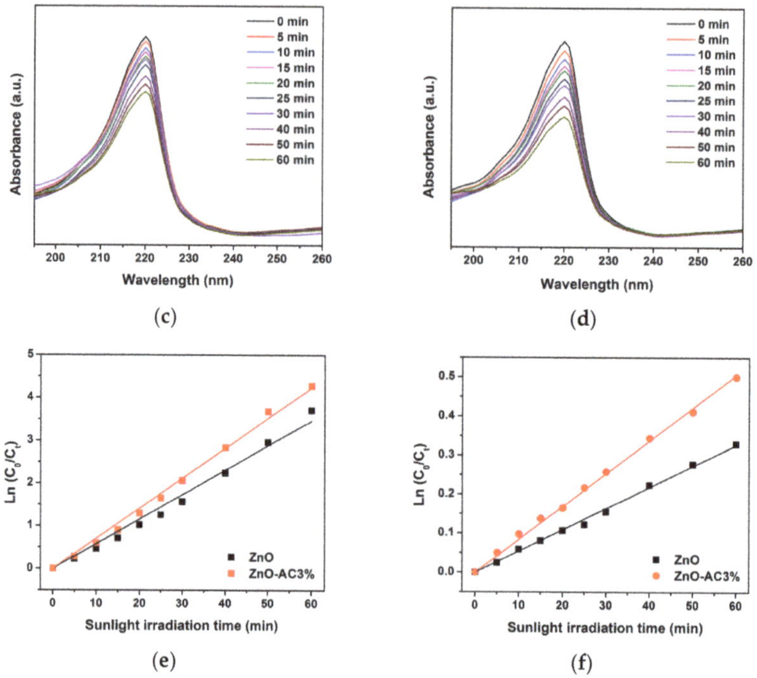

Figure 3. Photocatalytic degradation analysis under natural sunlight irradiation: absorbance of MB using (**a**) ZnO and (**b**) ZnO-AC3%, absorbance of CBR using (**c**) ZnO and (**d**) ZnO-AC3%, and degradation rate constant of (**e**) MB and (**f**) CBR degradation.

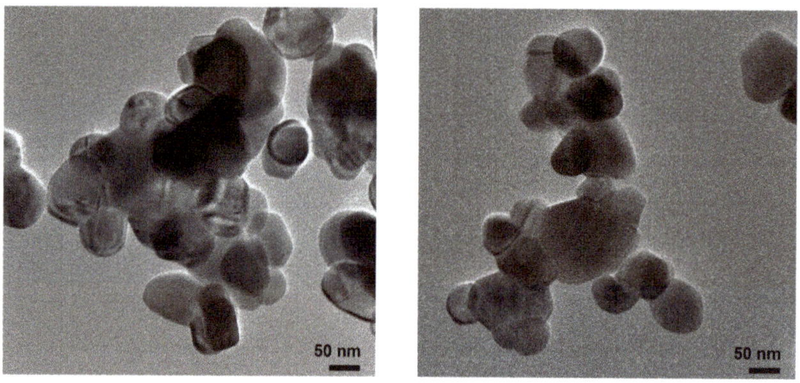

(**a**) (**b**)

Figure 4. TEM images of (**a**) ZnO and (**b**) ZnO-AC nanostructures.

The crystalline structures of AC, ZnO, and ZnO-AC were also evaluated through analysis of their XRD patterns (Figure 5). Carbon-related peaks were unclearly observed for AC. This indicated an amorphous structure for the AC sample. Regarding the unidentified peaks, there should be several mineral components in AC due to the use of biomass as a raw material. For ZnO and ZnO-AC, consistent diffraction peaks were evident with distinct patterns observed at 2θ of 31.8°, 34.6°, 36.4°, 47.6°, 56.9°, 63.0°, 66.6°, 68.1°, and 69.3° that correspond to the (100), (002), (101), (102), (110), (103), (200), (112), and (201) diffraction planes, respectively. These characteristic planes strongly suggested the presence of a ZnO-hexagonal-wurtzite structure in both photocatalysts, which aligned with the JCPDs no. 36-1451 standard.

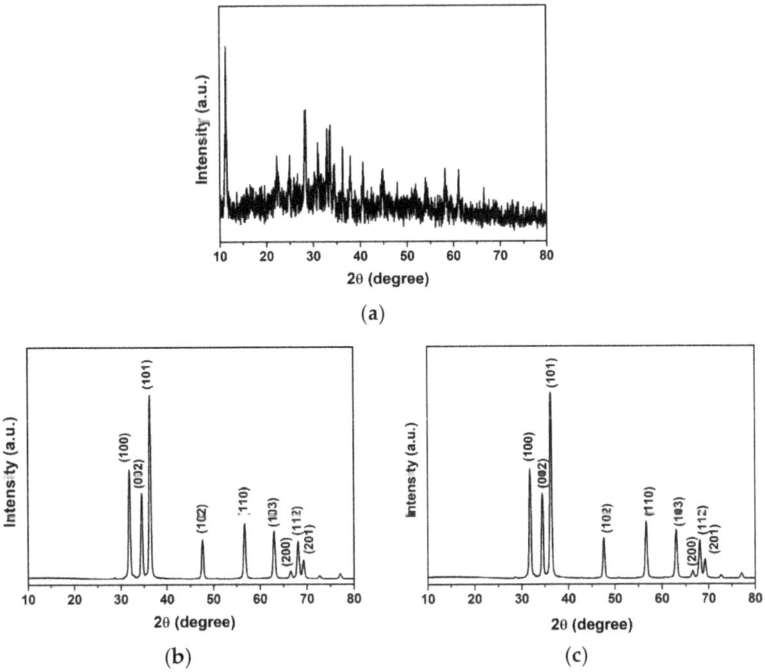

Figure 5. XRD patterns of (**a**) AC, (**b**) ZnO, and (**c**) ZnO-AC.

The findings from the Raman spectroscopy are presented in Figure 6. Figure 6a presented disordered and graphitic peaks of AC at 1349 and 1594 cm^{-1}, respectively, which implied a high order of the amorphous carbon structure. The Raman shift analysis of ZnO (Figure 6b) and ZnO-AC (Figure 6c) exhibited consistent peaks at 332, 440, 586, and 1159 cm^{-1} [25–28]. The peak at 332 cm^{-1} was assigned to the E_2(high)–E_2(low) mode, while the strong peak at 440 cm^{-1} corresponded to the E_2(high) modes due to the oxygen vibration in the ZnO matrix. Meanwhile, the two peaks at 586 and 1159 cm^{-1} correspond to the E_1(LO) and $2E_1$(LO) modes representing the multi-phonon process. These outcomes indicate the intrinsic nature of ZnO hexagonal wurtzite structures in ZnO and ZnO-AC samples, consistent with the XRD results. However, the peaks of AC were not observed in the ZnO-AC samples, which might be due to the low AC content in the ZnO-AC sample.

The physicochemical properties of ZnO and ZnO-AC samples were investigated using the N_2 adsorption–desorption method. Figure 7 demonstrates that the N_2 adsorption–desorption isotherm of ZnO and ZnO-AC features type H3 hysteresis due to the single-layer adsorption stage, which could be attributed to the ZnO solid structure. Moreover, the adsorption–desorption isotherm indicates that both photocatalysts have meso- or macroporous structures. The Barrett–Joyner–Halenda (BJH) adsorption analysis was conducted to determine ZnO and ZnO-AC porosities (Table 3). ZnO and ZnO-AC each demonstrated a mesoporous structure with pore diameters of 6.55 and 3.83 nm, respectively. The smaller pore diameter of ZnO-AC resulted in larger pore volume and higher BET surface area (S_{BET}) than those of ZnO, as calculated using the multipoint Brunauer–Emmett–Telle (BET) method.

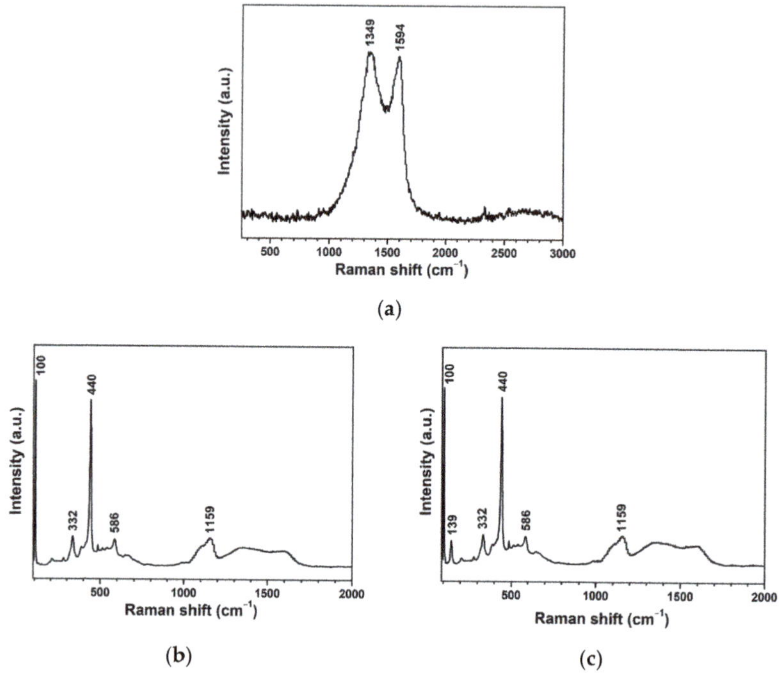

Figure 6. Raman spectroscopy analysis of (**a**) AC, (**b**) ZnO, and (**c**) ZnO-AC.

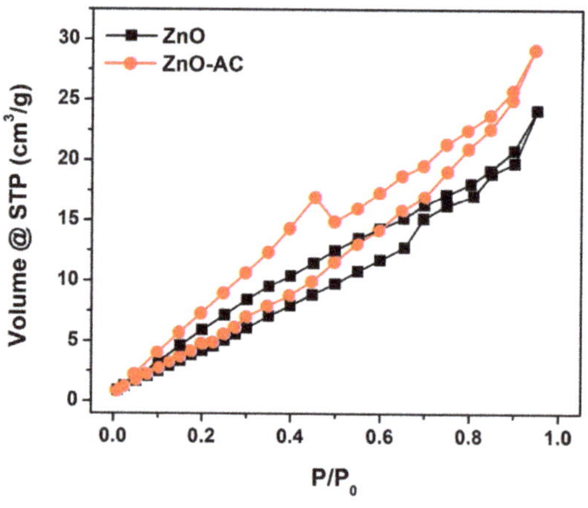

Figure 7. N_2 adsorption–desorption isotherms of ZnO and ZnO-AC.

Table 3. BJH porosities and BET surface areas of ZnO and ZnO-AC.

Sample	Pore Diameter (nm)	Pore Volume (cm^3/g)	S_{BET} (m^2/g)
ZnO	6.55	3.44×10^{-2}	27.08
ZnO-AC	3.83	4.40×10^{-2}	28.99

The FTIR spectra revealed a major peak at 490 cm^{-1}, indicating the Zn-O vibration of the ZnO matrix in the structural composition of both photocatalysts (Figure 8a) and the prevalence of ZnO structures in these examined samples. Meanwhile, the functional group analysis demonstrated that the wavenumber in the ZnO-AC sample ranged between 2000 and 1000 cm^{-1} (Figure 8b). The distinct vibrations at wavenumbers of 1738, 1367, and 1214 cm^{-1} represented characteristic peaks associated with C=O stretching, methyl groups, and C-H bending, respectively [29,30]. These findings are particularly significant as they suggest the potential adsorption of carbon on the ZnO surface, a phenomenon that can likely be attributed to the presence of AC in the ZnO-AC composite.

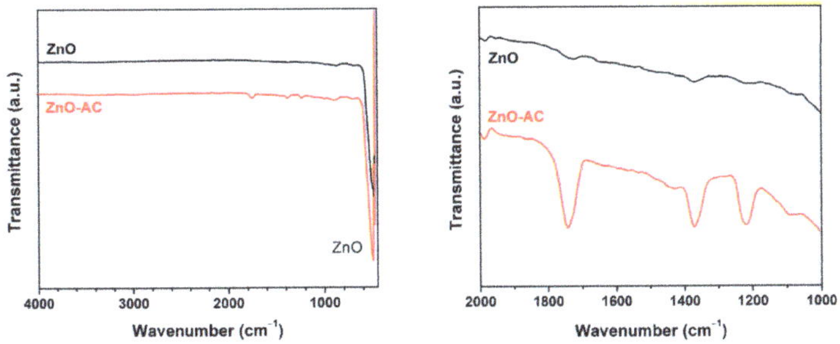

(**a**) (**b**)

Figure 8. FTIR analysis of ZnO and ZnO-AC using (**a**) full scanning, and (**b**) functional group scanning.

The binding energy (BE) of each sample was investigated using XPS to evaluate the ZnO-AC surface (Figure 9). In Figure 9a, the survey XPS spectra of ZnO and ZnO-AC identified chemical elements Zn, O, and C, reflecting comparable chemical compositions for the samples. The high-resolution Zn 2p core level in Figure 9b illustrates the distinct Zn $2p_{3/2}$ and Zn $2p_{1/2}$ peaks of ZnO exhibited at BEs of 1022.5 and 1045.6 eV, respectively, signifying the Zn^{2+} oxidation state [31]. Notably, there were no differences in these peaks between ZnO and ZnO-AC. The O 1s spectra of ZnO depicted peaks at BEs of 531.4 and 533.0 eV (Figure 9c), corresponding to O^{2-} ions in the metal oxide and compound groups, respectively [32]. The O^{2-} ions were associated with the bonding of Zn^{2+} ions in the ZnO structure, whereas the compound groups indicated the chemisorption on the ZnO surface responsible for the oxygenated carbon components in this case [33]. Meanwhile, a decrease in O^{2-} ions and an increase in the compound group were evident in the O 1s spectra of ZnO-AC, suggesting a stronger chemisorption effect of AC on ZnO structures. Figure 9d shows the C 1s spectra with peaks at BEs of 285.9 and 289.9 eV for ZnO, corresponding to C-O and C=O groups, respectively [11]. This might be due to residual levels of the starting ammonium bicarbonate. However, the relatively elevated intensity of these peaks in the ZnO-AC sample confirms the presence of AC on the ZnO surface. This detailed XPS analysis provides valuable insights into the surface compounds of ZnO-AC, which agree with the FTIR results.

The function of the adsorbing porous material of AC in caping surrounding toxic molecules near the ZnO-AC surface for improving photocatalytic mechanisms is assumed, as illustrated in Figure 10. In conventional photocatalysis, after electron–hole pairs are generated due to the incident light irradiation, there is a reaction in which dissolved oxygen (O_2) forms superoxide anions ($O_2^{\bullet-}$). Simultaneously, holes react with water (H_2O) or hydroxyl groups (OH^-), forming hydroxyl radicals ($^{\bullet}OH$). These reactive oxygen species (ROS) are continuously reacting with toxic molecules. In the case of ZnO-AC, toxic molecules were adsorbed by AC at a rapid rate and they reacted with ROS in the photocatalytic activity process for mineralization [34,35]. Thus, the presence of porous AC on ZnO surfaces plays a key role in the increasing photocatalytic activity rate. This mechanism can increase reaction density in the photocatalytic process and boost photocatalytic performance.

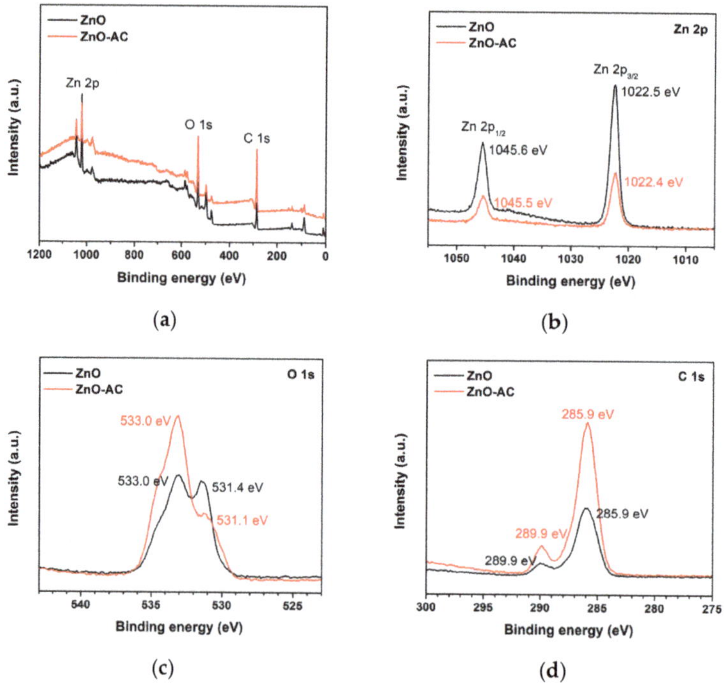

Figure 9. XPS analysis of ZnO and ZnO-AC: (**a**) full scan, (**b**) Zn 2p, (**c**) O 1s, and (**d**) C 1s.

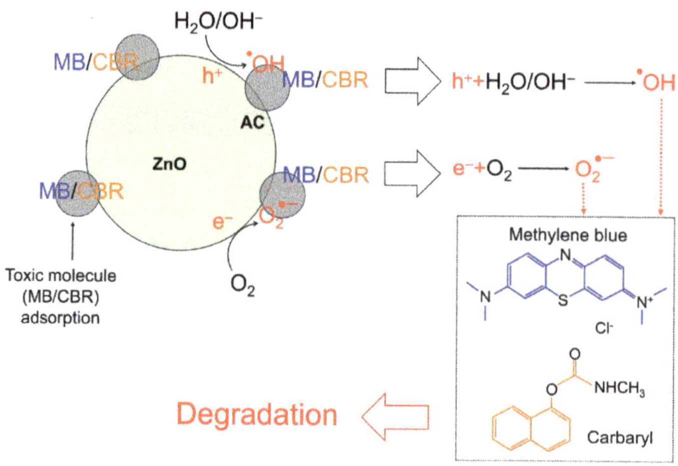

Figure 10. The mechanism of contaminant degradation using a ZnO-AC photocatalyst.

This study has briefly shown that aquatic weeds are of benefit as carbon-rich sources for transforming into carbon powder materials. The transforming process of Wh not only contributes to aquatic weed management but also serves as a viable carbon storage strategy. Employing a chemical activation process, carbon powder materials are activated to enhance surface properties, resulting in AC. This AC is then utilized to modify the ZnO photocatalyst, demonstrating the enhancement of photocatalysis under sunlight irradiation to degrade toxic molecules, including methylene blue and carbaryl. Thus, using water hyacinth as a carbon source for the transformation of AC can alternatively solve the problems of aquatic weed management and carbon storage strategies, and the application of AC to ZnO-AC photocatalysts can enhance photocatalysis.

4. Conclusions

Water hyacinth (Wh) is an aquatic weed considered a nuisance in agricultural and fishing activities. Therefore, this study proposed repurposing this plant into activated carbon (AC). First, the ZnO-AC was precipitated and applied as a photocatalyst for degrading methylene blue. For the preliminary photocatalytic test of methylene blue (MB) degradation, ZnO-AC recorded the highest degradation rate constant of 11.49×10^{-3} min^{-1}, which was double the value of pristine ZnO (5.55×10^{-3} min^{-1}). Furthermore, the photocatalytic degradation of MB and CBR under sunlight irradiation by ZnO-AC demonstrated degradation rate constants of 74.46×10^{-3} min^{-1} and 8.43×10^{-3} min^{-1}, respectively. To investigate the properties of ZnO-AC, several techniques were performed. ZnO-AC and ZnO exhibited similar results in morphology, crystalline structure, and Raman characteristics. However, ZnO-AC presented smaller pore diameters than those of ZnO, which enlarged the relevant surface area, and the presence of carbon-related groups implied the presence of AC on ZnO-AC surfaces. This can be attributed to the presence of AC on the ZnO surface, increasing the capture rate of surrounding toxic molecules and elevating the reaction density. This enhancement is thought to promote the degradation of toxic molecules. Therefore, using Wh as a carbon source for the transformation of AC can alternatively solve the problems of aquatic weed management and carbon storage strategies, and the application of AC to ZnO-AC photocatalysts can enhance photocatalysis.

Author Contributions: Conceptualization, S.K., T.R. and S.W. (Sutthipoj Wongrerkdee); methodology, S.K., T.R. and S.W. (Sutthipoj Wongrerkdee); validation, S.K., T.R., N.K. (Naphatson Khaodara) and S.W. (Sutthipoj Wongrerkdee); formal analysis, S.K., T.R. and S.W. (Sutthipoj Wongrerkdee); investigation, S.K., T.R., N.K. (Naphatson Khaodara), N.K. (Napat Kaewtrakulchai), K.M., S.W. (Sawitree Wongrerkdee) and S.W. (Sutthipoj Wongrerkdee); resources, S.K., N.K. (Napat Kaewtrakulchai), K.M., K.S., D.W., C.B. and S.W. (Sutthipoj Wongrerkdee); data curation, S.K., T.R., N.K. (Naphatson Khaodara) and S.W. (Sutthipoj Wongrerkdee); writing—original draft preparation, S.K., T.R., S.W. (Sawitree Wongrerkdee) and S.W. (Sutthipoj Wongrerkdee); writing—review and editing, S.K. and S.W. (Sutthipoj Wongrerkdee); visualization, N.K. (Naphatson Khaodara) and S.W. (Sutthipoj Wongrerkdee); supervision, S.K. and S.W. (Sutthipoj Wongrerkdee); project administration, S.W. (Sutthipoj Wongrerkdee); funding acquisition, S.K., C.B. and S.W. (Sutthipoj Wongrerkdee). All authors have read and agreed to the published version of the manuscript.

Funding: This work was funded by the Kasetsart University Research and Development Institute (Grant No. FF(S-KU)39.66) and the Faculty of Liberal Arts and Science, Kasetsart University Kamphaeng Saen Campus (Grant No. 205/2566). This work was also partially supported by Chiang Mai University.

Institutional Review Board Statement: Not applicable.

Informed Consent Statement: Not applicable.

Data Availability Statement: Data are contained within the article.

Acknowledgments: The authors would like to acknowledge the Baan Dinso Water Hyacinth Community Enterprise, Song Phi Nong District, Suphan Buri Province for supplying water hyacinth raw materials.

Conflicts of Interest: The authors declare no conflicts of interest.

References

1. Huang, C.W.; Chen, R.A.; Chen, W.Y.; Nguyen, V.H.; Lin, K.Y.A.; Lasek, J. Manipulating and Revealing the Roles of La and Zr Dopants into $ZnTiO_3$ Perovskite toward Heterogeneous Photocatalytic Degradation of Tetracycline Under Visible Light Irradiation. *Top. Catal.* **2023**, *66*, 34–40. [CrossRef]
2. Khalid, H.; Haq, A.; Naqvi, S.A.R.; Usman, M.; Bokhari, T.H. Enhancement of photocatalytic activity of Ba-doped CoO for degradation of Emamectin benzoate in aqueous solution. *Environ. Monit. Assess.* **2023**, *195*, 1245. [CrossRef] [PubMed]

3. Hassan, J.Z.; Raza, A.; Qumar, U.; Li, G. Recent advances in engineering strategies of Bi-based photocatalysts for environmental remediation. *Sustain. Mater. Technol.* **2022**, *33*, e00478. [CrossRef]
4. Sultana, K.A.; Hernandez Ortega, J.; Islam, M.T.; Dorado, Z.N.; Alvarado-Tenorio, B.; Galindo-Esquivel, I.R.; Noveron, J.C. Saccharide-Derived Zinc Oxide Nanoparticles with High Photocatalytic Activity for Water Decontamination and Sanitation. *Sustain. Chem.* **2023**, *4*, 321–338. [CrossRef]
5. Wongrerkdee, S.; Wongrerkdee, S.; Boonruang, C.; Sujinnapram, S. Enhanced Photocatalytic Degradation of Methylene Blue Using Ti-Doped ZnO Nanoparticles Synthesized by Rapid Combustion. *Toxics* **2023**, *11*, 33. [CrossRef] [PubMed]
6. Krobthong, S.; Rungsawang, T.; Wongrerkdee, S. Comparison of ZnO Nanoparticles Prepared by Precipitation and Combustion for UV and Sunlight-Driven Photocatalytic Degradation of Methylene Blue. *Toxics* **2023**, *11*, 266. [CrossRef] [PubMed]
7. Nagarajaiah, S.; Nanda, N.; Manjappa, P.; Nagabhushana, B.M.; Gadewar, M.; Rao, S.; Krishna, P.G. Evaluation of apoptosis in human breast cancer cell (MDA-MB-231) induced by ZnO nanoparticles synthesized using Piper betle leaf extract as bio-fuel. *Appl. Phys. A* **2023**, *129*, 461. [CrossRef]
8. Ramasubramanian, A.; Selvaraj, V.; Chinnathambi, P.; Hussain, S.; Ali, D.; Kumar, G.; Balaji, P.; Sagadevan, S. Enhanced photocatalytic degradation of methylene blue from aqueous solution using green synthesized ZnO nanoparticles. *Biomass Convers. Biorefin.* **2023**, *13*, 17271–17282. [CrossRef]
9. Yi, Y.; Guan, Q.; Wang, W.; Jian, S.; Li, H.; Wu, L.; Zhang, H.; Jiang, C. Recyclable Carbon Cloth-Supported ZnO@Ag_3PO_4 Core–Shell Structure for Photocatalytic Degradation of Organic Dye. *Toxics* **2023**, *11*, 70. [CrossRef]
10. Berehe, B.A.; Assen, A.H.; Kumar, A.S.K.; Ulla, H.; Duma, A.D.; Chang, J.Y.; Gedda, G.; Girma, W.M. Highly efficient visible light active ZnO/Cu-DPA composite photocatalysts for the treatment of wastewater contaminated with organic dye. *Sci. Rep.* **2023**, *13*, 16454. [CrossRef]
11. Hanif, M.A.; Akter, J.; Kim, Y.S.; Kim, H.G.; Hahn, J.R.; Kwac, L.K. Highly Efficient and Sustainable ZnO/CuO/g-C_3N_4 Photocatalyst for Wastewater Treatment under Visible Light through Heterojunction Development. *Catalysts* **2022**, *12*, 151. [CrossRef]
12. Zhang, Y.; Wang, M.; Li, G. Recent Advances in Aerobic Photo-Oxidation over Small-Sized IB Metal Nanoparticles. *Photochem* **2022**, *2*, 528–538. [CrossRef]
13. Mirikaram, N.; Pérez-Molina, Á.; Morales-Torres, S.; Salemi, A.; Maldonado-Hódar, F.J.; Pastrana-Martínez, L.M. Photocatalytic Perfomance of ZnO-Graphene Oxide Composites towards the Degradation of Vanillic Acid under Solar Radiation and Visible-LED. *Nanomaterials* **2021**, *11*, 1576. [CrossRef]
14. Kaewtrakulchai, N.; Faungnawakij, K.; Eiad-Ua, A. Parametric Study on Microwave-Assisted Pyrolysis Combined KOH Activation of Oil Palm Male Flowers Derived Nanoporous Carbons. *Materials* **2020**, *13*, 2876. [CrossRef]
15. Krobthong, S.; Wongrerkdee, S.; Pimpang, P.; Moungsrijun, S.; Sujinnapram, S.; Nilphai, S.; Rungsawang, T.; Wongrerkdee, S. ZnO Nanoparticles Coprecipitation with Aluminum and Copper Ions for Efficient Photocatalytic Degradation of Commercial Glyphosate. *Integr. Ferroelectr.* **2022**, *222*, 69–83. [CrossRef]
16. Abou Zeid, S.; Perez, A.; Bastide, S.; Le Pivert, M.; Rossano, S.; Remita, H.; Hautière, N.; Leprince-Wang, Y. Antibacterial and Photocatalytic Properties of ZnO Nanostructure Decorated Coatings. *Coatings* **2024**, *14*, 41. [CrossRef]
17. Irfan, M.; Hussain, H.; Saleem, B.; Saleem, M.; Shukrullah, S.; Legutko, S.; Petrů, J.; Naz, M.Y.; Pagáč, M.; Rahman, S.; et al. Evaluation of Ultrasonically ZnO Loading Effect on Photocatalytic Self-Cleaning, UV Protection and Antibacterial Activity of Plasma/Citric Acid-Activated Cotton Fabric. *Nanomaterials* **2022**, *12*, 2122. [CrossRef]
18. Sujinnapram, S.; Wongrerkdee, S. Synergistic effects of structural, crystalline, and chemical defects on the photocatalytic performance of Y-doped ZnO for carbaryl degradation. *J. Environ. Sci.* **2023**, *124*, 667–677. [CrossRef]
19. Wongrerkdee, S.; Krobthong, S. Synthesis, Characterization, and Photocatalytic Property of Ba-Doped ZnO Nanoparticles Synthesized Using Facile Precipitation. *Integr. Ferroelectr.* **2022**, *224*, 192–204. [CrossRef]
20. Sheik Mydeen, S.; Raj Kumar, R.; Sambathkumar, S.; Kottaisamy, M.; Vasantha, V.S. Facile Synthesis of ZnO/AC Nanocomposites using *Prosopis juliflora* for Enhanced Photocatalytic Degradation of Methylene Blue and Antibacterial Activity. *Optik* **2020**, *224*, 165426. [CrossRef]
21. Phophayu, S.; Pimpang, P.; Wongrerkdee, S.; Sujinnapram, S.; Wongrerkdee, S. Modified graphene quantum dots-zinc oxide nanocomposites for photocatalytic degradation of organic dyes and commercial herbicide. *J. Reinf. Plast. Compos.* **2020**, *39*, 81–94. [CrossRef]
22. Negash, A.; Mohammed, S.; Weldekirstos, H.d.; Ambaye, A.D.; Gashu, M. Enhanced photocatalytic degradation of methylene blue dye using eco-friendly synthesized rGO@ZnO nanocomposites. *Sci. Rep.* **2023**, *13*, 22234. [CrossRef]
23. Zeljković, S.; Balaban, M.; Gajić, D.; Vračević, S.; Ivas, T.; Vranković, D.; Jelić, D. Mechanochemically induced synthesis of N-ion doped ZnO: Solar photocatalytic degradation of methylene blue. *Green Chem. Lett. Rev.* **2022**, *15*, 869–880. [CrossRef]
24. Karthik, K.V.; Raghu, A.V.; Reddy, K.R.; Ravishankar, R.; Sangeeta, M.; Shetti, N.P.; Reddy, C.V. Green synthesis of Cu-doped ZnO nanoparticles and its application for the photocatalytic degradation of hazardous organic pollutants. *Chemosphere* **2022**, *287*, 132081. [CrossRef] [PubMed]
25. Kaur, M.; Kumar, V.; Singh, J.; Datt, J.; Sharma, R. Effect of Cu-N co-doping on the dielectric properties of ZnO nanoparticles. *Mater. Technol.* **2022**, *37*, 2644–2658. [CrossRef]

26. Ščajev, P.; Durena, R.; Onufrijevs, P.; Miasojedovas, S.; Malinauskas, T.; Stanionyte, S.; Zarkov, A.; Zukuls, A.; Bite, I.; Smits, K. Morphological and optical property study of Li doped ZnO produced by microwave-assisted solvothermal synthesis. *Mater. Sci. Semicond. Process.* **2021**, *135*, 106069. [CrossRef]
27. Russo, V.; Ghidelli, M.; Gondoni, P.; Casari, C.S.; Bassi, A.L. Multi-wavelength Raman scattering of nanostructured Al-doped zinc oxide. *J. Appl. Phys.* **2014**, *115*, 073508. [CrossRef]
28. Batterjee, M.G.; Nabi, A.; Kamli, M.R.; Alzahrani, K.A.; Danish, E.Y.; Malik, M.A. Green Hydrothermal Synthesis of Zinc Oxide Nanoparticles for UV-Light-Induced Photocatalytic Degradation of Ciprofloxacin Antibiotic in an Aqueous Environment. *Catalysts* **2022**, *12*, 1347. [CrossRef]
29. Gaitán-Alvarez, J.; Berrocal, A.; Mantanis, G.I.; Moya, R.; Araya, F. Acetylation of tropical hardwood species from forest plantations in Costa Rica: An FTIR spectroscopic analysis. *J. Wood Sci.* **2020**, *66*, 49. [CrossRef]
30. Algarni, T.S.; Al-Mohaimeed, A.M.; Al-Odayni, A.-B.; Abduh, N.A.Y. Activated Carbon/$ZnFe_2O_4$ Nanocomposite Adsorbent for Efficient Removal of Crystal Violet Cationic Dye from Aqueous Solutions. *Nanomaterials* **2022**, *12*, 3224. [CrossRef]
31. Tien, T.-M.; Chen, E.L. A Novel ZnO/Co_3O_4 Nanoparticle for Enhanced Photocatalytic Hydrogen Evolution under Visible Light Irradiation. *Catalysts* **2023**, *13*, 852. [CrossRef]
32. Lemraski, E.G.; Sharafinia, S. Kinetics, equilibrium and thermodynamics studies of Pb^{2+} adsorption onto new activated carbon prepared from Persian mesquite grain. *J. Mol. Liq.* **2016**, *219*, 482–492. [CrossRef]
33. Soliman, A.I.A.; Abdel-Wahab, A.M.A.; Abdelhamid, H.N. Hierarchical porous zeolitic imidazolate frameworks (ZIF-8) and ZnO@N-doped carbon for selective adsorption and photocatalytic degradation of organic pollutants. *RSC Adv.* **2022**, *12*, 7075–7084. [CrossRef] [PubMed]
34. Shi, H.; Shi, Q.; Li, J.; Li, G. Hollow sphere manganese–ceria solid solution enhances photocatalytic activity in tetracycline degradation. *New J. Chem.* **2023**, *47*, 21264–21269. [CrossRef]
35. Li, Z.; Xu, L.; Babar, Z.U.D.; Raza, A.; Zhang, Y.; Gu, X.; Miao, Y.X.; Zhao, Z.; Li, G. Fabrication of MXene-Bi_2WO_6 heterojunction by $Bi_2Ti_2O_7$ hinge for extraordinary LED-light-driven photocatalytic performance. *Nano Res.* 2023, *in press*. [CrossRef]

Disclaimer/Publisher's Note: The statements, opinions and data contained in all publications are solely those of the individual author(s) and contributor(s) and not of MDPI and/or the editor(s). MDPI and/or the editor(s) disclaim responsibility for any injury to people or property resulting from any ideas, methods, instructions or products referred to in the content.

toxics

MDPI

Article

Complete Photooxidation of Formaldehyde to CO_2 via Ni-Dual-Atom Decorated Crystalline Triazine Frameworks: A DFT Study

Zhao Lu [1,2] and Zhongliao Wang [3,*]

1 HKUST Shenzhen-Hong Kong Collaborative Innovation Research Institute, Shenzhen 515100, China; zhaolu@ust.hk
2 Research and Development Center, Shenzhen Foundation Engineering Co., Ltd., Shenzhen 515100, China
3 Anhui Province Industrial Generic Technology Research Center for Alumics Materials, Anhui Province Key Laboratory of Pollutant Sensitive Materials and Environmental Remediation, School of Physics and Electronic Information, Huaibei Normal University, Huaibei 235000, China
* Correspondence: wangzl@chnu.edu.cn

Abstract: Formaldehyde (CH_2O) emerges as a significant air pollutant, necessitating effective strategies for its oxidation to mitigate adverse impacts on human health and the environment. Among various technologies, the photooxidation of CH_2O stands out owing to its affordability, safety, and effectiveness. Nitrogen-rich crystalline triazine-based organic frameworks (CTFs) exhibit considerable potential in this domain. Nevertheless, the weak and unstable CH_2O adsorption hinders the overall oxidation efficiency of CTF. To address this limitation, we incorporate single and dual Ni atoms into nitrogen-rich CTFs by density functional theory (DFT) calculations, resulting in CTF-Ni and CTF-2Ni. This strategic modification significantly enhances the adsorption capability of CH_2O. Notably, this synergy between Ni dual atoms activates CH_2O by strong chemical adsorption, thereby reducing the energy barrier of CH_2O oxidation and achieving the complete oxidation of CH_2O to CO_2. Moreover, the introduction of dual-atom Ni over CTF ameliorates visible and near-infrared light absorption and facilitates photoexcited charge transfer and separation. Finally, the underlying mechanisms of complete CH_2O oxidation over CTF-2Ni are proposed. This work offers novel insights into the rational design of photocatalysts for CH_2O oxidation through the integration of Ni dual atoms into CTFs.

Keywords: CTFs; dual atoms; formaldehyde oxidation; DFT calculation; structural design

1. Introduction

Indoor formaldehyde (CH_2O) pollution, primarily resulting from construction and renovation materials, poses significant threats to human health [1–3]. Consequently, employing green and eco-friendly methods for the adsorption, transformation, or degradation of CH_2O is crucial for attaining indoor air quality standards and protecting human health [4–6]. Among various CH_2O remediation technologies, the photocatalytic oxidation of CH_2O, with its green, economical, and efficient characteristics, emerges as a promising means to mitigate the hazards of CH_2O [7,8]. Numerous inorganic semiconductors, such as TiO_2 [9–12], $BiVO_4$ [13], g-C_3N_4 [14], ZnO [15], single-atom catalysts [16], and metal–organic frameworks (MOFs) [17,18], have been applied in the research on CH_2O photooxidation. Despite these materials displaying certain levels of CH_2O degradation performance, the efficient degradation of CH_2O over inorganic semiconductors is hindered by the severe recombination of photogenerated charges [19–22]. Additionally, their complex structures make the elucidation of mechanisms behind CH_2O adsorption and oxidation challenging [23,24].

Covalent organic frameworks (COFs) with well-defined units, high surface areas, and abundant porous structure, have garnered extensive attention for enhancing CH_2O oxidation performance and elucidating oxidation mechanisms [25–27]. The composition of COFs primarily consists of light non-metallic elements, wherein the atoms are interconnected through covalent bonds to ensure the coordination saturation of these atoms [28]. While the high surface area of COFs facilitates enhancing the adsorption capacity for CH_2O, the interaction between CH_2O molecules and COFs is predominantly physical adsorption, which is driven mainly by weak van der Waals forces [29–32]. This leads to difficulties in the activation and transformation of adsorbed CH_2O molecules with adsorption easily reaching saturation. Under certain conditions, such as high temperature and pressure, these physically adsorbed CH_2O molecules may be released, leading to secondary pollution. Common modification methods for COFs, such as heterojunction composite and ligand adjustment, improve their oxidation capabilities to a certain extent [32–35]. However, these schemes primarily enhance the light absorption and charge separation without fundamentally addressing the adsorption mode of CH_2O molecules to form stable chemical adsorption, thus failing to achieve the deep activation of CH_2O molecules and reduce the energy barrier for their oxidative decomposition [36].

Considering the abundant d-orbital electrons in transition metals, which facilitate bonding with gas molecules in comparison to pristine organic COFs, they are more prone to forming chemical adsorption with CH_2O molecules [37,38]. Therefore, incorporating transition metals into COFs to enhance the chemical adsorption of CH_2O molecules represents a viable strategy. The exceptional catalytic performance of COFs can be achieved by loading single atoms, which maximizes the preservation of their high surface area, porous structure, and atomic utilization efficiency of transition metals [39]. Although single-atom loading can improve adsorption capacity and molecule activation, the catalytic transformation of small gas molecules by single atoms is generally incomplete with activity still in need of enhancement [40]. Thus, embedding dual-atom catalysts in COFs presents a substantial opportunity to surpass the constraints of single-atom catalysts. However, the precise design and successful construction of dual-atom catalyst loading sites on COFs remain significant experimental challenges with few successes [16]. Thus, the strategic design of dual-atom photocatalysts becomes imperative to direct experimental synthesis and unveil the underlying mechanisms of CH_2O photooxidation.

In this study, nickel was introduced in both single and dual atoms into a covalent triazine framework (CTF) enriched with pyridinic nitrogen, resulting in the CTF-Ni and CTF-2Ni variants, and their structural, optical, and catalytic characteristics were investigated theoretically. The impact of nickel integration on CTF's structure was initially explored through X-Ray Diffraction (XRD) and Fourier Transform Infrared (FT-IR) spectroscopy. The influence of nickel on the optical characteristics was then investigated using Ultraviolet-Visible Diffuse Reflectance Spectroscopy (UV-Vis DRS) and X-ray Absorption Spectroscopy (XAS). The effect of nickel on the internal electric field was examined via dipole moment analysis, and its impacts on the conduction band, valence band, and frontier orbitals were assessed using the band structure, density of states (DOS), and HOMO–LUMO analyses. Charge transfer and separation were analyzed through the work function and charge density difference. The rate-determining step (RDS) and energy barriers in the CH_2O oxidation process were evaluated by changes in the free energy of adsorbed intermediates. This research aims to serve as a benchmark for the future development of dual-atom photocatalysts in the photooxidation of CH_2O.

2. Computational Details

Density functional theory (DFT) simulations were executed using the Vienna ab initio Simulation Package (VASP), utilizing the projector augmented-wave (PAW) method. The exchange-correlation functional was approached through the generalized gradient approximation (GGA) as proposed by Perdew, Burke, and Ernzerhof (PBE). For surface calculations, the Brillouin zone sampling was conducted with a mesh of $2 \times 2 \times 1$ K-points.

The energy cutoff was established at 500 eV. Structural optimizations were pursued until the convergence criteria for energy and force were achieved at less than 1×10^{-5} eV and 0.02 eV/Å, respectively. To prevent interactions among periodic surface models, a vacuum spacing of 15 Å was implemented. The consideration of van der Waals (vdW) interactions was enhanced by incorporating the DFT-D3 method with zero damping, as developed by Grimme. The computation of Gibbs free energy, ΔG, was formulated as the sum of the electronic energy change (ΔE), zero-point energy change (ΔE_{ZPE}), and the entropy change (ΔS) times temperature (T), explicitly $\Delta G = \Delta E + \Delta E_{ZPE} - T\Delta S$. The calculation of zero-point energies for both isolated and adsorbed intermediates was based on frequency analysis. Vibrational frequencies and entropies for gas-phase molecules were conducted by the VASPKIT code [41]. Spectral analyses, including vibration and excited state spectra, were conducted with the CP2K-2023.2 software package, employing the PBE functional for system description. The electronic structure was assessed through unrestricted Kohn–Sham DFT within the Gaussian and plane waves (GPW) approach, using Goedecker–Teter–Hutter (GTH) pseudopotentials and the triple-zeta valence with a two polarization functions (TZV2P-MOLOPT-GTH) basis set for all elements. A plane-wave cutoff of 400 Ry was applied, and geometries were optimized employing the Broyden–Fletcher–Goldfarb–Shanno (BFGS) algorithm. The self-consistent field (SCF) method's density matrix convergence was set to 1×10^{-5} Hartree with the force convergence criterion established at 4.5×10^{-4} Bohr per Hartree. The excited states and their corresponding spectra were conducted employing the Multiwfn suite [42].

3. Results and Discussion

The distinct structure of CTF reveals the alternating arrangement and connection of triazine ring (TR) and pyridazine (Figure 1a). For a comprehensive investigation into the efficacy and specific mechanisms of the photocatalytic CH_2O oxidation by single and dual atoms, we constructed CTF-Ni with Ni single atoms (31.1 wt% Ni) and CTF-2Ni with Ni dual atoms (47.4 wt% Ni). The Ni single atom is loaded onto an N atom of TR and an N atom of the connected pyridazine, forming simultaneous bonds with both N atoms (Figure 1b). In contrast, the other Ni in the Ni dual atoms only bonds with another N atom within the same pyridazine and the N phase on another TR connected to pyridazine (Figure 1c). To elucidate the specific mechanism of single- or dual-atom photocatalytic oxidation of CH_2O, a single-layer structure was employed for simulation calculations of the three aforementioned materials (Figure 1d–f). The structural diagrams highlight that loading Ni atoms onto the CTF induces a reduction in the pore size. The pore diameter of the pure CTF is 14.1 Å, whereas after loading Ni single atoms (CTF-Ni), the pore diameter decreases to 12.3 Å. With the continued loading of Ni atoms to form CTF-2Ni, the pore diameter further decreases to 10.5 Å, marking a 25% reduction compared to the pure CTF. The XRD simulation calculations reveal that the primary peak of CTFs were observed within the 10° range. Specifically, CTF exhibits three robust peaks at 5.90, 7.00, and 9.15°, respectively. It is worth noting that these three strong peaks of CTF-Ni containing Ni single atoms are also located at 5.90, 7.00 and 9.15°, reflecting the pattern of pure CTF. Similarly, the three prominent peaks of CTF-2Ni align entirely with those of CTF. The XRD results indicate that the incorporation of Ni atoms does not alter the lattice parameters of CTFs, affirming the retention of a highly ordered structure. The simulated structures demonstrate complete uniformity in lattice parameters with a = b = 14.57 Å and c = 15.00 Å for all three types.

In the following analysis, the bond lengths in three CTF materials are investigated to assess how single or dual atoms affect the chemical bonds within the CTF (Table 1). According to previous research findings, the C–N bond length within the isolated TR group is 1.327 Å [25]. However, within the TR group embedded in CTF, the C2–N1 bond length expands to 1.347 Å. This expansion signifies that the TR group is influenced by pyridazine, specifically revealing a repulsive interaction between the N atoms in pyridazine and those within the TR group. This leads to an extension of the C2–N1 bond length within the TR group in CTF. Subsequently, with the loading of Ni single atoms onto CTF, the bond lengths of N1–C2 and C3–N4 experience an increase. This is attributed to the formation of covalent bonds between Ni and N, driven by the overlapping of electron clouds, inducing atomic shifts. Additionally, the larger size of Ni atoms and the generally lower zero-point vibration energy associated with heavier atoms contribute to their reduced susceptibility to vibration during covalent bond formation. Lighter atoms, being more prone to thermal vibrations, may undergo greater deflection during bond formation. Consequently, the N atom shifts toward the Ni atom, resulting in increased bond lengths for N1–C2 and C3–N4. However, the bond length of N4–N5 decreases when a single Ni atom is loaded, which shows that the N atom bonded to Ni on pyridazine moves toward another N on pyridazine under the influence of Ni. However, when Ni atoms continued to be added to CTF-Ni to form a Ni diatomic structure, it was found that the bond lengths of N1–Ni6 and N4–Ni6 both decreased. This shows that the dual-atoms configuration of Ni is more stable than a single Ni atom. Similarly, the N4–N5 bond length in CTF-2Ni only increases slightly, indicating that the addition of a single Ni atom has made the N4–N5 bond of pyridazine in a stable state. This means that dual atoms do not significantly affect the bond lengths within the CTF. In summary, the introduction of Ni atoms will affect the bond length inside CTF to a certain extent, thereby significantly changing the light absorption of CTF and affecting the photooxidation ability of CTF for CH_2O pollution.

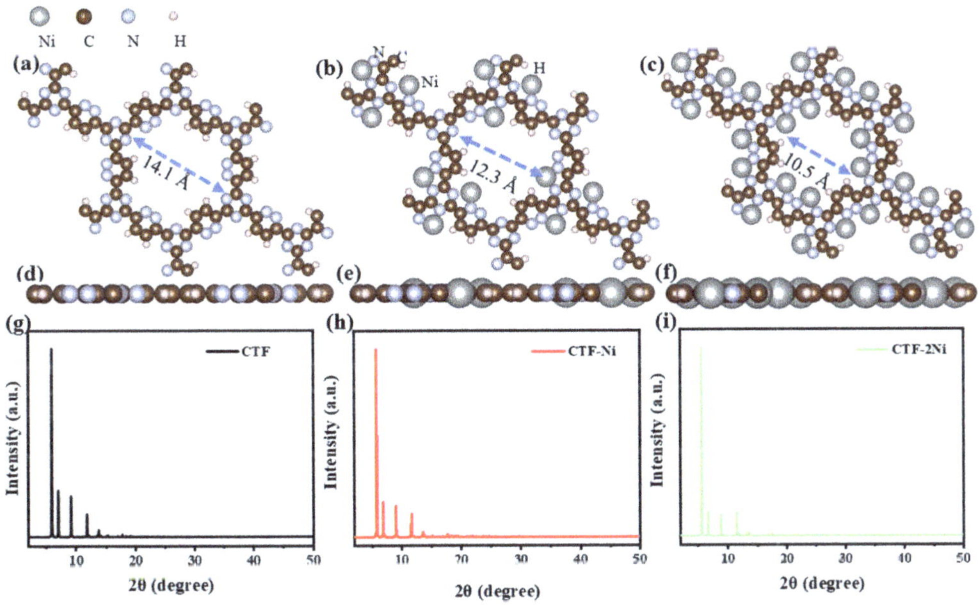

Figure 1. Top view of crystal structure of (**a**) CTF, (**b**) CTF-Ni and (**c**) CTF-2Ni. Side view of crystal structure of (**d**) CTF, (**e**) CTF-Ni and (**f**) CTF-2Ni. Simulated XRD pattern of (**g**) CTF, (**h**) CTF-Ni and (**i**) CTF-2Ni.

The infrared spectroscopy relies on the vibrational motion of atoms within molecules, encompassing modes like stretching, bending, and twisting. Atoms within a molecule engage in relative vibrational motions, and these vibrations closely correlate with the molecule structure and bond characteristics. These principles allow for using DFT simulations to describe electronic structural and molecular vibrations within a quantum mechanical framework. In the simulated infrared spectrum of CTF, characteristic absorption peaks of TR are observed at wave numbers 1331 and 1476 cm^{-1} (Figure 2a). Specifically, the 1331 cm^{-1} peak corresponds to the stretching vibration mode of C–N in TR, while the 1476 cm^{-1}

peak corresponds to the stretching vibration mode of C=N in TR. For CTF-Ni with the incorporation of a single Ni atom, characteristic absorption peaks belonging to TR are retained. The 1345 cm^{-1} peak corresponds to the stretching vibration mode of C–N in TR, while the 1487 cm^{-1} peak corresponds to the stretching vibration mode of C=N in TR. However, in comparison to CTF, the peak intensity of the TR-associated absorption peak in CTF-Ni is significantly weakened, indicating that the presence of Ni single atoms strongly influences the TR vibration in CTF. Similarly, in CTF-2Ni, the characteristic absorption peak associated with TR is notably weakened. However, with the introduction of single Ni atoms, a new peak at 629 cm^{-1} emerges in CTF-Ni, representing the bending vibration peak of Ni–N. Notably, as the Ni loading increases, the bending vibration peak at 632 cm^{-1} belonging to Ni–N in CTF-2Ni becomes much stronger than that in CTF-Ni.

Table 1. Bond length of CTF, CTF-Ni and CTF-2Ni.

Bond Length (Å)	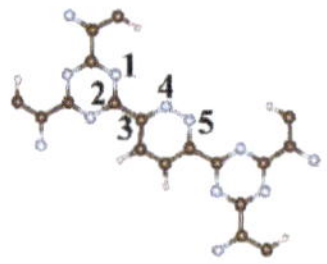	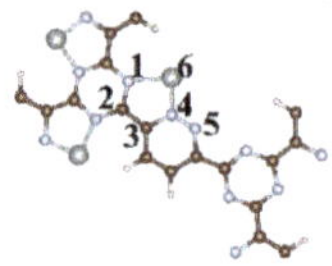	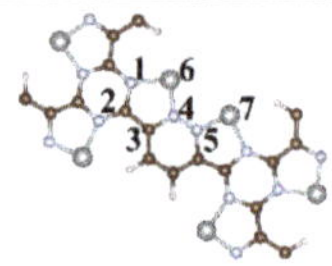
N1–C2	1.347	1.378	1.390
C2–C3	1.524	1.465	1.453
C3–N4	1.357	1.399	1.398
N4–N5	1.335	1.324	1.326
N1–Ni6	-	2.124	1.946
N4–Ni6	-	1.813	1.804
Ni6–Ni7	-	-	3.446

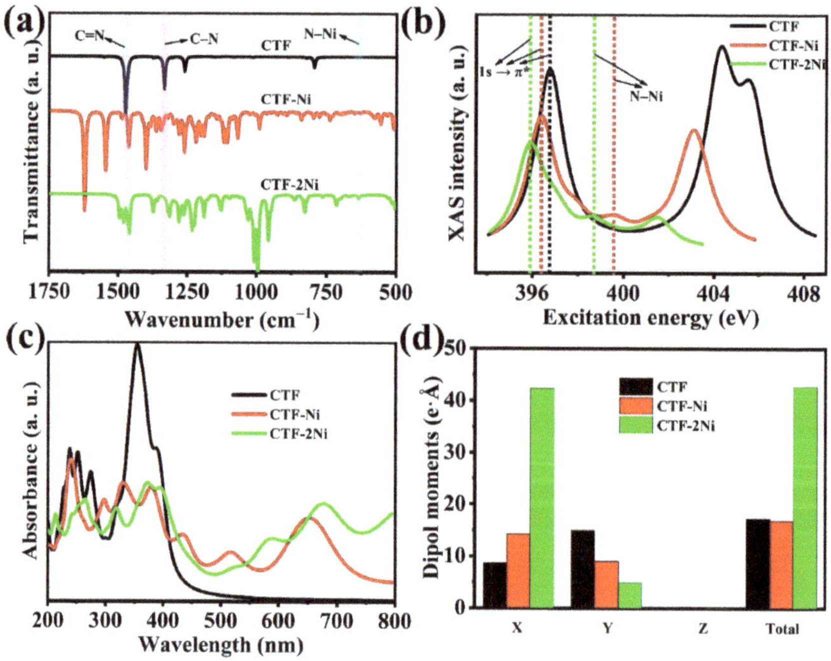

Figure 2. (**a**) Simulated IR spectra of CTF, CTF-Ni and CTF-2Ni. (**b**) Simulated synchrotron X-ray absorption spectra of the N K edge of CTF, CTF-Ni and CTF-2Ni. (**c**) Simulated UV-Vis spectra of CTF, CTF-Ni and CTF-2Ni. (**d**) Dipole moments on different components of CTF, CTF-Ni and CTF-2Ni.

XAS revolves around the absorption of inner shell electrons in materials when the energy of incident X-rays aligns with the energy level of specific element inner-shell electrons. DFT provides a precise description of these core electrons, rendering it a potent tool for simulating XAS. Given the lone pair of electrons on N, facilitating its effective combination with metal Ni to form a Ni single/dual-atom structure, XAS simulation calculations are executed for the N sites in CTFs materials (Figure 2b). This aims to elucidate the chemical environment surrounding the N sites in different CTFs. In CTF, a substantial absorption peak at 396.76 eV corresponds to 1s $\rightarrow \pi^*$ of N. Specifically, the 1s orbital of N serves as the inner orbital, while the π orbital represents the antibonding orbital linked to the π bond. The XAS absorption edge manifests when the incident X-ray energy aligns with the electronic transition from the 1s orbital to the π orbital of the N atom. Upon the introduction of Ni atoms, the 1s $\rightarrow \pi^*$ absorption peak of N in CTFs starts shifting toward lower energy. This shift signifies a reduction in the oxidation state of the N atom, indicating the acquisition of electrons by the N atom when bonding with Ni. The appearance of a peak at 399.56 eV in CTF-Ni, absent in CTF, is attributed to a newly formed electronic excited state resulting from electron transfer from N to Ni during their bonding. Notably, the peak in CTF-2Ni, positioned at 398.57 eV, is shifted toward lower energy compared to CTF-Ni. This shift indicates a lower oxidation state of N on the N–Ni bond in CTF-2Ni, promoting the stability of the N–Ni bond and enabling Ni to consistently participate in catalytic reactions.

Utilizing DFT for electron transition calculations and employing the Kubo–Greenwood formula to compute the UV-Vis absorption spectrum, we investigated the impact of Ni atom loading on the light absorption properties of CTF (Figure 2c). Initially, CTF exhibits two prominent absorption peaks at 210–299 nm and 308–450 nm, respectively. The peaks at 210–299 nm may be associated with intramolecular electron transfer, while those in the 308–450 nm range, within the blue light region, may be linked to diverse electronic excitation processes. Notably, the absorption edge of CTF is around 450 nm, indicating an insufficient absorption of visible light. Upon loading Ni atoms onto CTF to form CTF-Ni, there is a notable extension of the absorption range to 450–750 nm. The absorption edge of CTF-Ni reaches 750 nm. This extension may be attributed to the interaction between the 3D orbital of the Ni atom and the orbital of the N atom, causing an adjustment in the energy level structure of the entire system. Simultaneously, the formation of new chemical bonds, such as Ni-N bonds, leads to charge transfer processes, which are observable in the UV-Vis spectrum. Continuing to load Ni atoms into CTF-Ni to create CTF-2Ni with Ni dual atoms, it is observed that the absorption edge of CTF-2Ni extends to 1000 nm, nearly reaching the infrared region. Consistently loading Ni atoms onto CTF to form Ni single/dual atoms enhances the light absorption characteristics of CTF.

The dipole moment magnitudes of the three single-layer CTFs materials were computed using DFT to illustrate the asymmetry of the positive and negative charge distribution within the CTFs materials (Figure 2d). Firstly, the dipole moment has an impact on the absorption spectrum of electromagnetic radiation. Molecules with larger dipole moments are generally more prone to absorbing light in the visible or near-infrared spectral range, contributing to enhanced photocatalytic activity. Notably, CTF-2Ni exhibits the largest dipole moment, suggesting that it could possess the broadest absorption edge. This observation aligns with the findings from the previously calculated UV-Vis spectrum. Secondly, a larger dipole moment is typically associated with an uneven charge distribution within CTFs. CTF-2Ni with the largest dipole moment possesses the strongest built-in electric field, thus facilitating the transfer of photoexcited charge. This implies that during the photocatalytic reaction process, CTF-2Ni may possess the lowest recombination efficiency of photogenerated carriers.

To achieve a more profound understanding of the influence of loaded Ni single/dual atoms on the electronic structure of CTFs, the band structure and state density of the three simulated CTF materials were calculated. After a detailed analysis of the energy band structure diagrams for the three CTF materials, it was found that all of them have indirect band gaps. Notably, CTF has the widest band gap of 1.30 eV (Figure 3a–c). It is clear that as the concentration of Ni atoms in CTF increases, the band gap decreases. Specifically, CTF-2Ni has the smallest band gap at 1.16 eV. This result indicates that low-energy light is sufficient to excite electrons in CTF-2Ni from the HOMO to the LUMO, facilitating their participation in the oxidation of CH_2O. Simultaneously, the reduced band gap in CTF-2Ni corresponds to an expanded light absorption edge, indicating that CTF-2Ni can capture sunlight more efficiently. Incorporating Ni single atoms into CTF shifts the Fermi level (E_f) from the valence band to the conduction band, resulting in the formation of n-type degenerate semiconductors (Figure 3d,e). This observation indicates that by incorporating Ni single atoms, CTF transitions toward a metal-like semiconductor, thereby improving its light absorption efficiency. Notably, the Fermi level of CTF-2Ni extends further into the conduction band compared to that of CTF-Ni, indicating that CTF-2Ni possesses a higher charge transfer efficiency. Furthermore, the specific effects of Ni loading on the density of states of CTFs are examined (Figure 3d–f). The valence band of CTF is primarily contributed by the 2s and 2p orbitals of N, whereas the conduction band of CTF is contributed by the 2p orbitals of C and N. This phenomenon arises from the distinct electronic structures of C and N. It is noteworthy that in CTF-Ni, the 3d orbital of Ni significantly contributes to the composition of the valence band owing to the coordinate covalent bond between N 2p and Ni 3d orbitals. With the further formation of the diatomic CTF-2Ni, the contribution of Ni 3d orbitals to the valence band becomes more obvious.

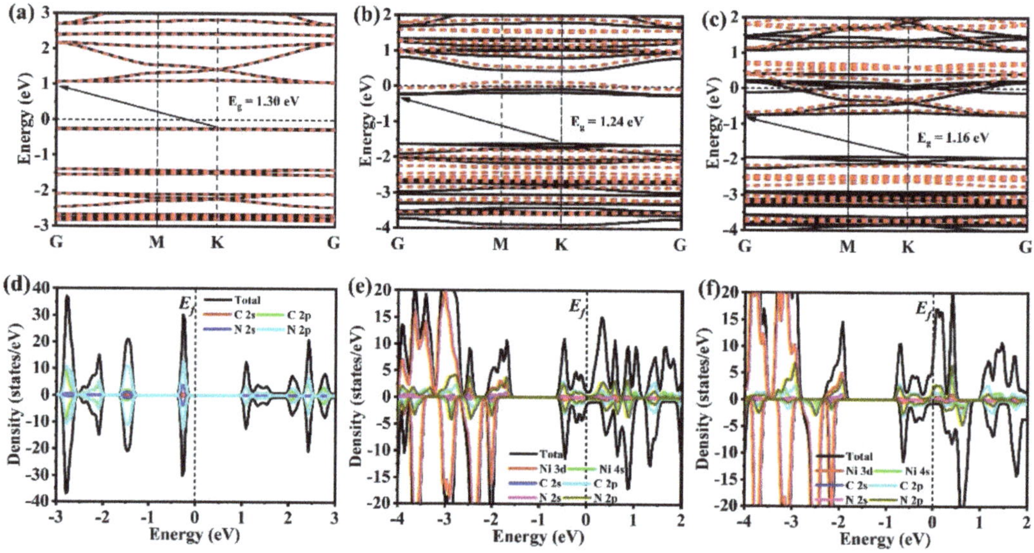

Figure 3. Band structure of (**a**) CTF, (**b**) CTF-Ni and (**c**) CTF-2Ni. Density of states (DOS) of (**d**) CTF, (**e**) CTF-Ni and (**f**) CTF-2Ni.

To demonstrate the effect of loading Ni single or dual atoms on the electronic structure of CTFs, 2D top views of the HOMO and LUMO bands for three CTF materials are displayed. For pristine CTF, the HOMO energy mainly locates on the nitrogen atoms in pyridazine and adjacent carbon atoms (Figure 4a). Nitrogen on triazine rings also slightly contributes to the HOMO bands, aligning with the density of states of CTF. The LUMO band distribution covers carbon and nitrogen in both triazine and pyridazine (Figure 4d). Introducing Ni single atoms into CTF modifies the HOMO and LUMO bands significantly, reflecting the strong interaction of the metallic Ni with CTF (Figure 4b,e). Part of the HOMO bands localizes on single Ni atoms, and the rest is on triazine. Notably, a large part of the LUMO bands centers on Ni atoms. CTF-2Ni presents notable changes in its

HOMO and LUMO bands relative to CTF-Ni, further underscoring that Ni can effectively tune frontier orbitals and active sites (Figure 4c,f). In CTF-2Ni, Ni atoms are the main contributors to both HOMO and LUMO bands. This highlights the critical role of dual-atom Ni in CTF-2Ni for the modulation of electrical structure and active sites for CH_2O oxidation.

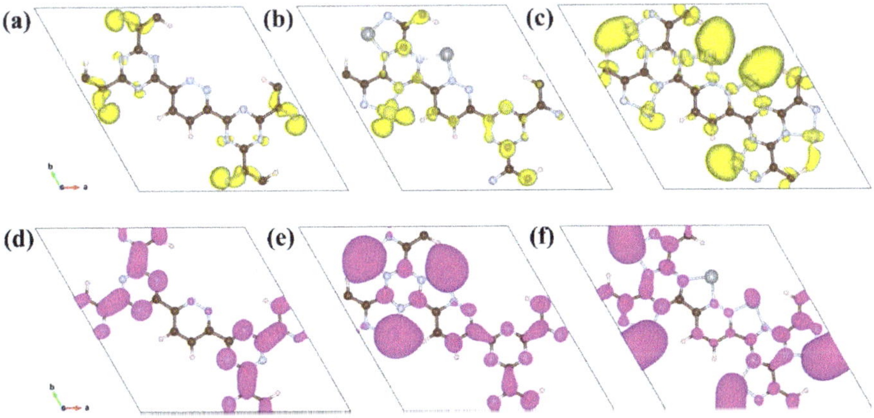

Figure 4. Top view of the HOMO of (**a**) CTF, (**b**) CTF-Ni, (**c**) CTF-2Ni and the LUMO of (**d**) CTF, (**e**) CTF-Ni, (**f**) CTF-2Ni.

Catalytic formaldehyde oxidation (CFO) calculation results reveal that the rate-determining step (RDS) for CFO on CTF is the adsorption of CHO on CTF (* + CH_2O = *CH_2O) with an energy barrier of 2.57 eV (Figure 5a). The incorporation of Ni atoms significantly enhances the adsorption energy of CTF for CH_2O molecules. The negative adsorption free energy of CH_2O on CTF-Ni and CTF-2Ni suggests that the introduction of Ni enables CTF to spontaneously adsorb CH_2O from the indoor or external environment (Figure 5b,c). The adsorption free energy of CTF-2Ni shows a stronger negativity, indicating an improved ability to adsorb CH_2O. Furthermore, the intermediates during the CH_2O oxidation on CTF, CTF-Ni and CTF-2Ni are displayed to illustrate the significance of Ni for the oxidation process of CH_2O and decreased energy barrier (Table 2). Initially, CH_2O is adsorbed on CTF through O4–N5 and C3–N6 bonds with the two N atoms on the pyridazine, respectively. However, the C3–O4 bond in CH_2O is stretched to 1.451 Å, which is not conducive to the catalytic oxidation of CH_2O to CO_2 in CTF. Furthermore, the low adsorption capacity of CTF causes the adsorbed CO to desorb easily, resulting in incomplete CH_2O oxidation (Figure 5a). However, incomplete oxidation of CH_2O molecules into CO remains detrimental to human health. The integration of Ni modifies the rate-determining step in the CFO process by chemically linking the oxygen in CH_2O to the Ni atom, enabling Ni-enriched CTF to adsorb CH_2O molecules spontaneously (Figure 5b,c). Moreover, with CTF-2Ni having twice the Ni content of CTF-Ni, CTF-2Ni can provide more active sites and strong adsorption capability. In addition, with the introduction of Ni, the RDS of CFO changes into the conversion of *CO to *COOH (*CO + H_2O + h^+ = *COOH + H^+). Finally, *COOH is completely oxidized to CO_2 under the influence of photogenerated holes. Furthermore, the reaction energy barriers corresponding to the introduction of Ni single atoms and double atoms on CTF are 0.89 and 0.63 eV, respectively. This result indicates that diatomic Ni is thermodynamically more favorable for oxidizing CH_2O molecules to CO_2 than a Ni single atom. In summary, Ni diatoms not only facilitate the adsorption and complete oxidation of formaldehyde molecules but also significantly reduce the reaction energy barrier for CH_2O oxidation.

$$* + CH_2O = *CH_2O \quad (1)$$

$$*CH_2O + h^+ = *CHO + H^+ \quad (2)$$

$$*CHO + h^+ = *CO + H^+ \quad (3)$$

$$*CO + H_2O + h^+ = *COOH + H^+ \quad (4)$$

$$*COOH + h^+ = *CO_2 + H^+ \quad (5)$$

$$*CO = * + CO \quad (6)$$

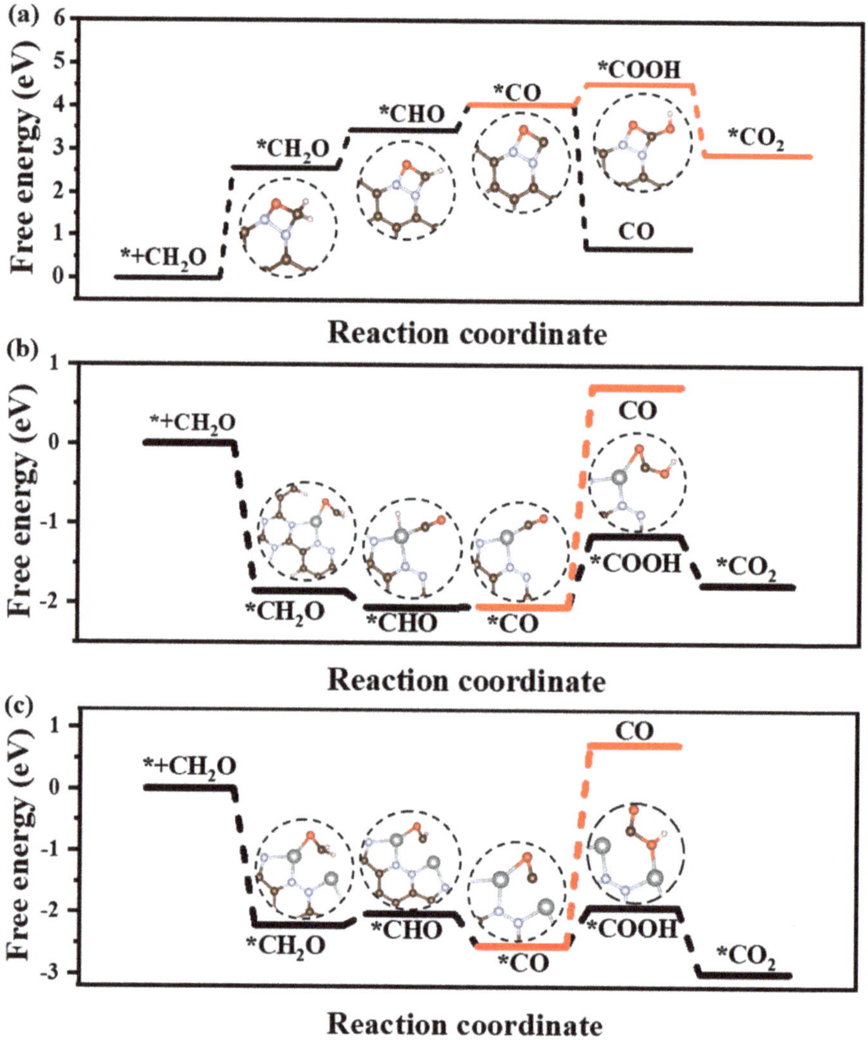

Figure 5. Free-energy diagrams and key intermediates during the CH_2O oxidation for the CH_2O oxidation to CO or CO_2 on the (**a**) CTF, (**b**) CTF-Ni and (**c**) CTF-2Ni.

Table 2. Bond length and angle distribution of CH_2O of CH_2O, CTF, CTF-Ni and CTF-2Ni.

Structure	Bond Length (Å)						Bond Angle (°)		
	C3–H1	C3–H2	C3–O4	N5–O4	N6–C3	Ni7–O4	H1–C3–H2	H1–C3–O4	H2–C3–O4
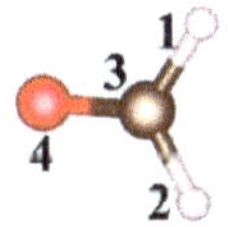	1.118	1.118	1.215	-	-	-	116.15	121.94	121.911.1
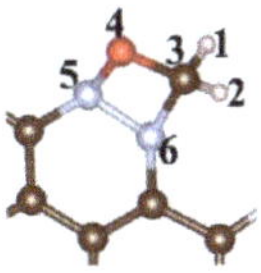	1.103	1.095	1.451	1.573	1.466	-	111.84	111.51	113.59
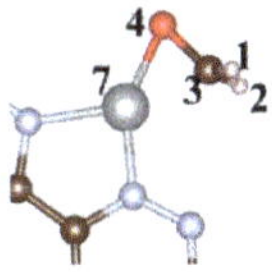	1.105	1.105	1.338	-	-	1.857	115.45	118.59	118.47
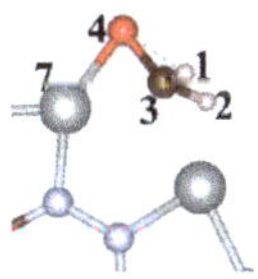	1.11	1.191	1.343	-	-	1.863			

To further analyze the direction and quantity of electron transfer in the three CTFs after adsorbing CH_2O, Bader charge analysis was conducted. After CH_2O adsorption on CTF, 0.2 electrons transfers from CH_2O to CTF (Figure 6a,d). Upon loading Ni onto CTF, CH_2O chemically bonds with Ni on CTF-Ni. However, the electron transfer direction is altered, and 0.48 electrons transfers from CTF-Ni to CH_2O (Figure 6b,e). CTF-2Ni transfers the most electrons to CH_2O with 0.65 electrons transfers from CTF-2Ni to CH_2O (Figure 6c,f). These results demonstrate that the introduction of Ni atoms not only changes the direction of charge transfer between CTF and CH_2O but also alters the quantity of transferred electrons. During adsorption, the more electrons that transfer between the substrate and the adsorbed substance, the more favorable the adsorption. The Bader charge analysis shows that CH_2O is most readily adsorbed on CTF-2Ni, whereas CTF offers the least favorable conditions for CH_2O adsorption. The values of the d band centers of CTF-Ni and CTF-2Ni also give the reason why CTF-2Ni has strong adsorption of CH_2O. The d band center of diatomic Ni moves downward compared to single-atom Ni, which results in fewer electrons filling the antibonding orbital of Ni 3d (Figure 3e,f). This is a critical factor regarding why diatomic Ni has a stronger ability to adsorb CH_2O than single-atom Ni. Interestingly, the loading of Ni atoms alters the direction of electron transfer, which is key for the adsorption and activation of CH_2O. In CTF, electron transfer from CH_2O to CTF creates an inherent electric field (IEF) oriented from CH_2O toward CTF. This IEF, however, drives photoexcited holes from CH_2O to CTF, hindering CH_2O oxidation. The introduction of Ni reverses the direction of IEF, thus facilitating more photoexcited holes to engage in the oxidation processes of CH_2O.

The work functions of the CTF (001), CTF-Ni (001), and CTF-2Ni (001) surfaces are calculated at 5.4, 4.1, and 3.6 eV, respectively, demonstrating that the incorporation of Ni leads to a reduction in work function values (Figure 7a–c). Moreover, CTF-2Ni with the smallest work function may donate more electrons to adsorbed CH_2O and elevate CH_2O oxidation efficiency. This effect primarily arises from Ni acting as an electron donor to CTF, resulting in an upward shift of the Fermi level. The surface electrostatic potentials of the three CTFs offer further insights into the mechanism behind the electron reversal phenomenon triggered by the addition of Ni atoms (Figure 7d–f). The potential at the N-N bond position of CTF is low (Figure 7d). Introducing Ni leads to electron transfer from Ni to the N-N bridge, elevating the potential at the Ni site (Figure 7e,f). However, the electrostatic potential at the N–N bridge and Ni site are opposite, leading to a reversed electron transfer after CH_2O is adsorbed. As the number of Ni atoms increases, the overall potential of the CTF decreases further. The electrostatic potential of the N-N bridge in pyridazine is notably lower in CTF-2Ni compared to CTF-Ni, accounting for increased electron transfer from CTF-2Ni to adsorbed CH_2O.

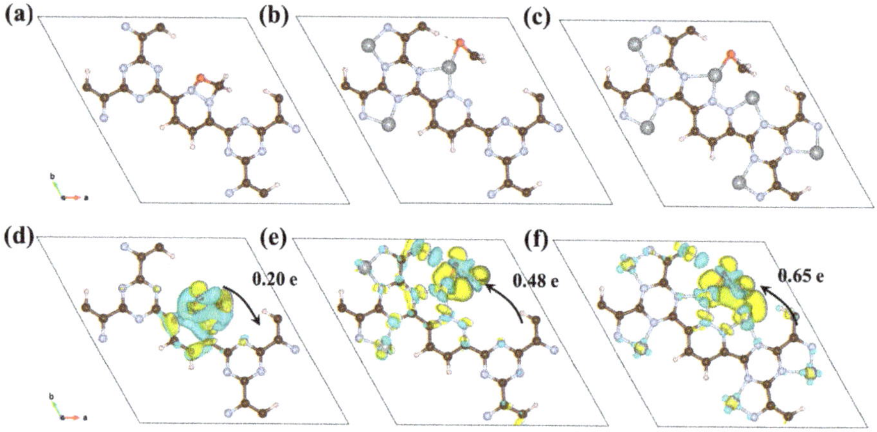

Figure 6. Top view of the optimized (**a**) CTF, (**b**) CTF-Ni, (**c**) CTF-2Ni. Top view of the charge density difference of (**d**) CTF, (**e**) CTF-Ni, and (**f**) CTF-2Ni adsorbed with CH_2O with an isosurface of 2×10^{-3} e/Å^3 (charge accumulation is marked in yellow, while reduction is in cyan).

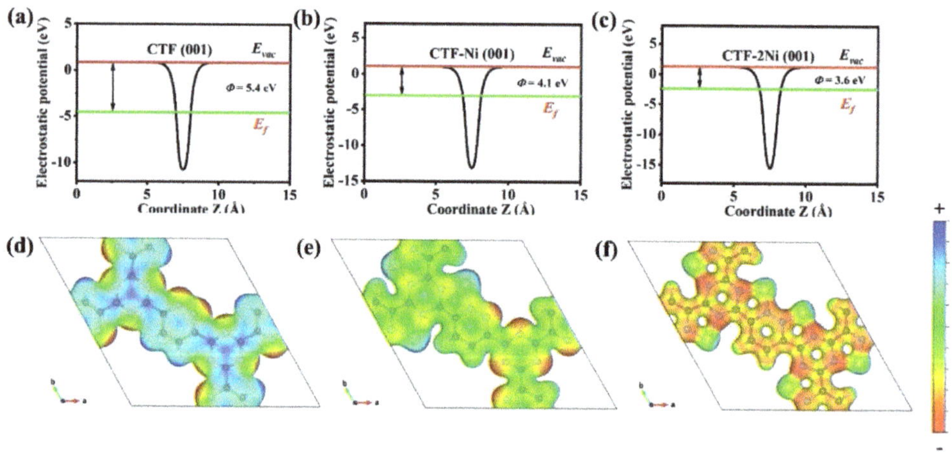

Figure 7. Electrostatic potentials of (**a**) CTF (001) surface, (**b**) CTF-Ni (001) surface and (**c**) CTF-2Ni (001) surface. Electrostatic potential diagram of (**d**) CTF, (**e**) CTF-Ni and (**f**) CTF-2Ni.

4. Conclusions

In summary, this work explores the photooxidation of CH_2O over pristine CTF and CTF embedded with single and dual Ni configurations. First, incorporating Ni into CTF slightly reduces the band gap and induces a modest redshift of absorption edge, thereby facilitating visible and near-infrared light absorption. In addition, the introduction of Ni lowers the work function and elevates the Fermi level, consequently reversing photoexcited charge transfer compared with pristine CTF and driving the hole from the CTF-2Ni to CH_2O to be involved in photooxidation processes. Moreover, the introduction of Ni intensifies the dipole moment of CTF-2Ni and induces a strong IEF, boosting photoexcited charge separation. More importantly, the strong chemical adsorption and increased charge transfer of CH_2O over CTF-2Ni endow CH_2O with sufficient activation, thus decreasing the energy barrier of CH_2O photooxidation and completely converting CH_2O to CO_2. This work provides a reference for designing dual-atom covalent organic frameworks for the complete photooxidation of CH_2O.

Author Contributions: Writing—original draft preparation, data curation, investigation, validation, formal analysis, writing—review and editing, visualization, Z.L.; conceptualization, methodology, software, resources, supervision, project administration, Z.W.; funding acquisition, Z.L. and Z.W. All authors have read and agreed to the published version of the manuscript.

Funding: This work was supported by the Project of Hetao Shenzhen-Hong Kong Science and Technology Innovation Cooperation Zone (HZQB-KCZYB-2020083), Shenzhen Science and Technology Program (No. KCXFZ20211020163816023), Shenzhen Zhongdi Construction Engineering Co., Ltd. cooperation project (22100245), Key Research Project in Natural Sciences for Higher Education Institutions by the Ministry of Education (2022AH050396).

Institutional Review Board Statement: Not applicable.

Informed Consent Statement: Not applicable.

Data Availability Statement: All the data are available within the manuscript.

Conflicts of Interest: Author Zhao Lu was employed by the company (Research and Development Center, Shenzhen Foundation Engineering Co., Ltd.). The remaining authors declare that the research was conducted in the absence of any commercial or financial relationships that could be construed as a potential conflict of interest.

References

1. Gushit, J.S.; Mohammed, S.U.; Moda, H.M. Indoor Air Quality Monitoring and Characterization of Airborne Workstations Pollutants within Detergent Production Plant. *Toxics* **2022**, *10*, 419. [CrossRef]
2. Jiang, C.; Li, D.; Zhang, P.; Li, J.; Wang, J.; Yu, J. Formaldehyde and volatile organic compound (VOC) emissions from particleboard: Identification of odorous compounds and effects of heat treatment. *Build. Environ.* **2017**, *117*, 118. [CrossRef]
3. Talaiekhozani, A.; Rezania, S.; Kim, K.-H.; Sanaye, R.; Amani, A.M. Recent advances in photocatalytic removal of organic and inorganic pollutants in air. *J. Clean. Prod.* **2021**, *278*, 123895. [CrossRef]
4. Wang, Z.; Fan, J.; Cheng, B.; Yu, J.; Xu, J. Nickel-based cocatalysts for photocatalysis: Hydrogen evolution, overall water splitting and CO_2 reduction. *Mater. Today Phys.* **2020**, *15*, 100279. [CrossRef]
5. Guo, J.; Lin, C.; Jiang, C.; Zhang, P. Review on noble metal-based catalysts for formaldehyde oxidation at room temperature. *Appl. Surf. Sci.* **2019**, *475*, 237. [CrossRef]
6. Mohammadifard, Z.; Saboori, R.; Mirbagheri, N.S.; Sabbaghi, S. Heterogeneous photo-Fenton degradation of formaldehyde using MIL-100(Fe) under visible light irradiation. *Environ. Pollut.* **2019**, *251*, 783. [CrossRef] [PubMed]
7. Wu, M.; Ye, J.; Cheng, B.; Yu, J.; Zhang, L. Pt/$NiCo_2O_4$ nanowire arrays on Ni foam as catalysts for efficient formaldehyde oxidation at room temperature. *Appl. Surf. Sci.* **2023**, *608*, 155183. [CrossRef]
8. Low, W.; Boonamnuayvitaya, V. Enhancing the photocatalytic activity of TiO_2 co-doping of graphene-Fe^{3+} ions for formaldehyde removal. *J. Environ. Manag.* **2013**, *127*, 142. [CrossRef] [PubMed]
9. Chen, M.; Wang, H.; Chen, X.; Wang, F.; Qin, X.; Zhang, C.; He, H. High-performance of Cu-TiO_2 for photocatalytic oxidation of formaldehyde under visible light and the mechanism study. *Chem. Eng. J.* **2020**, *390*, 124481. [CrossRef]
10. Bathla, A.; Kukkar, D.; Heynderickx, P.M.; Younis, S.A.; Kim, K.-H. Removal of gaseous formaldehyde by portable photocatalytic air purifier equipped with bimetallic Pt@Cu-TiO_2 filter. *Chem. Eng. J.* **2023**, *469*, 143718. [CrossRef]
11. Khanmohammadi, M.; Garmarudi, A.B.; Elmizadeh, H.; Koochi, M.B. Spectrophotometric evaluation of the photocatalytic degradation of formaldehyde by Fe_2O_3–TiO_2 nano hybrid. *J. Ind. Eng. Chem.* **2014**, *20*, 1841. [CrossRef]
12. Song, Y.W.; Kim, S.E.; Park, J.C. Experiment on the formaldehyde removal performance of TiO_2 coating agent for finishing materials. *J. Photochem. Photobiol. A* **2024**, *447*, 115229.
13. Li, Y.-W.; Li, S.-Z.; Zhao, M.-B.; Ma, W.-L. Decahedral $BiVO_4$/tubular g-C_3N_4 S-scheme heterojunction photocatalyst for formaldehyde removal: Charge transfer pathway and deactivation mechanism. *Sep. Purif. Technol.* **2023**, *327*, 124966. [CrossRef]

14. Zheng, M.-W.; Lin, Y.-R.; Liu, S.-H. TiO_2@g-C_3N_4/SiO_2 with superior visible-light degradation of formaldehyde for indoor humidity control coatings. *Mater. Today Sustain.* **2023**, *24*, 100496. [CrossRef]
15. Rezaee, A.; Rangkooy, H.; Khavanin, A.; Jafari, A.J. High photocatalytic decomposition of the air pollutant formaldehyde using nano-ZnO on bone char. *Environ. Chem. Lett.* **2014**, *12*, 353. [CrossRef]
16. Chen, M.; Wang, H.; Wang, J.; Sun, M.; Hu, Y.; Zhao, X.; Zhou, Y. Efficient degradation of formaldehyde based on DFT-screened metal-doped C_3N_6 monolayer photocatalysts: Performance evaluation and mechanistic insights. *Phys. Chem. Chem. Phys.* **2023**, *25*, 25353. [CrossRef] [PubMed]
17. He, J.; Yu, X.; Luan, X.; Li, H.; Shah, S.J.; Su, W.; Jia, Z.; Chen, J.; Zhou, L.; Deng, J.; et al. Facile construction of tourmaline@MOFs composites via micron-scale electrical polarization to enhance photodegradation of gaseous formaldehyde. *Chem. Eng. J.* **2024**, *479*, 147828. [CrossRef]
18. Huang, Q.; Hu, Y.; Pei, Y.; Zhang, J.; Fu, M. In situ synthesis of TiO_2@NH_2-MIL-125 composites for use in combined adsorption and photocatalytic degradation of formaldehyde. *Appl. Catal. B Environ.* **2019**, *259*, 118106. [CrossRef]
19. Yi, Y.; Guan, Q.; Wang, W.; Jian, S.; Li, H.; Wu, L.; Zhang, H.; Jiang, C. Recyclable Carbon Cloth-Supported ZnO@Ag_3PO_4 Core-Shell Structure for Photocatalytic Degradation of Organic Dye. *Toxics* **2023**, *11*, 70. [CrossRef]
20. Ye, J.; Wang, L.; Zhu, B.; Cheng, B.; He, R. Light-enhanced metal-support interaction for synergetic photo/thermal catalytic formaldehyde oxidation. *J. Mater. Sci. Technol.* **2023**, *167*, 74. [CrossRef]
21. Li, Z.; Liu, W.; Chen, C.; Ma, T.; Zhang, J.; Wang, Z. Transforming the Charge Transfer Mechanism in the In_2O_3/CdSe-DETA Nanocomposite from Type-I to S-Scheme to Improve Photocatalytic Activity and Stability during Hydrogen Production. *Acta Phys. Chim. Sin.* **2023**, *39*, 202208030.
22. Bourgeois, P.-A.; Puzenat, E.; Peruchon, L.; Simonet, F.; Chevalier, D.; Deflin, E.; Brochier, C.; Guillard, C. Characterization of a new photocatalytic textile for formaldehyde removal from indoor air. *Appl. Catal. B Environ.* **2012**, *128*, 171. [CrossRef]
23. Hu, Q.; Liu, K.; Ye, J.; Ming, L.; Xu, J.; Cao, S. Hierarchical Pt/$NiCo_2O_4$ nanosheets decorated carbon nanofibers for room-temperature catalytic formaldehyde oxidation. *Appl. Surf. Sci.* **2023**, *623*, 157012. [CrossRef]
24. Belhadj, H.; Hamid, S.; Robertson, P.K.J.; Bahnemann, D.W. Mechanisms of Simultaneous Hydrogen Production and Formaldehyde Oxidation in H_2O and D_2O over Platinized TiO_2. *ACS Catal.* **2017**, *7*, 4753. [CrossRef]
25. Wang, J.; Wang, Z.; Zhang, J.; Dai, K. Single-layer crystalline triazine-based organic framework photocatalysts with different linking groups for H_2O_2 production. *Chin. J. Struct. Chem.* **2023**, *42*, 100202. [CrossRef]
26. Xie, F.; Bie, C.; Sun, J.; Zhang, Z.; Zhu, B. A DFT study on Pt single atom loaded COF for efficient photocatalytic CO_2 reduction. *J. Mater. Sci. Technol.* **2024**, *170*, 87. [CrossRef]
27. Segura, J.L.; Royuela, S.; Mar Ramos, M. Post-synthetic modification of covalent organic frameworks. *Chem. Soc. Rev.* **2019**, *48*, 3903. [CrossRef]
28. Francis Kurisingal, J.; Kim, H.; Choe, J.H.; Hong, C.S. Covalent organic framework-based catalysts for efficient CO_2 utilization reactions. *Coord. Chem. Rev.* **2022**, *473*, 214835. [CrossRef]
29. Lu, Z.; Xu, X.; Wang, Z. Highly crystalline carbon nitride by covalent remedy for CO_2 photoreduction. *Sep. Purif. Technol.* **2024**, *330*, 125457. [CrossRef]
30. Zhang, Y.; Cao, L.; Bai, G.; Lan, X. Engineering Single Cu Sites into Covalent Organic Framework for Selective Photocatalytic CO_2 Reduction. *Small* **2023**, *19*, 2300035. [CrossRef]
31. Wang, J.; Zhu, W.; Meng, F.; Bai, G.; Zhang, Q.; Lan, X. Integrating Dual-Metal Sites into Covalent Organic Frameworks for Enhanced Photocatalytic CO_2 Reduction. *ACS Catal.* **2023**, *13*, 4316. [CrossRef]
32. Do, K.H.; Kumar, D.P.; Rangappa, A.P.; Lee, J.; Yun, S.; Kim, T.K. Design and synthesis of a covalent organic framework bridging CdS nanoparticles and a homogeneous cobalt–bipyridine cocatalyst for a highly efficient photocatalytic CO_2 reduction. *J. Mater. Chem. A* **2023**, *11*, 8392. [CrossRef]
33. Sun, K.; Qian, Y.; Jiang, H.L. Metal-Organic Frameworks for Photocatalytic Water Splitting and CO_2 Reduction. *Angew. Chem. Int. Ed.* **2023**, *62*, 202217565. [CrossRef] [PubMed]
34. Yue, Y.; Cai, P.; Xu, K.; Li, H.; Chen, H.; Zhou, H.C.; Huang, N. Stable Bimetallic Polyphthalocyanine Covalent Organic Frameworks as Superior Electrocatalysts. *J. Am. Chem. Soc.* **2021**, *143*, 18052. [CrossRef] [PubMed]
35. Babu, H.V.; Bai, M.G.M.; Rajeswara Rao, M. Functional pi-Conjugated Two-Dimensional Covalent Organic Frameworks. *ACS Appl. Mater. Interfaces* **2019**, *11*, 11029. [CrossRef] [PubMed]
36. Lu, Z.; Wang, Z. S-scheme $CuWO_4$@g-C_3N_4 core-shell microsphere for CO_2 photoreduction. *Mater. Sci. Semicond. Process.* **2023**, *153*, 107177. [CrossRef]
37. Ye, J.; Cheng, B.; Yu, J.; Ho, W.; Wageh, S.; Al-Ghamdi, A.A. Hierarchical Co_3O_4-NiO hollow dodecahedron-supported Pt for room-temperature catalytic formaldehyde decomposition. *Chem. Eng. J.* **2022**, *430*, 132715. [CrossRef]
38. Qiu, X.F.; Zhu, H.L.; Huang, J.R.; Liao, P.Q.; Chen, X.M. Highly Selective CO_2 Electroreduction to C_2H_4 Using a Metal-Organic Framework with Dual Active Sites. *J. Am. Chem. Soc.* **2021**, *143*, 7242. [CrossRef] [PubMed]
39. Zhou, Y.; Chen, L.; Sheng, L.; Luo, Q.; Zhang, W.; Yang, J. Dual-metal atoms embedded into two-dimensional covalent organic framework as efficient electrocatalysts for oxygen evolution reaction: A DFT study. *Nano Res.* **2022**, *15*, 7994. [CrossRef]
40. Zhu, Z.; Tao, H.; Zhang, R.; Gan, L.; Zhou, Y. Synergistic effect of diatomic materials on efficient formaldehyde sensing and degradation. *J. Mater. Chem. A* **2024**, *12*, 419. [CrossRef]

41. Wang, V.; Xu, N.; Liu, J.-C.; Tang, G.; Geng, W.-T. VASPKIT: A user-friendly interface facilitating high-throughput computing and analysis using VASP code. *Comput. Phys. Commun.* **2021**, *267*, 108033. [CrossRef]
42. Lu, T.; Chen, F. Multiwfn: A multifunctional wavefunction analyzer. *J. Comput. Chem.* **2012**, *33*, 580. [CrossRef] [PubMed]

Disclaimer/Publisher's Note: The statements, opinions and data contained in all publications are solely those of the individual author(s) and contributor(s) and not of MDPI and/or the editor(s). MDPI and/or the editor(s) disclaim responsibility for any injury to people or property resulting from any ideas, methods, instructions or products referred to in the content.

toxics

MDPI

Review

Selective Adsorption of Hazardous Substances from Wastewater by Hierarchical Oxide Composites: A Review

Wenjun Tu and Weiquan Cai *

Guangzhou Higher Education Mega Center, School of Chemistry and Chemical Engineering, Guangzhou University, 230 Wai Huan Xi Road, Guangzhou 510006, China; 19972083102@163.com
* Correspondence: cccaiwq@gzhu.edu.cn; Tel.: +86-20-39366905

Abstract: Large volumes of wastewater containing toxic contaminants (e.g., heavy metal ions, organic dyes, etc.) are produced from industrial processes including electroplating, mining, petroleum exploitation, metal smelting, etc., and proper treatment prior to their discharge is mandatory in order to alleviate the impacts on aquatic ecosystems. Adsorption is one of the most effective and practical methods for removing toxic substances from wastewater due to its simplicity, flexibility, and economics. Recently, hierarchical oxide composites with diverse morphologies at the micro/nanometer scale, and the combination advantages of oxides and composite components have been received wide concern in the field of adsorption due to their multi-level structures, easy functionalization characteristic resulting in their large transport passages, high surface areas, full exposure of active sites, and good stability. This review summarizes the recent progress on their typical preparation methods, mainly including the hydrothermal/solvothermal method, coprecipitation method, template method, polymerization method, etc., in the field of selective adsorption and competitive adsorption of hazardous substances from wastewater. Their formation processes and different selective adsorption mechanisms, mainly including molecular/ion imprinting technology, surface charge effect, hard-soft acid-base theory, synergistic effect, and special functionalization, were critically reviewed. The key to hierarchical oxide composites research in the future is the development of facile, repeatable, efficient, and scale preparation methods and their dynamic adsorption with excellent cyclic regeneration adsorption performance instead of static adsorption for actual wastewater. This review is beneficial to broaden a new horizon for rational design and preparation of hierarchical oxide materials with selective adsorption of hazardous substances for wastewater treatment.

Keywords: hierarchical oxide composites; preparation method; hazardous pollutants; selective adsorption; selective adsorption mechanism

1. Introduction

With industrial development, wastewater treatment has become one of the most serious environmental problems. The major pollutants, including metal ions, dyes,and other toxic organics, in the effluents seriously harm biodiversity, ecosystem functions, and aquatic systems. Some methods, including ion exchange, chemical precipitation, adsorption, membrane separation, and electrochemistry treatment, are applied for wastewater treatment. Among them, adsorption has aroused widespread concern because of its convenient and flexible operation, high efficiency, and friendly and regenerative characteristics. Furthermore, some pollutants are valuable and can be reused after separation from the pollution system. Therefore, it is necessary and urgent to rationally design and prepare novel materials with special selectivity towards specific pollutants for different situations.

In the past years, many hierarchical oxides with multi-level structures and diverse morphologies resulting in improved physicochemical and surface properties and potential applications including photocatalysis [1], ion detection [2], membrane separation [3], and adsorption [4] have been reported. The construction of their structures is usually considered to be a process of self-assembly from a low-dimensional structure to a high-dimensional structure. Commonly, hierarchical oxide has the advantages of a high surface area, ease of modification, special application structure, and various morphologies. Especially, the well-structured hierarchical structure with an interconnected porous network facilitates the movement of pollutants to the exposing active sites located on the pore walls, resulting in better diffusion and adsorption process. Therefore, with rational design, preparation, and modification of functional groups, hierarchical oxide composites show great potential for selective adsorption of specific pollutants from wastewater.

Applications of hierarchically structured metal oxides, including cobalt oxide, iron oxide, and ceria, for the removal of As(V) and Cr(VI) ions in water [5], metal oxide heterostructures for arsenic removal from contaminated water [6], MgO with micro-nanostructures and composites of nano-MgO-based materials focusing on their composition and application [7], and different hierarchical nanostructures of TiO_2 for energy and environmental applications [8] have been concisely reviewed. Thereafter, rationally designing appropriate hierarchical oxide composites for effective removal of hazardous pollutants has been the top strategy for wastewater treatment. Herein, recent research progress on their development and selective adsorption of hazardous substances from wastewater, together with their selective adsorption mechanisms, was presented to deepen their understanding and promote their applications in wastewater treatment. Especially, our focus is on the effects of their hierarchical structures with diverse morphologies instead of hierarchical porous structures with multi-level pores, including macro-, meso-, and micropores.

2. Preparation of Hierarchical Oxide Composites with Selective Adsorption Abilities

2.1. *Hydrothermal/Solvothermal Method*

The hydrothermal/solvothermal method is widely used to prepare hierarchical materials via dissolution and recrystallization processes in a sealed pressure vessel. It has the advantages of a relatively uniform distribution of particle sizes, poor particle aggregation, complete grain development, and easy control of the morphology in a specific direction. It is worth noting that polyols are usually used as solvents, such as ethylene glycol (EG) [9] and triethylene glycol [10], and under solvothermal conditions, the mixed water-polyol solvent plays a key role in controlling the self-assembly of oxide nanoparticles to form a hierarchical morphology. For example, Liu et al. [11] adopted a modified hydrothermal process and followed heat treatment to fabricate an octahedral $ZnO/ZnFe_2O_4$ composite using $FeCl_3 \cdot 6H_2O$ and $ZnCl_2$ as the metal resources, CH_3COONa as the assistant, and EG as the solvent. Figure 1 shows that the octahedron composite presents particle sizes of about 200 nm composed of three elements (Zn, Fe, and O) uniformly. The $ZnO/ZnFe_2O_4$ can adsorb malachite green (MG) with a maximum adsorption capacity as high as 4983.0 $mg \cdot g^{-1}$ from mixed dye solutions including methyl orange (MO) and rhodamine B (RhB) via ion exchange. Due to that, its pore volume and pore sizes can be adjusted by different heat treatments. Hierarchical rod-like CuO with a remarkable affinity for hazardous dichromate adsorption properties was also hydrothermally synthesized, and its inherent self-assemble structure and high surface area allowed the creation of effective adsorption sites. However, it is interesting that rod-shaped CuO nanomaterials could not be obtained in the absence of EG [12]. EG was also used as a solvent in our lab to prepare Fe_3O_4 microspheres with an average diameter of about 500 nm. After modification with chitosan, the composite showed more than 90% removal efficiency for Cr(VI) in the presence of various cations, including Cd^{2+}, Cu^{2+}, Zn^{2+}, and Ni^{2+}, or anions, including Cl^-, $C_2O_4^{2-}$, SO_4^{2-}, and HPO_4^{2-} [13]. During the hydrothermal process, the growth of nanoparticles was achieved by the condensation of hydroxyl groups in alcohols, resulting in the local formation of trace water.

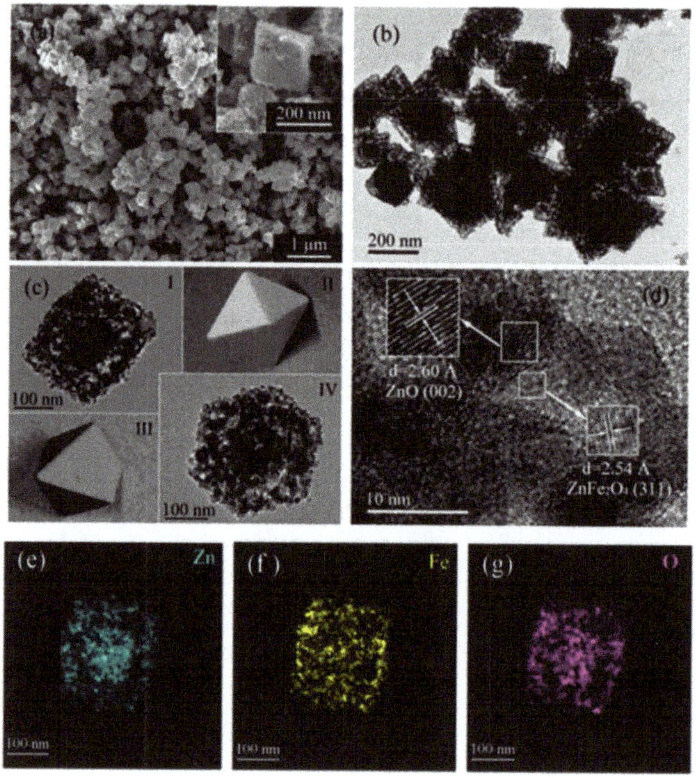

Figure 1. (**a**) FESEM (inset is a high solution image of a particle), (**b**,**c**) FETEM, (**d**) HRTEM images of the $ZnO/ZnFe_2O_4$ nanoparticles; EDS elemental mapping images to the corresponding area: (**e**) Zn, (**f**) Fe, and (**g**) O [11]. (In (**c**), images I and IV are the TEM scannograms from different perspectives of a same octahedral particle, photo-illustrated as images II and III, respectively).

Hydrothermal time is an important factor that significantly affects the formation of hierarchical structures. Wu et al. [14] used a similar method to prepare porous α-MoO_3 for selective removal of Pb^{2+} in a solution containing Cu^{2+}, Zn^{2+}, $Cr^{3+,}$ and Cd^{2+} through the formation of lead molybdate by MoO_3 and Pb^{2+} as follows:

$$MnO_3 + H_2O + Pb^{2+} \rightarrow PbMnO_4 + 2H^+ \quad (1)$$

The SEM with different times shows that the α-MoO_3 nucleuses can gradually aggregate to form flower-like microspheres with diameters of 2.5–4.5 μm composed of massive nanobelts. Then, they grew on capillaries to obtain hand-like nanosheet arrays along the preferential orientation.

Some researchers used hierarchical oxides of rare earth elements, including La-, Ce-, to remove pollutants. For example, Chen et al. [15] prepared a three-dimensional (3D) graphene-La_2O_3 composite composed of graphene and La_2O_3 particles, which were prepared in advance and then heated together. La- could bind to graphene through La-O, and the composite achieved 100% removal efficiency for phosphate through La-O-P in the presence of Cl^-, ${SO_4}^{2-}$, and ${NO_3}^-$. Rotzetter et al. [16] also proved that La_2O_3 could adsorb phosphate as follows:

$$La_2O_3 + 2{HPO_4}^{2-} + H_2O \rightarrow 2LaPO_4 + 4OH^- \quad (2)$$

Likewise, Sun et al. [17] used La- to synthesize lanthanum molybdenum oxide with an average diameter of about 270 nm, and it showed almost removal efficiency of 100% for various mixed dyes with multi-sulfonic groups. Because La- center lacks electrons, it can attract electron-rich groups such as sulfonic groups. Furthermore, Tong et al. [18] reported a cerium oxide doped by two-dimensional (2D) molybdenum disulfide nanosheets. CeO_2 nanoparticles (3–5 nm) could grow on 2D MoS_2, and the maximum adsorption capacity of MoS_2/CeO_2 nanohybrids for Pb^{2+} is 333 $mg \cdot g^{-1}$ at a pH value of 2.0. It also showed an amazing selective adsorption for Pb^{2+} from the coexisting ions of K^+, Na^+, Ca^{2+}, Mg^{2+}, Mn^{2+}, Cr^{3+}, Co^{2+}, Ni^{2+}, Zn^{2+}, Cu^{2+}, and Cd^{2+} due to the coordination effect between S and Pb.

The development of TiO_2-based composites for pollution removal has also attracted the attention of researchers. For example, Lee et al. [19] reported a hydrothermal preparation of flower-like TiO_2-graphene oxide (GO). Nonmetallic oxides GO (100 nm in diameter) with oxygen-containing functional groups was used as a supporting matrix to induce the self-assembled growth of 3D flowerlike TiO_2 nanoparticles. The GO-TiO_2 showed the following adsorption order: $Zn^{2+} > Cd^{2+} > Pb^{2+}$ due to ionic radius or electronegativity effects. Similarly, Xie et al. [20] designed a simple method to synthesize layered protonated titanate hierarchical microspheres with average diameters of 2 μm and an extremely high specific surface area of 450 $m^2 \cdot g^{-1}$, and the self-assembly process was achieved through aggregation, crystallization, and growth processes (Figure 2). This material shows specific selectivity for methylene blue (MB) compared to MO, cresol red, and RhB, and the two key factors affecting adsorption capability are the electrostatic interaction and steric structure of the organic dyes.

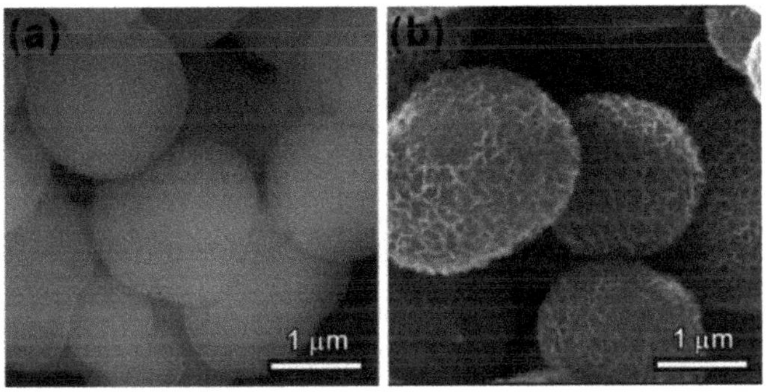

Figure 2. SEM images of the products at the early stage of the reaction: (**a**) 2 h, and (**b**) 6 h [20].

Using waste biomass or carbon as a dispersion carrier of metal oxides to prepare composites could reduce their agglomeration, further enhancing their adsorption performance via a synergistic effect. For example, durian shell fibers modified with Cu-Al bimetallic oxide (Cu-Al/DBF) with hierarchical structure were hydrothermally synthesized, and the Cu-Al/DBF was used to remove ammonia nitrogen from wastewater. Its maximum adsorption capacity was 18.04 $mg \cdot g^{-1}$ fitted from the Langmuir isotherm due to the synergistic effect of biomass carbon, alumina, and copper oxide [21]. Furthermore, the magnetic Fe_3O_4@C hollow microspheres obtained by a solvothermal method coupled with an annealing strategy showed an adsorption capacity of 197.2 $mg \cdot g^{-1}$ for Cr(VI). Especially, the magnetic Fe_3O_4 core can provide easily separated characteristics by magnetic separation, and the carbon layers can effectively prevent the aggregation of the magnetic Fe_3O_4 nanoparticles [22].

Deep eutectic solvents (DESs), as a type of ionic liquid analog, are regarded as an alternative to conventional solvents in the fields of material chemistry. Li et al. [23] reported uniformly distributed MgO microcubes based on DES consisting of $MgCl_2 \cdot 6H_2O$ and urea with a molar ratio of 1:2 via a solvothermal method and the following calcination process.

Their adsorption capacities for anionic dyes Congo red (CR), Amaranth, and Indigo carmine were 666.7, 43.74, and 54.32 mg·g^{-1}, respectively, at 25 °C, relating to electrostatic attraction and hydrogen bonding.

2.2. Coprecipitation Method

Coprecipitation is a method of adding precipitant into a mixed metal salt solution to precipitate two or more cations together to form a precipitate. Its typical application is the preparation of magnetic Fe_3O_4, and this process achieves the transformation from zero-dimensional nanoparticles to 3D microspheres using Fe^{2+} and Fe^{3+} as the cores for forming 3D core-shell material. Wang et al. [24] designed a core-shell magnetic calcium silicate/GO; the average diameter of Fe_3O_4 nanoparticles coated by calcium silicate was about 10 nm. The composite showed selective adsorption for acridine orange via electrostatic, hydrophobic, and π-π interactions. Zavareh et al. [25] also reported a Cu-chitosan/Fe_3O_4 nanocomposite to selectively adsorb phosphate in the presence of chloride, nitrate, and sulfate, whose concentrations are 10 times higher than phosphate due to complex formation between Cu(II) and phosphate; the mean size of magnetite nanoparticles was between 20 and 30 nm.

This method can be extended to prepare other composite metal oxides. For example, Rahman et al. [26] synthesized Ag_2O_3-ZnO nanocones for selective Co(II) adsorption under the interference of Cd(II), Co(II), Cr(III), Cu(II), Fe(III), Ni(II), and Zn(II). The growth mechanism of nanocones could be described as a process of self-assembled nucleation and subsequent aggregation. The final product shows a 3D nanocone composed of nanoparticles, and the mean length and cross-section (center) of the nanocones were approximately 1.42 mm and 0.53 mm, respectively. Similarly, Marwani et al. [27] prepared a CdO coordinated Fe_2O_3 composite of aggregated nanofiber with an average width size of 70 nm using cadmium and ferric salts as metal resources and NaOH as precipitant, respectively. The selective research indicated that Pb(II) showed the highest distribution coefficient (K_d) value of 1.21×10^5 mL·g^{-1} among the coexisting cations, including Cd(II), Co(II), Cu(II), Cr(III), Cr(VI), Fe(III), Ni(II), and Zn(II). As another example, hierarchical Ti_3C_2@FeOOH nanocomposites for antimony-contaminated wastewater treatment were developed via in-situ anchored FeOOH into the interlamination of alkali-treated Ti_3C_2 nanosheets, and they showed excellent adsorption capacities for both Sb(V) and Sb(III), with the maximum adsorption capacities of 93.22 and 111.50 mg·g^{-1}, respectively, based on isotherm analysis. They also showed superior selectivity, especially for Sb(III), irrelevant to foreign anions [28]. A flower-like AlOOH/AlFe intermetallic composite with a high adsorption rate and adsorption capacity for As(V) removal up to 200 mg·g^{-1} was prepared via direct precipitation of a bimetallic Al/Fe nanopowder with water at 60 °C, indicating that this micro/nanostructure with a specific surface area of 247.1 m^2·g^{-1} facilitated deliverability [29].

Hierarchical MgO is also being shown applications for removing various pollutants. For example, an EG-assisted precipitation route was developed to prepare 3D flower-like MgO samples at room temperature. The Mg^{2+} coordinates the -OH groups of EG to form an alkoxide coordination complex in a basic medium with different divalent cations to form sheets arranged in an ordered way to construct microspheres. The MgO showed the highest adsorption capacity of 574.71 mg·g^{-1} for phosphate [30]. Then, a 3D nest-like porous magnetic MgO hybrid (Fe_3O_4/MgO) with a specific surface area of 135.2 m^2·g^{-1}, uniform mesochannels of 5–35 nm, and easily separated characteristics (Figure 3) was controllably synthesized based on a seed-induced precipitation process containing a suspension of Fe_3O_4 nanoparticles with an average diameter of 200 nm and following calcination of the precursor. The Fe_3O_4/MgO had a good removal performance for simultaneous removal of 12 polycyclic aromatic hydrocarbons (PAHs) and Cd^{2+} with fast adsorption (~0.25 h) and high removal efficiencies (>70% for PAHs and >80% for Cd^{2+}, respectively) [31].

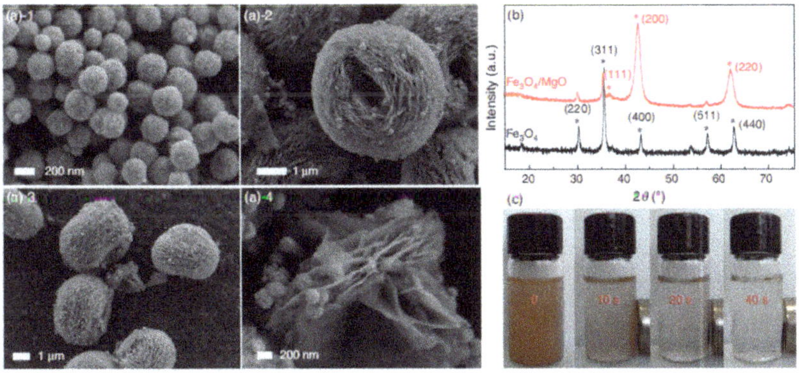

Figure 3. SEM images of Fe_3O_4 (**(a)-1**), Fe_3O_4/MgO (**(a)-2–(a)-4**), XRPD patterns of Fe_3O_4, Fe_3O_4/MgO (**b**) and the performance of magnetic separation (**c**) [31].

2.3. Template Method

Template synthesis is a process in which materials with a nanostructure, unique morphology, and low cost are used as templates, and related materials are deposited into their holes or surfaces by chemical or physical methods. Then, the template is removed to obtain the desired hierarchical materials. Templates include soft templates and hard templates. Surfactants such as polyvinylpyrrolidone (PVP), cetyl trimethyl ammonium bromide (CTAB), polyethylene oxide-polypropylene oxide-polyethylene oxide (P123), etc. are used as soft templates to influence the morphologies and pore structures of the metal oxides. Li et al. [32] reported a PVP-assisted preparation method of hollow Fe_3O_4 with a mean diameter of 300 nm. Firstly, a hollow Fe_3O_4 precursor (Figure 4a,b) was prepared using $FeCl_3 \cdot 6H_2O$ assisted by sodium citrate and urea in deionized water; then, PVP was added following a hydrothermal process. Then, the PVP was removed by a 2-aminoterephthalic acid solution of ethanol to obtain hollow Fe_3O_4. Figure 4c depicts Fe_3O_4 with NH_2-MIL-101(Fe) doped with inner hollow structures, while Figure 4d shows the layers belonging to NH_2-MIL-101(Fe). The as-prepared adsorbent showed selective adsorption for phosphates in the presence of Cl^-, F^-, Br^-, NO_3^-, and SO_4^{2-}.

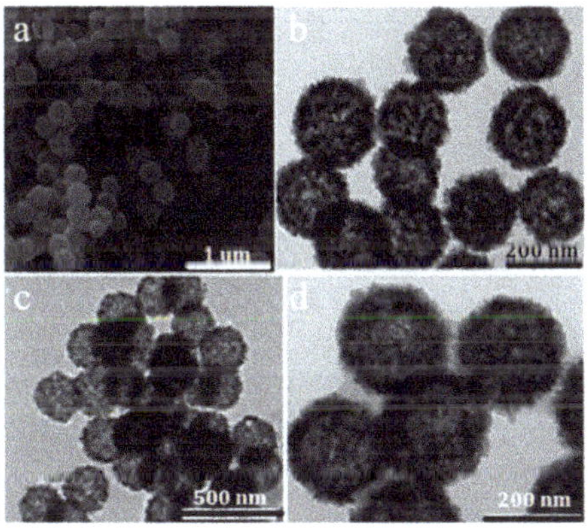

Figure 4. SEM (**a**) and TEM (**b**) images of hollow Fe_3O_4 precursor, and TEM (**c**,**d**) images of the as-prepared hollow porous magnetic $Fe_3O_4@NH_2$-MIL-101(Fe) [32].

Cai et al. [33] designed hollow $C@SiO_2$ nanoparticles using CTAB as the template. Their mean sizes were approximately 500 nm, while the sizes of the core and the shell were 450 nm and 50 nm, respectively. CTAB was removed by ethanol and HCl solutions, resulting in a porous structure with a high specific surface of 150.57 $m^2 \cdot g^{-1}$. The adsorption capacity of C@SiO_2 nanoparticles for Cr(VI) was 90.53 $mg \cdot g^{-1}$ in the presence of Cu(II), Cd(II), Zn(II), and Ni(II), which just declined 10% compared with no interfering cations. Furthermore, anions like Cl^- and NO_3^- showed no influence on Cr(VI) adsorption, while SO_4^{2-} and HPO_4^{2-} could slightly affect it due to the similar hydration degree with $HCrO_4^-$. It was found that adding an appropriate amount of CTAB can change the morphology, pore properties of hierarchical MgO and its adsorption behavior for phosphate in a CTAB-assisted solvothermal route. Especially, the gardenia flower-like MgO with the highest specific surface area of 336.54 $m^2 \cdot g^{-1}$ and a total pore volume of 0.843 $cm^3 \cdot g^{-1}$ showed the highest adsorption capacity of 348.32 $mg \cdot g^{-1}$ for phosphate with a short equilibrium time of 4 h [34].

Hierarchical structures also include mesoporous materials, which possess several levels of structure via soft template modification. For example, Sarafraz et al. [35] used P123 as the template to prepare phosphonic functional groups-modified mesoporous silica microspheres for uranium selective adsorption, and P123 was removed via a boiled mixture of methanol and HCl. Similarly, Yang et al. [36] used P123 to synthesize phosphoric mesoporous silica with ion-imprinting technology for selective removal of uranium. The above researches showed that the phosphorous group is beneficial for selectively capturing uranium.

Unlike soft templates, hard templates, such as carbon, carbonate, and silica, can retain their morphologies in the sample. For example, Zhang et al. [37] prepared the porous Al_2O_3 microspheres/acrylic ester resin hybrids for selective absorption of oil and organic solvent via a microwave polymerization process by using Al_2O_3 spheres as modifiers. The Al_2O_3 microspheres with an average size of 2.0 μm were obtained using glucose as the carbon source of the carbon template from a combined hydrothermal and sintering processes, followed by surface modification with the silane coupling agent KH 570 (Figure 5) to enhance their hydrophobicity and reactivity. Pervaiz also reported a sacrificial carbonate template coupled with organic ligands and polymers for preparing cobalt iron oxide microspheres via a soft prototype route. They showed a high specific surface area of 786 $m^2 \cdot g^{-1}$ and a high magnetism of 63 $emu \cdot g^{-1}$, and were excellent adsorbents and catalysts for the oxidative-dissociation of RhB and 4-Nitrophenol from water at room temperature [38]. In another example, Zhang et al. [39] synthesized mesoporous δ-Bi_2O_3 using SBA-15 silicas prepared by different silicon sources as hard templates, which were later removed by a 2 M NaOH solution. The δ-Bi_2O_3 showed a selective adsorption capacity of 2.21 $mmol \cdot g^{-1}$ for I^- in the presence of Cl^-.

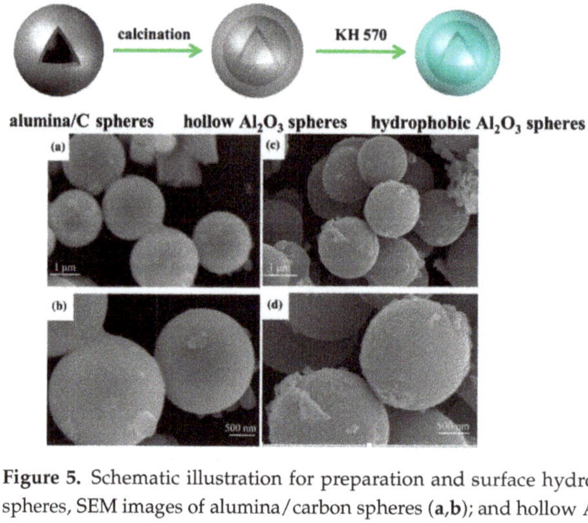

Figure 5. Schematic illustration for preparation and surface hydrophobic modification of Al_2O_3 spheres, SEM images of alumina/carbon spheres (**a**,**b**); and hollow Al_2O_3 microspheres (**c**,**d**) [37].

2.4. Polymerization

Polymerization is usually used to combine polymers with metal oxides, and some core-shell structures are modified by polymers to obtain special functionality. For example, Kliangsuwan et al. reported a hierarchical composite which was incorporated a nanocomposite of zinc oxide and carbon foam embedded in a magnetic molecularly imprinted polymer (ZnO@CF@Fe_3O_4-SiO_2-NH_2@MIP) for extracting sulfonamides (Figure 6). The foam nanocomposite helped to improve the adsorption performance of sulfonamides; the molecularly imprinted polymer (MIP) provided highly specific recognition cavities for three sulfonamides, and the magnetic material enabled its simple and rapid separation after adsorption and desorption. This developed strategy determined sulfonamides in milk and water, with extraction recoveries between 84.3 and 96.2% [40]. In another example, Zhang et al. [41] prepared a Fe_3O_4/PANI/MnO_2 core-shell hybrid with a diameter of 300 nm, and the coating thicknesses of polyaniline (PANI) and MnO_2 shells could be controlled by determining the polymerization time and $KMnO_4$ amount, respectively. It was found that the superior adsorption capacity of this hybrid for Cd(II), Zn(II), Pb(II) and Cu(II) was attributed to the synergetic effect between PANI and MnO_2. Gu et al. [42] also designed an amino functionalized Fe_3O_4@SiO_2 core-shell structure with a mean diameter of 320 nm by one-pot co-condensation. The Fe_3O_4@SiO_2 modified by amino silane contained one N atom, and it showed the adsorption amount for multiple ions solution containing Cr(VI), Cu(II), Ni(II), Zn(II) and Cd(II) at the same time due to the abundant amino and hydroxyl groups of the adsorbent. Compared with that without interference ions, it is noteworthy that its adsorption capacity for Cr(VI) did not decrease. In addition, Hwang et al. [43] prepared a porous phenol resin containing lithium manganese oxide (LMO) through polycondensation and carbonization for selective adsorption of lithium. Poly(vinyl alcohol) (PVA) and hexamethylenetetramine were chosen as the stabilizer and the curing agent, respectively. Finally, the LMO microspheres with a rough surface were obtained. Hierarchical porous, magnetic Fe_3O_4@carbon nanofibers (Fe_3O_4@CNFs) based on polybenzoxazine precursors have been synthesized by a combination of electrospinning and in situ polymerization at 250 °C. The fibers with an average diameter of 130 nm were comprised of carbon fibers with embedded Fe_3O_4 nanocrystals and showed a high specific surface area of 1885 $m^2 \cdot g^{-1}$ and a pore volume of 2.3 $cm^3 \cdot g^{-1}$. The Fe_3O_4@CNFs showed efficient adsorption properties for organic dyes in water and excellent magnetic separation performance [44].

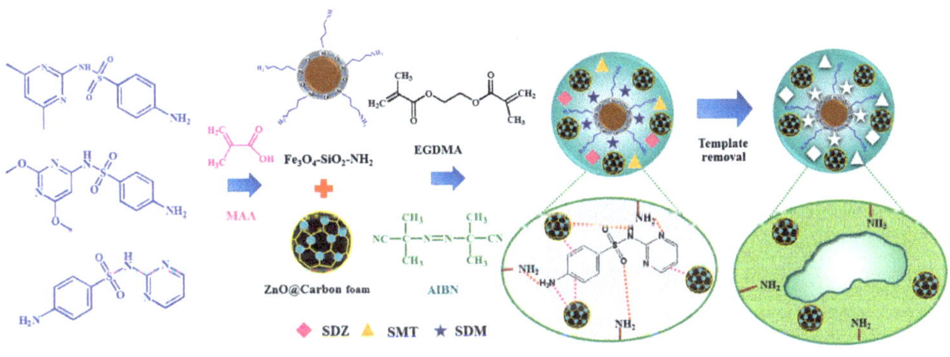

Figure 6. The preparation procedure of the proposed hierarchical nanocomposite ZnO@CF@Fe_3O_4 SiO_2-NH_2@MIP adsorbent for the extraction of sulfonamides [40].

2.5. Other Preparation Methods

Other preparation methods, mainly sol-gel, electrospinning, reflux, thermal decomposition, and chemical bath deposition, could also be used to prepare hierarchical oxide composites for pollutant removal via adsorption. Sol-gel method: hierarchical magnetic graphene oxide-titanate nanocomposites (MGO@TNs) with a high specific surface area of

193.4 $m^2 \cdot g^{-1}$ and magnetite nanoparticles anchored on them were prepared via a modified sol-gel and subsequent alkaline hydrothermal process. When 3 $g \cdot L^{-1}$ MGO@TNs was used for removing Pb(II) from stimulated realistic battery wastewater, safe discharge with a concentration lower than 0.05 $mg \cdot L^{-1}$ could be achieved due to ion exchange and surface complexation [45]. Electrospinning method: Min et al. [46] reported an electrospinning chitosan/Fe-Mn nanofibrous composite (Fe-Mn@CS NF) to remove trace As(III) from water, and its concentration decreased from 550 $\mu g \cdot L^{-1}$ to less than 1.2 $\mu g \cdot L^{-1}$ while using 0.5 $g \cdot L^{-1}$ Fe-Mn@CS NF. The presence of F^- or SO_4^{2-} showed a negligible impact on As(III) removal, while PO_4^{3-} impeded its adsorption via competing for adsorption sites. Reflux method: Solanki et al. [47] reported a 3D flower-like Fe_3O_4 architecture decorated with SALDETA moieties via refluxing the reaction mixture (Figure 7). This magnetical composite showed excellent adsorption capacity of 415.5 $mg \cdot g^{-1}$, faster kinetics of 8 min, rapid separation of 40 s, facile regeneration of 5 min, and good reusability of 5 runs for Pb^{2+} ions resulting from its hierarchical structure, immobilized functional groups, and chelation property. Electrostatic self-assembly method: to solve the disadvantages of easily distorted and aggregated into other uncontrolled morphologies for 2D lamellar-like graphene, resulting in a remarkable decline in performance, 3D macroporous reduced GO-Fe_3O_4 nanocomposites were synthesized via an electrostatic self-assembly method. They showed high adsorption capacities, rapid adsorption rates for Cr(VI), and easy magnetic separation for reusability. Interestingly, the Fe_3O_4 nanoparticles serve as stabilizers for separating graphene nanosheets from aggregation, while the graphene nanosheets favor hindering them from agglomeration and enabling their good distribution on the surface of graphene [48]. Thermal decomposition method: ferromagnetic 3D flower-like γ-Fe_2O_3 particles with an adsorption capacity of 102.7 $mg \cdot g^{-1}$ for CR were prepared by a simple direct thermal decomposition method using cheap and nontoxic ferric nitrate as an iron source and CTAB as a structure-directing agent. With increasing amounts of CTAB, the morphology of γ-Fe_2O_3 particles was transformed from 1D to 3D. Especially, the sample obtained by adding 15% CTAB showed a complete flower-like structure with smooth petals [49]. Chemical bath deposition method: hierarchical NiO hollow architectures (HPHAs) assembled from nanoflakes with a thickness of about 8 nm were synthesized via a one-pot facile chemical bath deposition method and the following calcination process. The HPHAs showed the maximum adsorption capacity of 490.2 $mg \cdot g^{-1}$ for CR from the Langmuir equation due to the synergistic effect of porous structure, large specific surface area, and the electrostatic attraction of NiO with CR molecules [50].

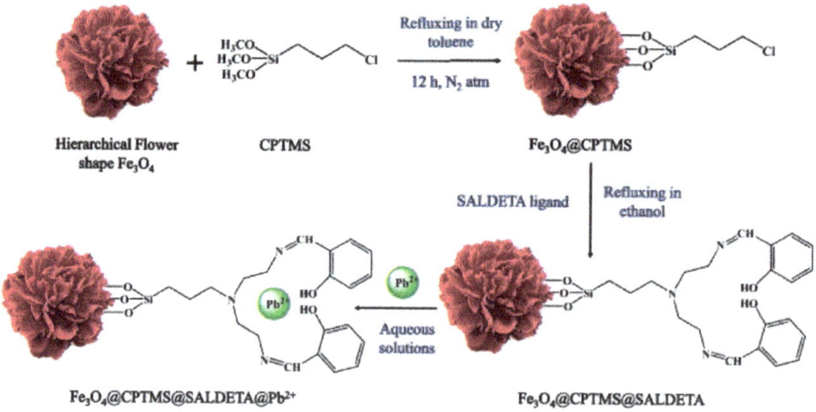

Figure 7. Schematic diagram of the surface modification of 3D hierarchical Fe_3O_4 structures and their adsorption for Pb^{2+} ions (CPTMS and SALDETA represent 3-chloropropyltrimethoxysilane and salicylaldehyde-diethylene triamine, respectively) [47].

2.6. Non-Powder Adsorption Materials

All of the above adsorption materials are powders, and thus it is difficult to separate them after adsorption unless external forces, such as a magnetic field, are used. Non-powder materials such as membranes and aerogels with a 3D structure could solve this problem, and it is interesting that GO is popular to assist in their preparation. For example, Zhao et al. [51] synthesized a 3D aerogel based on GO modified by positively charged polyetherimide (PEI) via a sol-gel method. The GO/PEI aerogel with a tunable surface charge at different pH values was formed after freeze drying and was stable in acidic and basic aqueous solutions. It showed high adsorption capacities of 249.6 $mg \cdot g^{-1}$ for MB at pH 10.5 and 3331.0 $mg \cdot g^{-1}$ at pH 2.0 for anionic MO, respectively. Likewise, Rahmani et al. [52] prepared a N-doped reduced GO aerogel with a 3D inter-connected network via a hydrothermal method. This aerogel exhibited excellent selective adsorption performance for oil pollutants because of the coordination ability of N. Li et al. [53] designed a GO membrane functionalized with phenanthroline diamide (GO-PDA) through a modified sol-gel method.

Furthermore, SiO_2 aerogel is a type of material that could be tailored in terms of its specific features and surface chemistry as an adsorbent for pollutant removal. For example, Lamy-Mendes et al. [54] synthesized methyltrimethoxysilane (MTMS)-based carbon nanostructures—silica aerogels via a two-step acidbase-catalyzed sol-gel process, and they can remove various organic compounds and drugs, and achieve adsorption capacities of 200 $mg \cdot g^{-1}$ for xylene and 170 $mg \cdot g^{-1}$ for toluene, respectively. The addition of co-precursors containing carbon nanomaterials and/or amine groups was a valuable tool to alter their properties, thus enhancing their adsorption performance. Zhang et al. [55] successfully fabricated flexible and hierarchical TiO_2-SiO_2 nanofibrous mats with superior adsorption efficiency and recyclable performance for methyl blue removal via sol-gel metnod and the following calcination process, and they could maintain integrated morphology after bending to curvature in which the radius is 0.6 mm, indicating that the brittleness of inorganic oxides was successfully overcome.

Membrane adsorption is an efficient and easily segregated method for pollutant removal. Park et al. [56] reported a chitosan-coated iron oxide nanocomposite immobilized hydrophilic poly(vinylidene) fluoride membrane (Chi@Fe_2O_3-PVDF) to remove Cr(VI) with adsorption capacities of 14.45 $mg \cdot g^{-1}$ in a batch system, and 14.10 $mg \cdot g^{-1}$ in a continuous in-flow system, respectively. Importantly, its removal efficiency was not changed significantly in the presence of competing ions, including Cl^-, NO_3^-, SO_4^{2-}, and PO_4^{3-}. Our previous work [57] reported a γ-AlOOH/PVA membrane via a sol-gel method, and it presented a good adsorption capacity for Cr(VI), due to electrostatic interaction. The adsorption results of coexisting anions (HCO_3^-, HPO_4^-, $C_2O_4^{2-}$, F^-, Cl^-, and SO_4^{2-}) indicated that the composite membrane showed a highly selective adsorption capacity for Cr(VI) while only HCO_3^-, SO_4^{2-}, and HPO_4^{2-} could slightly cause interference due to the similar radii with $HCrO_4^-$. Moreover, both Sun et al. [58] and Tan et al. [59] used the sol-gel method with PVA as an adjuvant to form membranes. It seems that this sol-gel method, assisted by PVA, has the ability to make each component form a uniform 3D network structure. Hierarchical hybrid nanocomposite MgO@PES-PDA membranes were also constructed by coating MgO nanoparticles on the PES-PDA derived from co-deposition of polydopamine (PDA) and PEI under mild basic conditions, and such membranes showed a rapid capture and high destructive adsorption capacity for paraoxon toxin (up to 92% within 40 min). Especially the smaller MgO-coated PES-PDA membrane with relatively high MgO content showed the best destructive absorption ability [60].

The advantages and disadvantages of the different preparation methods for preparation of hierarchical oxides were compared in Table 1.

Table 1. Comparison of different methods for the preparation of hierarchical oxides.

Methods	Advantages	Disadvantages
Hydrothermal/ solvothermal method	Beneficial to crystal nucleation and growth, easy to control morphology.	Slight aggregation, not easy to control in a closed space; relatively long reaction time; high reaction temperature; and high pressure.
Coprecipitation	Usually involving metal salts and precipitants, mild reaction conditions usually occur at room temperature.	Only be used to prepare metal oxides; similar precipitation equilibrium constants of metal salts.
Template	Precise control of the size, morphology, and structure, aggregation reduction of nanoparticles.	Structure destruction while removing the template, limits of the reaction conditions for the template.
Polymerization	Combination of polymer and metal oxide to form composites with more functional groups.	Complex preparation process, easy aggregation resulting in destroyed morphology.
Sol-gel	Easy separation, beneficial to form a 3D network, and homogeneous mixing of components at the molecular level.	Long reaction time for several days, making it easy to introduce anionic impurities.

3. Selective Adsorption Mechanism

3.1. Molecular/Ion Imprinting Technology

Molecularly imprinted technology (MIT) was first used in biological detection. In the field of adsorption, ion imprinting technology (IIP), which is similar to MIP but uses ions as a template, has received extensive attention. MIT and IIP can generate specific binding sites for molecules and ions in the adsorbent with a non-biological strategy [61]. The adsorbent after removing the template showed extremely selective adsorption for target molecules or ions. Liang et al. [62] prepared a Cr(VI) ion-imprinted polymer using four organic compounds as chemical additives with different functions. Core-shell Fe_3O_4@SiO_2 was introduced for easy separation, and GO can prevent Fe_3O_4@SiO_2 from aggregation. The adsorbent reached adsorption equilibrium within 5 min for Cr(VI), and its adsorption capacity increased from 182.77 to 301.89 mg·g^{-1} compared with the sample without using IIP. Furthermore, its selectivity coefficient for Cr(VI) was high in the presence of Cu(II), Cd(II), Cr(III), Ni(II), SO_4^{2-} and NO_3^- indicating its enhanced selective adsorption performance for Cr(VI) by using IIP. Compared with cations, its adsorption was significantly affected by anions because of their similar charge and ionic radii with Cr(VI). Figure 8 showed that the template ions left identifiable sites in the material after washing by acidified thiourea solution; the memory effect enabled the adsorbent to adsorb Cr(VI) faster and selectively. In another example, Zhang et al. [63] used MIP to selectively adsorb 2-aminopyridine based on magnetic chitosan and β-cyclodextrin, and then a methanol-acetic acid solution was chosen to remove the template. Compared with the adsorbent obtained without template, the adsorption capacity of the adsorbent obtained with template increased three times, and it also showed a higher selectivity for 2-aminopyridine in the presence of its structural analogues or coexistent ions (Na^+, K^+, Mg^{2+}, Ca^{2+}, Cl^-, and SO_4^{2-}) due to the recognition effect. Similar reports were also reported [64–66], and the key of MIP or IIP is to add the adsorbate to the adsorbent as the template. After removing the template, a highly selective adsorbent for this adsorbate can be obtained.

3.2. Electrostatic Interaction

3.2.1. Surface Charge Effect

Because many pollutants are charged, selective adsorption can be achieved by choosing a material with appropriate opposite charges [67]. Chen et al. [68] designed a Fe_3O_4@NH_2@PEI core/shell composite, and the average size of the Fe_3O_4@NH_2 cores was 40 nm, while the average shell thickness was 40 nm. NH_2- groups act as a bridge for PEI to be loaded on the magnetic core. Due to the fact that PEI is positively charged, it can selectively adsorb anionic dyes from a mixed solution of cationic and anionic dyes; the adsorption can reach equilibrium within 30 min. Likewise, Huang et al. [69] prepared a similar Fe_3O_4@Tb/AMP core-shell composite with a mean size of 10 nm. It showed similarly selective adsorption performance for anionic dyes because adenosine 5′-monophosphate monohydrate (AMP) was positively charged in the solution. Sarkar et al. [70] synthesized a more flexible amine-functionalized reduced graphene oxide-carbon nanotube hybrid (rGO-CNT-PPD) with an average width of 20 nm. As shown in Figure 9, rGO-CNT-PPD can selectively adsorb cationic dyes in an acidic environment, while it shows selective adsorption performance for anionic dyes in a neutral environment because of the change in surface charge.

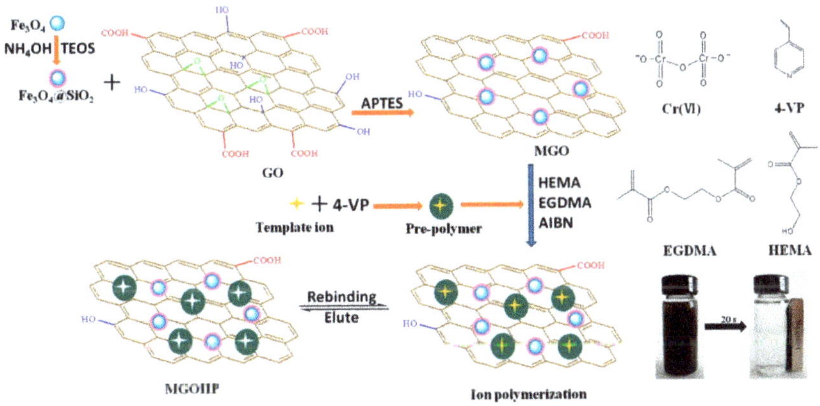

Figure 8. Schematic of synthesis route of Cr(VI) imprinted polymer (the meanings of the corresponding abbreviations are as follows: tetraethoxysilane (TEOS), 3-aminopropyltriethoxysilane (APTES), 4-vinyl pyridine (4-VP), 2-hydroxyethyl methacrylate (HEMA), ethylene glycol dimethacrylate (EGDMA), N,N-azoisobisbutyronitrile (AIBN), ion-imprinted polymer (MGOIIP)) [62].

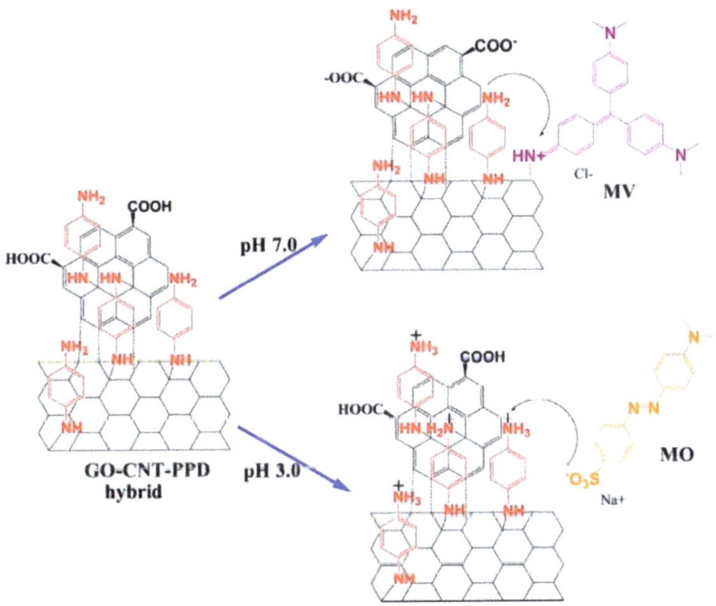

Figure 9. Effect of pH on adsorption of rGO–CNT–PPD hybrid for methyl violet (MV) and MO [70].

3.2.2. Hard-Soft-Acid-Base Theory

Hard-soft-acid-base theory (HSAB) claims that acids and bases are divided into hard and soft categories depending on their different properties. "Hard acid first binds to hard base, soft acid first binds to soft base" is one of the most important empirical rules to explain adsorption selectivity. Ashour et al. [71] prepared citric acid and l-cysteine-modified Fe_3O_4 microspheres (CA@Fe_3O_4 and Cys@Fe_3O_4, respectively) to selectively adsorb rare earth ions (RE^{3+}). The coexistence cations (RE^{3+}, Mg^{2+}, Ca^{2+}, and Ni^{2+}) adsorption experiment showed that O and N in CA@Fe_3O_4 and Cys@Fe_3O_4 are hard Lewis base atoms, and thus they can selectively complex with RE^{3+} ions, which are hard Lewis acids. The authors used separation factor (SF) based on K_d to describe selective ability, and further research showed that the adsorbent had better selectivity for Gd^{3+} and Nd^{3+} than La^{3+} and Y^{3+}, indicating that a moderate ion radius was beneficial to adsorption. In another example, Wu et al. [72] developed 3D S-impregnated nano-MnO_2 nanorods with lengths ranging from 300 to 800 nm for selective removal of Pd^{2+}. The competitive adsorption experiment showed that the S atom is a soft base that can bind with a soft acid, Pd^{2+}. However, interfering ions Ni^{2+}, Cu^{2+}, Zn^{2+}, and Co^{2+} are hard Lewis acids, so they cannot be adsorbed. Fu et al. [73] also prepared a 2D porous Fe_3O_4/Poly($C_3N_3S_3$) nanocomposite for selective removal of Pb^{2+} and Hg^{2+}; the average size of Fe_3O_4 was about 7.5 nm while the pore size of the poly($C_3N_3S_3$) network ranged from 5 to 500 nm. The competitive adsorption experiment showed that Hg^{2+} as a soft acid, can be easily bound to S, while Pb^{2+} and Cu^{2+} as intermediate acids have a medium affinity for the nitrogen groups. On the contrary, Mg^{2+} and Ca^{2+} are hard acids, so Fe_3O_4/Poly($C_3N_3S_3$) has the worst adsorption selectivity for them.

3.3. *Synergistic Effect*

Different from other adsorption mechanisms, the synergistic effect refers to the combined work of each component, resulting in selective adsorption ability. Chen et al. [74] prepared a PPy/TiO_2 core-shell composite that can selectively adsorb heavy metal ions. The SEM images showed that TiO_2 was the core while polypyrrole (PPy) was the shell. It was found that the selective adsorption order was as follows: $Zn^{2+} > Pb^{2+} \gg Cu^{2+}$, and the result cannot be explained by any known theory. The adsorption mechanism is a synergistic effect, which can be briefly described in Figure 10. PPy can conduct a doping-dedoping process, through which TiO_2 displays a selective adsorption performance towards Zn^{2+}, Pb^{2+}, and Cu^{2+}. However, the reason for the higher adsorption tendency of TiO_2 to Zn^{2+} than Cu^{2+} is unknown.

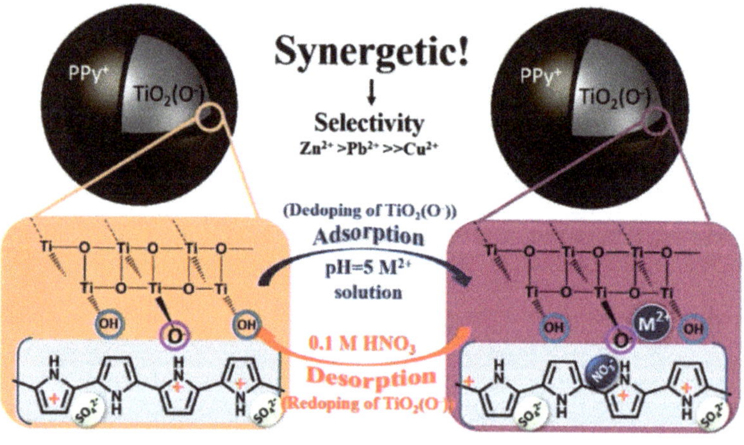

Figure 10. Synergistic adsorption between PPy and TiO_2 in the PPy/TiO_2 composite of selective adsorption for heavy metal ions [74].

Zhang et al., presented a hydrothermal synthesis of hierarchical rod-like mesoporous Mg-Al bimetallic oxides (Mg/Al-BOs) with a high specific surface area of 472.4 $m^2 \cdot g^{-1}$ and high adsorption capacity, selectivity, and reusability for U(VI) uptake via both surface complexation and electrostatic interaction [75]. The oxygen-containing groups on the surface of Mg/Al-BOs play significant roles in the U(VI) adsorption in addition to the electrostatic attractions, for which the complexation process could be described as follows:

$$\equiv M{-}O^- + H_2O \rightarrow \equiv M{-}OH + OH^- \quad (3)$$

$$\equiv M{-}OH + U(VI) \rightarrow \equiv M{-}O{-}U(VI) + H^+ \quad (4)$$

Bian et al. [76] also found a synergistic effect between ferrite and TiO_2. They synthesized a series of core-shell MFe_2O_4-TiO_2 (M=Mn, Fe, Zn, Co, or Ni) to selectively adsorb toxic UO^{2+} in the presence of Rb^+, Sr^{2+}, Cr^{3+}, Mn^{2+}, Ni^{2+}, Zn^{2+}, and Cd^{2+}. The shell MFe_2O_4 increased the average diameter of the core TiO_2 from 2.5 nm to hundreds of nanometers. The adsorption mechanism was embodied as follows: at first, M^{2+} ions act as a mediator to make holes from MFe_2O_4 to TiO_2; then, TiO_2 on the surface can adsorb all cations with hydroxyl radicals except UO^{2+}; after that, the remaining UO^{2+} is adsorbed in the interface of MFe_2O_4-TiO_2 through holes, leading to selective adsorption. Hierarchical Mg-Al bimetallic oxide/straw fiber (Mg-Al/SF) was hydrothermally synthesized and incorporated with the in situ growth method.

Hierarchical Mg-Al bimetallic oxide/straw fber (Mg-Al/SF) was hydrothermally synthesized and incorporated with in the situ growth method. The synergistic combination of surface adsorption from biomass fiber and chemical adsorption from alumina and magnesium oxide made Mg-Al/SF a promising adsorbent for enhanced phosphorus removal with the maximum adsorption capacity of 89.37 mg/g [77].

3.4. Special Functionalization

Surface functionalization is an efficient method to reduce the agglomeration of the powder adsorbents, and it can also improve their stability and adsorption performance. Some selective adsorptions are achieved by special functionalized adsorbents aimed at the adsorbate, such as specific ion exchange, special substances, and specific chemical bonds. Kera et al. [78] prepared a PPy-PANI/Fe_3O_4 core-shell nanocomposite with mean sizes of 50–100 nm for selective adsorption of Cr(VI). The protonated N atoms on the surface of PPy-PANI/Fe_3O_4 can adsorb Cr(VI) in acidic conditions, and realize selective ion exchange of Cr(VI) with innocuous Cl^-. Therefore, neither cation nor anion could affect the adsorption, and the maximum adsorption capacity can reach 434.78 $mg \cdot g^{-1}$. Futher, Fe_3O_4 nanoparticles were functionalized by a modified mussel-inspired method with dopamine (DA) and (3-aminopropyl)triethoxysilane (KH550), thus obtaining core-shell Fe_3O_4/poly(DA+KH550) hybrids for MB. The surface of the hybrid is negatively charged at a higher pH, resulting in interaction with MB molecules via electrostatic attraction, hydrogen bonding, and π–π stacking, and it showed a maximum adsorption capacity of 400.00 $mg \cdot g^{-1}$ for MB, easy separation, and excellent reusability characteristics because of the high specific surface area and abundant active adsorption sites [79].

Similarly, Wu et al. [80] prepared a $La(OH)_3/Fe_3O_4$ core-shell nanocomposite with a maximum adsorption capacity of 83.5 $mg \cdot g^{-1}$ for phosphate, and the adsorption process was described as follows:

$$La - OH + H_2PO_4{}^- \leftrightarrow La - H_2PO_4 + OH^- \quad (5)$$

$$La = OH + HPO_4{}^{2-} \leftrightarrow La = HPO_4 + 2OH^- \quad (6)$$

$$La \equiv OH + PO_4{}^{3-} \leftrightarrow La \equiv PO_4 + 3OH^- \quad (7)$$

It seems that harmless –OH groups on La can selectively exchange phosphorus groups in the actual wastewater containing ions of various concentrations. This result was the same as the previous research about phosphate removal via La-based materials, including lanthanum oxides or lanthanum hydroxides [81].

Some substances or groups have a specific adsorption affinity for certain toxic metals. For example, Cs^+, a widely present rare metal element ion in nuclear radiation wastewater, can be specifically adsorbed by Prussian Blue (PB), which is analogous because it has specific recognition ability for Cs^+ [82,83]. PB can form specific lattices with Cs^+ because its lattice parameter and hydrated ion radius of $^{137}Cs^+$ are well matched. U(VI), similar to Cs^+, can preferentially form complexes with phosphorous groups [35,36,84]. As for fluorine-containing pollutants, it is a good choice to choose fluorine-containing functional groups to construct adsorbents due to the fluorine-fluorine (F-F) interaction [85,86]. Table 2 summarizes the advantages and disadvantages of some current adsorption selection mechanisms. It was found that adsorbents with good performance can be prepared by combining the properties of the adsorbates and materials to be used.

Table 2. Comparison of different selective mechanisms.

Mechanism	Advantage	Disadvantage	Selectivity
Molecularly imprinted polymer or ion imprinting technology	Adsorption of a specific adsorbate by introducing a corresponding template, not easily affected by external factors	Complex preparation method, only adsorption for single adsorbate, bad morphology	Excellent
Surface charge effect	Adsorption of many adsorbates with the same charge, wide scope of application	Affected by pH value, influenced by coexisting ions with the same charge	Good
Hard-soft-acid-base theory	Suitable for atomically modified adsorbents	Limited scope of application, influenced by amphoteric coexisting ions	Good
Synergistic effect	No modification results in good morphology; the simple preparation of adsorbent	Speculative and unclear adsorption mechanism, limited scope of application	Good
Special functionalization	Adsorption of a specific adsorbate, not affected by external factors	Useless to multiple pollutants, limited scope of application	Excellent

4. Conclusions

This review mainly focuses on the synthesis methodologies of hierarchical oxide composites, their selective adsorption performance, and adsorption mechanisms for hazardous substances from the aqueous matrix. Their greatest advantage for the removal of pollutants is that they can combine the advantages of the physico-chemical characteristics of all the components and their multi-level structures with diverse morphologies that realize the effect of "1 + 1 > 2". However, there still exist some problems for most of the functionalized hierarchical oxides that influence their selective adsorption performance, such as the agglomeration of magnetic Fe_3O_4 modified by polymers. Future research should focus on retaining their original hierarchical morphology with good dispersibility. Secondly, although many studies have reported their applications for removing various pollutants, including heavy metal ions, organic dyes, etc. Most of them are not suitable for mass-scale preparation for large-scale applications. Therefore, it is of increasing interest to explore high-efficiency methods for controllable-scale fabrication of hierarchical oxide composites. Thirdly, though surface functionalization has been proven to be an effective method to improve the adsorption performance of hierarchical oxides, the amounts of their functional groups for binding pollutants are strongly dependent on the functionalization method, whereas the functionalization process of most reported functionalized adsorbents is cumbersome and time-consuming. Fourthly, simple wet chemical routes, including hydrothermal/solvothermal methods, coprecipitation methods, and sol-gel methods are largely limited in utility due to the low pore density of agglomeration and hence the low specific surface area. To simultaneously avoid these shortcomings, offer monodispersed structures, and achieve high functionality, soft templates and self-assembled composites need to be further explored in the near future. Fifthly, most studies are static adsorption for the specific model pollutants prepared in the laboratory, and they almost have no practical application potential. Finally, it is quite difficult for the hierarchical oxide composites to in-situ monitor their adsorption processes for pollutants, and thus their thorough removal mechanisms for pollutants are unclear in detail. Therefore, future studies should pay attention to the in-situ adsorption process and the dynamic selective adsorption of the hierarchical oxide composites for actual industrial wastewater. In short, this review contains systematically gathered information using highly related references on selective adsorption of hazardous substances from wastewater by hierarchical oxide composites, and thus can serve as a starting point for research in the related field.

Author Contributions: Writing—original draft preparation, W.T.; writing—review and editing, supervision, project administration and funding acquisition, W.C. All authors have read and agreed to the published version of the manuscript.

Funding: This work was funded by the National Natural Science Foundation of China (No.22378088), the Natural Science Foundation of Guangdong Province (No.2024A1515011239), and the Scientific Research Project of Guangzhou University (JC2023007).

Institutional Review Board Statement: Not applicable.

Informed Consent Statement: Not applicable.

Data Availability Statement: The data underlying this article are available in the corresponding references.

Acknowledgments: We thank the National Natural Science Foundation of China, the Natural Science Foundation of Guangdong Province and Guangzhou University for funding this paper.

Conflicts of Interest: The authors declare no conflict of interest.

References

1. Mu, X.; Xu, Q.; Xie, Y.; Ma, Y.; Zhang, Z.; Shen, Z.; Guo, Y.; Yu, J.; Ajmal, S.; Zhang, W.; et al. Hollow cubic TiO_2 Loaded with Copper and Gold Nanoparticles for Photocatalytic CO_2 reduction. *J. Alloys Compd.* **2024**, *980*, 173589. [CrossRef]
2. Elfiky, M.; Salahuddin, N.; Matsuda, A. Green fabrication of 3D Hierarchical Blossom-Like Hybrid of Peeled Montmorillonite-Zno for In-Vitro Electrochemical Sensing of Diltiazem Hydrochloride Drug. *Mat. Sci. Eng. C Mater.* **2020**, *111*, 110773. [CrossRef] [PubMed]
3. Jayachitra, R.; Lincy, V.; Prasannan, A.; Jebaranjitham, J.N.; Sangaraju, S.; Hong, P.D. Tailored Fabrication of Biodegradable Polymer/Fe_3O_4 doped WO_3 Nano Star-based Porous Membrane with Enhanced Photo Fentonic Activity for Environmental. *Environ. Res.* **2024**, *248*, 118262. [CrossRef] [PubMed]
4. Zhou, J.; Cai, W.; Yang, Z.; Xia, Q.; Chen, J.; Fan, J.; Du, C. N,N-dimethylformamide Assisted Facile Hydrothermal Synthesis of Boehmite Microspheres for Highly Effective Removal of Congo Red from Water. *J. Colloid Interf. Sci.* **2021**, *583*, 128–138. [CrossRef]
5. Hu, J.S.; Zhong, L.S.; Song, W.G.; Wan, L.J. Synthesis of Hierarchically Structured Metal Oxides and Their Application in Heavy Metal Ion Removal. *Adv. Mater.* **2008**, *20*, 2977–2982. [CrossRef]
6. Chen, L.; Xin, H.; Fang, Y.; Zhang, C.; Zhang, F.; Cao, X.; Zhang, C.; Li, X. Application of Metal Oxide Heterostructures in Arsenic Removal from Contaminated Water. *J. Nanomat.* **2014**, *2014*, 793610. [CrossRef]
7. Duan, Z.J.; Li, X.Q.; Dong, B. Recent Development in the Environmental Application of Nano sized MgO. *Bull. Mater. Sci.* **2022**, *45*, 204. [CrossRef]
8. Reghunath, S.; Pinheiro, D.; Devi, K.R.S. A Review of Hierarchical Nanostructures of TiO_2: Advances and Applications. *Appl. Surf. Sci. Adv.* **2021**, *3*, 100063. [CrossRef]
9. Lü, H.X.; Wang, X.M.; Yang, J.Q.; Xie, Z.H. One-step Synthesis of CDTA Coated Magnetic Nanoparticles for Selective Removal of Cu(II) from Aqueous Solution. *Int. J. Biol. Macromol.* **2015**, *78*, 209–214. [CrossRef] [PubMed]
10. Helal, A.S.; Mazario, E.; Mayoral, A.; Decorse, P.; Losno, R.; Lion, C.; Ammar, S.; Hémadi, M. Highly Efficient and Selective Extraction of Uranium from Aqueous Solution Using a Magnetic Device: Succinyl-β-cyclodextrin-APTES@maghemite Nanoparticles. *Environ. Sci. Nano* **2018**, *5*, 158–168. [CrossRef]
11. Liu, J.; Zeng, M.; Yu, R.H. Surfactant-free Synthesis of Octahedral $ZnO/ZnFe_2O_4$ Heterostructure with Ultrahigh and Selective Adsorption Capacity of Malachite Green. *Sci. Rep.* **2016**, *6*, 25074. [CrossRef] [PubMed]
12. Banerjee, S.; Sarkar, S.; Sarkar, S.; Patra, A.K. Rational Design of Self-assembled Copper Oxide Nanoparticles into Hierarchical Nanorods with High-surface-area for Environmental Remediation of Wastewater. *Inorg. Chem. Commun.* **2024**, *160*, 111925. [CrossRef]

13. Jiang, Y.H.; Cai, W.Q.; Tu, W.J.; Zhu, M.Y. Facile Cross-link Method to Synthesize Magnetic Fe_3O_4@SiO_2-chitosan with High Adsorption Capacity toward Hexavalent Chromium. *J. Chem. Eng. Data* **2019**, *64*, 226–233. [CrossRef]
14. Wu, Y.Y.; Cheng, X.L.; Zhang, X.F.; Xu, Y.M.; Gao, S.; Zhao, H.; Huo, L.H. High Efficient and Selective Removal of Pb^{2+} through Formation of Lead Molybdate on α-MoO_3 Porous Nanosheets Array. *J. Colloid Interf. Sci.* **2017**, *491*, 80–88. [CrossRef] [PubMed]
15. Chen, M.L.; Huo, C.B.; Li, Y.K.; Wang, J.H. Selective Adsorption and Efficient Removal of Phosphate from Aqueous Medium with Grapheme-lanthanum Composite. *ACS Sustain. Chem. Eng.* **2016**, *4*, 1296–1302. [CrossRef]
16. Rotzetter, A.C.C.; Kellenberger, C.R.; Schumacher, C.M.; Mora, C.; Grass, R.N.; Loepfe, M.; Luechinger, N.A.; Stark, W.J. Combining Phosphate and Bacteria Removal on Chemically Active Filter Membranes Allows Prolonged Storage of Drinking Water. *Adv. Mater.* **2013**, *25*, 6057–6063. [CrossRef] [PubMed]
17. Sun, M.; Ma, Y.Y.; Tan, H.Q.; Yan, J.; Zhang, H.Y.; Shi, H.F.; Wang, Y.H.; Li, Y.G. Lanthanum Molybdenum Oxide as a New Platform for Highly Selective Adsorption and Fast Separation of Organic Dyes. *RSC Adv.* **2016**, *6*, 90010–90017. [CrossRef]
18. Tong, S.S.; Deng, H.X.; Wang, L.; Huang, T.; Liu, S.H.; Wang, J.Y. Multi-functional Nanohybrid of Ultrathin Molybdenum Disulfide Nanosheets Decorated with Cerium Oxide Nanoparticles for Preferential Uptake of Lead (II) Ions. *Chem. Eng. J.* **2018**, *335*, 22–31. [CrossRef]
19. Lee, Y.C.; Yang, J.W. Self-assembled Flower-like TiO_2 on Exfoliated Graphite Oxide for Heavy Metal Removal. *J. Ind. Eng. Chem.* **2012**, *18*, 1178–1185. [CrossRef]
20. Xie, S.F.; Zheng, B.J.; Kuang, Q.; Wang, X.; Xie, Z.X.; Zheng, L.S. Synthesis of Layered Protonated Titanate Hierarchical Microspheres with Extremely Large Surface Area for Selective Adsorption of Organic Dyes. *CrystEngComm* **2012**, *14*, 7715–7720. [CrossRef]
21. Yuan, J.; Zhu, Y.; Wang, J.; Liu, Z.; Wu, J.; Zhang, T.; Li, P.; Qiu, F. The Application of the Modifed Durian Biomass Fiber as Adsorbent for Effective Removal of Ammonia Nitrogen. *J. Iran. Chem. Soc.* **2022**, *19*, 435–445. [CrossRef]
22. Wu, X.P.; Cheng, L.; Song, C.S.; Zhang, Y.Z.; Shi, X.J.; Li, X.Y.; Lin, P.; Wang, S.L.; Wang, P.; Xu, L.B.; et al. Constructing Carbon-coated Fe_3O_4 Hierarchical Microstructures with a Porous Structure and Their Excellent Cr(VI) Ion Removal Properties. *J. Mater. Sci-Mater. El.* **2021**, *32*, 20844–20855. [CrossRef]
23. Li, N.; Dang, H.; Chang, Z.; Zhao, X.; Zhang, M.; Li, W.; Zhou, H.; Sun, C. Synthesis of Uniformly Distributed Magnesium Oxide Micro-/nanostructured Materials with Deep Eutectic Solvent for Dye Adsorption. *J. Alloys Compd.* **2019**, *808*, 151571. [CrossRef]
24. Wang, H.; Chen, Y.N.; Wei, Y.M. A Novel Magnetic Calcium Silicate/graphene Oxide Composite Material for Selective Adsorption of Acridine Orange from Aqueous Solutions. *RSC Adv.* **2016**, *6*, 34770–34781. [CrossRef]
25. Zavareh, S.; Behrouzi, Z.; Avanes, A. Cu (II) Binded Chitosan/Fe_3O_4 Nanocomomposite as a New Biosorbent for Efficient and Selective Removal of Phosphate. *Int. J. Biol. Macromol.* **2017**, *101*, 40–50. [CrossRef] [PubMed]
26. Rahman, M.M.; Khan, S.B.; Marwani, H.M.; Asiri, A.M. Selective Divalent Cobalt Ions Detection Using Ag_2O_3-ZnO Nanocones by ICP-OES Method for Environmental Remediation. *PLoS ONE* **2014**, *9*, e114084. [CrossRef] [PubMed]
27. Marwani, H.M.; Lodhi, M.U.; Khan, S.B.; Asiri, A.M. Selective Extraction and Determination of Toxic Lead Based on Doped Metal Oxide Nanofiber. *J. Taiwan Inst. Chem. Eng.* **2015**, *51*, 34–43. [CrossRef]
28. Wan, K.; Wang, Y.; Liu, C.; Wei, C.; Lv, S.; Tang, X.; Fang, T.; Zhao, J.; Wei, G.; Qi, P.; et al. Facile Synthesis of Hierarchical Ti_3C_2@FeOOH Nanocomposites for Antimony Contaminated Wastewater Treatment: Performance, Mechanisms, Reutilization, and Sustainability. *Chem. Eng. J.* **2022**, *450*, 138038. [CrossRef]
29. Svarovskaya, N.; Bakina, O.; Glazkova, E.; Rodkevich, N.; Lerner, M.; Vornakova, V.; Chzhou, V.; Naumova, L. Synthesis of Novel Hierarchical Micro/nanostructures AlOOH/AlFe and Their Application for As(V) Removal. *Environ. Sci. Pollut. Res.* **2022**, *29*, 1246–1258. [CrossRef] [PubMed]
30. Ahmed, S.; Pan, J.; Ashiq, M.N.; Li, D.; Tang, P.; Feng, Y. Ethylene Glycol-assisted Fabrication and Superb Adsorption Capacity of Hierarchical Porous Flower-like Magnesium Oxide Microspheres for Phosphate. *Inorg. Chem. Front.* **2019**, *6*, 1952–1961. [CrossRef]
31. Tan, D.; Jin, J.; Guo, C.; Dhanjai; Chen, J. Magnetic Magnesium Oxide Composites for Rapid Removal of Polycyclic Aromatic Hydrocarbons and Cadmium Ions from Water. *Environ. Chem.* **2020**, *17*, 479–487. [CrossRef]
32. Li, Y.; Xie, Q.Y.; Hu, Q.; Li, C.P.; Huang, Z.J.; Yang, X.J.; Guo, H. Surface Modifcation of Hollow Magnetic Fe_3O_4@NH_2-MIL-101(Fe) Derived from Metal-organic Frameworks for Enhanced Selective Removal of Phosphates from Aqueous Solution. *Sci. Rep.* **2016**, *6*, 30651. [CrossRef] [PubMed]
33. Cai, W.Q.; Gu, M.M.; Jin, W.; Zhou, J.B. CTAB-functionalized C@SiO_2 Double-shelled Hollow Microspheres with Enhanced and Selective Adsorption Performance for Cr(VI). *J. Alloys Compd.* **2019**, *777*, 1304–1312. [CrossRef]
34. Ahmed, S. CTAB-assisted Fabrication of Hierarchical Flower-like Magnesium Oxide Adsorbent for Enhanced Removal Performance towards Phosphate. *J. Magne. Alloys* **2023**, *11*, 3231–3240. [CrossRef]
35. Sarafraz, H.; Minuchehr, A.; Alahyarizadeh, G.; Rahimi, Z. Synthesis of Enhanced Phosphonic Functional Groups Mesoporous Silica for Uranium Selective Adsorption from Aqueous Solutions. *Sci. Rep.* **2017**, *7*, 11675. [CrossRef] [PubMed]
36. Yang, S.; Qian, J.; Kuang, L.J.; Hua, D.B. Ion-imprinted Mesoporous Silica for Selective Removal of Uranium from Highly Acidic and Radioactive Effluent. *ACS Appl. Mater. Inter.* **2017**, *9*, 29337–29344. [CrossRef] [PubMed]
37. Zhang, T.; Yue, X.J.; Yang, D.Y.; Guo, Q.; Qiu, F.X.; Li, Z.D. Hybridization of Al_2O_3 Microspheres and Acrylic Ester Resins as a Synergistic Absorbent for Selective Oil and Organic Solvent Absorption. *Appl. Organomet. Chem.* **2018**, *32*, e4244. [CrossRef]

38. Pervaiz, E.; Thomas, T.; Afzal, M.J.; Yang, M. Template Synthesis of $CoFe_2O_4$ Extended Surface Microspheres for Efficient Water Decontamination and Absorption of Electromagnetic Waves: Twin Behavior. *Mate. Res. Express* **2019**, *6*, 075506. [CrossRef]
39. Zhang, L.P.; Jarouiec, M. SBA-15 Templating Synthesis of Mesoporous Bismuth Oxide for Selective Removal of Iodide. *J. Colloid Interf. Sci.* **2017**, *501*, 248–255. [CrossRef] [PubMed]
40. Kliangsuwan, A.; Phonchai, A.; Bunkoed, O. A Magnetic Molecularly Imprinted Polymer Hierarchical Composite Adsorbent Embedded with a Zinc Oxide Carbon Foam Nanocomposite for the Extraction of Sulfonamides. *Microchem. J.* **2022**, *179*, 107443. [CrossRef]
41. Zhang, J.; Han, J.; Wang, M.G.; Guo, R. Fe_3O_4/PANI/MnO_2 Core-shell Hybrids as Advanced Adsorbents for Heavy Metal ions. *J. Mater. Chem. A* **2017**, *5*, 4058–4066. [CrossRef]
42. Gu, M.M.; Cai, W.Q.; Ma, T. Preparation of Amino Modified Fe_3O_4@SiO_2 Composite by One-pot Co-Condensation Method with Enhanced Adsorption Performance towards Cr(VI). *Chin. J. Inorg. Chem.* **2018**, *34*, 1293–1302.
43. Hwang, C.W.; Park, S.G.; Hwang, T.S. Synthesis of Lithium Ion Selective Porous Phenolic Microsphere Adsorbents with Lithium Manganese Oxide (LMO) by Template and Their Lithium Ion Adsorption Properties. *Macromol. Res.* **2015**, *23*, 313–319. [CrossRef]
44. Si, Y.; Ren, T.; Li, Y.; Ding, B.; Yu, J. Fabrication of Magnetic Polybenzoxazine-based Carbon Nanofibers with Fe_3O_4 Inclusions with a Hierarchical Porous Structure for Water Treatment. *Carbon* **2012**, *50*, 5176–5185. [CrossRef]
45. Yang, X.; Guo, N.; Yu, Y.; Li, H.; Xia, H.; Yu, H. Synthesis of magnetic graphene oxide-titanate composites for efficient removal of Pb(II) from wastewater: Performance and mechanism. *J. Environ. Manag.* **2020**, *256*, 109943. [CrossRef] [PubMed]
46. Min, L.; Ma, Y.; Zhang, B.; He, D.; Chen, J.; Li, X.; Wang, S.; Chi, Y. Electrospinning Chitosan/Fe-Mn Nanofibrous Composite for Efficient and Rapid Removal of Arsenite from Water. *Toxics* **2024**, *12*, 230. [CrossRef] [PubMed]
47. Solanki, K.; Sharma, S.; Rana, P.; Kaushik, B.; Yadav, S.; Dixit, R.; Birdar, A.V.; Gupta, A.; Sharma, R.K. A Structurally Engineered Flower Shaped Magnetic Hierarchical Sorbent for Rapid and Selective Uptake of Pb^{2+} Ions from Water Samples. *Mater. Chem. Front.* **2023**, *7*, 4482–4496. [CrossRef]
48. Liu, Y.; Zhang, Z.; Sun, X.; Wang, T. Design of Three-dimensional Macroporous Reduced Graphene Oxide–Fe_3O_4 Nanocomposites for the Removal of Cr(VI) from Wastewater. *J. Porous Mat.* **2019**, *26*, 109–119. [CrossRef]
49. Zhang, B.; Zhou, L.H.; Zhao, S.; Asuha, S. Direct synthesis of 3D Flower-Like Maghemite Particles and Their Properties. *J. Alloys Compd.* **2020**, *817*, 152802. [CrossRef]
50. Hu, H.; Deng, C.; Sun, M.; Zhang, K.; Wang, M.; Xu, J.; Le, H. Facile Template-free Synthesis of Hierarchically Porous NiO Hollow Architectures with High-efficiency Adsorptive Removal of Congo Red. *J. Porous Mat.* **2019**, *26*, 1743–1753. [CrossRef]
51. Zhao, Q.; Zhu, X.Y.; Chen, B.L. Stable Graphene Oxide/poly(ethyleneimine) 3D Aerogel with Tunable Surface Charge for High Performance Selective Removal of Ionic Dyes from Water. *Chem. Eng. J.* **2018**, *334*, 1119–1127. [CrossRef]
52. Rahmani, Z.; Rashidi, A.M.; Kazemi, A.; Samadi, M.T.; Rahmani, A.R. N-doped Reduced Graphene Oxide Aerogel for The Selective Adsorption of Oil Pollutants from Water: Isotherm and Kinetic Study. *J. Ind. Eng. Chem.* **2018**, *61*, 416–426. [CrossRef]
53. Li, F.H.; Yang, Z.; Weng, H.Q.; Chen, G.; Lin, M.Z.; Zhao, C. High Efficient Separation of U(VI) and Th(IV) from Rare Earth Elements in Strong Acidic Solution by Selective Sorption on Phenanthroline Diamide Functionalized Graphene Oxide. *Chem. Eng. J.* **2018**, *332*, 340–350. [CrossRef]
54. Lamy-Mendes, A.; Lopes, D.; Girão, A.V.; Silva, R.F.; Malfait, W.J.; Durães, L. Carbon Nanostructures—Silica Aerogel Composites for Adsorption of Organic Pollutants. *Toxics* **2023**, *11*, 232. [CrossRef] [PubMed]
55. Zhang, J.; Mensah, A.; Narh, C.; Hou, X.; Cai, Y.; Qiao, H.; Wei, Q. Fabrication of Fexible TiO_2-SiO_2 Composite Nanofibers with Variable Structure as Efficient Adsorbent. *Ceram. Int.* **2020**, *46*, 3543–3549. [CrossRef]
56. Park, J.E.; Shin, J.H.; Oh, W.; Choi, S.J.; Kim, J.; Kim, C.; Jeon, J. Removal of Hexavalent Chromium(vi) from Wastewater Using Chitosan-coated Iron Oxide Nanocomposite Membranes. *Toxics* **2022**, *10*, 98. [CrossRef] [PubMed]
57. Luo, L.; Cai, W.Q.; Zhou, J.B.; Li, Y.Z. Facile Synthesis of Boehmite/PVA Composite Membrane with Enhanced Adsorption Performance towards Cr(VI). *J. Hazard. Mater.* **2016**, *318*, 452–459. [CrossRef] [PubMed]
58. Sun, D.S.; Meng, M.J.; Lu, Y.; Hu, B.; Yan, Y.S.; Li, C.X. Porous Nanocomposite Membranes Based on Functional GO with Selective Function for Lithium Adsorption. *New J. Chem.* **2018**, *42*, 4432–4442. [CrossRef]
59. Tan, P.; Hu, Y.Y.; Bi, Q. Competitive Adsorption of Cu^{2+}, Cd^{2+} and Ni^{2+} from an Aqueous Solution on Graphene Oxide Membranes. *Colloid Surface. A* **2016**, *509*, 56–64. [CrossRef]
60. Wei, X.; Li, C.; Wang, C.; Li, N.S.; Wu, J.; Guo, M. Rapid and Destructive Adsorption of Paraoxon-ethyl Toxin via a Selfdetoxifying Hybrid Electrospun Nanofibrous Membrane. *Chem. Eng. J.* **2018**, *351*, 31–39. [CrossRef]
61. Liu, Y.; Meng, X.G.; Liu, Z.C.; Meng, M.J.; Jiang, F.P.; Luo, M.; Ni, L.; Qiu, J.; Liu, F.F.; Zhong, G.X. Preparation of Two-Dimensional Ion-imprinted Polymer Based on Graphene Oxide/SiO_2 Composite for Selective Adsorption of Nickel Ions. *Langmuir* **2015**, *31*, 8841–8851. [CrossRef] [PubMed]
62. Liang, Q.W.; Geng, J.J.; Luo, H.J.; Fang, W.; Yin, Y.W. Fast and Selective Removal of Cr(VI) from Aqueous Solutions by a Novel Magnetic Cr(VI) Ion-Imprinted Polymer. *J. Mol. Liq.* **2017**, *248*, 767–774. [CrossRef]
63. Zhang, W.; Zhu, Z.L.; Zhang, H.; Qiu, Y.L. Selective Removal of the Genotoxic Compound 2-Aminopyridine in Water Using Molecularly Imprinted Polymers Based on Magnetic Chitosan and B-Cyclodextrin. *Int. J. Environ. Res. Public Health* **2017**, *14*, 991. [CrossRef] [PubMed]

64. Kong, D.L.; Wang, N.; Qiao, N.; Wang, Q.; Wang, Z.; Zhou, Z.Y.; Ren, Z.Q. Facile Preparation of Ion-Imprinted Chitosan Microspheres Enwrapping Fe_3O_4 and Graphene Oxide by Inverse Suspension Cross-linking for Highly Selective Removal of Copper(II). *ACS Sustain. Chem. Eng.* **2017**, *5*, 7401–7409. [CrossRef]
65. Ahmed, M.A.; Abdelbar, N.M.; Mohamed, A.A. Molecular Imprinted Chitosan-TiO_2 Nanocomposite for the Selective Removal of Rose Bengal from Wastewater. *Int. J. Biol. Macromol.* **2018**, *107*, 1046–1053. [CrossRef] [PubMed]
66. Etemadi, M.; Samadi, S.; Yazd, S.S.; Jafari, P.; Yousefi, N.; Aliabadi, M. Selective Adsorption of Cr(VI) Ions from Aqueous Solutions Using Cr^{6+}-imprinted Pebax/chitosan/GO/APTES Nanofibrous Adsorbent. *Int. J. Biol. Macromol.* **2017**, *95*, 725–733. [CrossRef] [PubMed]
67. Yu, B.; Li, Z.; Cong, H.L.; Li, G.L.; Peng, Q.H.; Yang, C.F. Synthesis and Application of Sulfonated Polystyrene/ferrosoferric Oxide/diazoresin Nanocomposite Microspheres for Highly Selective Removal of Dyes. *Mater. Des.* **2017**, *135*, 333–342. [CrossRef]
68. Chen, B.; Liu, Y.; Chen, S.J.; Zhao, X.S.; Yue, W.L.; Pan, X.J. Nitrogen-rich Core/shell Magnetic Nanostructures for Selective Adsorption and Separation of Anionic Dyes from Aqueous Solution. *Environ. Sci. Nano* **2016**, *3*, 670–681. [CrossRef]
69. Huang, W.; Xu, J.Z.; Lu, D.K.; Deng, J.J.; Shi, G.Y.; Zhou, T.S. Rational Design of Magnetic Infinite Coordination Polymer Core-shell Nanoparticles as Recyclable Adsorbents for Selective Removal of Anionic Dyes from Colored Wastewater. *Appl. Surf. Sci.* **2018**, *462*, 453–465. [CrossRef]
70. Sarkar, C.; Bora, C.; Dolui, S.K. Selective Dye Adsorption by pH Modulation on Amine Functionalized Reduced Graphene Oxide-carbon Nanotube Hybrid. *Ind. Eng. Chem. Res.* **2014**, *53*, 16148–16155. [CrossRef]
71. Ashour, R.M.; El-sayed, R.; Abdel-Magied, A.F.; Abdel-khalek, A.A.; Ali, M.M.; Forsberg, K.; Uheida, A.; Muhammed, M.; Dutta, J. Selective Separation of Rare Earth Ions from Aqueous Solution Using Functionalized Magnetite Nanoparticles: Kinetic and Thermodynamic Studies. *Chem. Eng. J.* **2017**, *327*, 286–296. [CrossRef]
72. Wu, S.J.; Xie, M.J.; Zhang, Q.; Zhong, L.J.; Chen, M.H.; Huang, Z.J. Isopentyl-sulfide-impregnated Nano-MnO_2 for the Selective Sorption of Pd(II) from the Leaching Liquor of Ores. *Molecules* **2017**, *22*, 1117. [CrossRef] [PubMed]
73. Fu, W.; Huang, Z.Q. One-pot Synthesis of a Two-dimensional Porous Fe_3O_4/poly($C_3N_3S_3$) Network Nanocomposite for the Selective Removal of Pb(II) and Hg(II) from Synthetic Wastewater. *ACS Sustain. Chem. Eng.* **2018**, *6*, 14785–14794. [CrossRef]
74. Chen, J.; Yu, M.T.; Wang, C.Y.; Feng, J.T.; Yan, W. Insight into the Synergistic Effect on Selective Adsorption for Heavy Metal Ions by a Polypyrrole/TiO_2 Composite. *Langmuir* **2018**, *34*, 10187–10196. [CrossRef] [PubMed]
75. Zhang, J.; Yin, X.; Ye, Z.; Chen, L.; Liu, L.; Wang, X.; Zhu, Y.; Fujita, T.; Wei, Y. Synthesis of Novel Hierarchical Rod-like Mg-Al Bimetallic Oxides for Enhanced Removal of Uranium (VI) from Wastewater. *Chemosphere* **2022**, *308*, 136546. [CrossRef] [PubMed]
76. Bian, L.; Nie, J.N.; Jiang, X.Q.; Song, M.X.; Dong, F.Q.; Shang, L.Q.; Deng, H.; He, H.C.; Belzile, N.; Chen, Y.W.; et al. Selective Adsorption of Uranyl and Potentially Toxic Metal Ions at the Core-shell MFe_2O_4-TiO_2 (M=Mn, Fe, Zn, Co, or Ni) Nanoparticles. *J. Hazard. Mater.* **2019**, *365*, 835–845. [CrossRef] [PubMed]
77. Yuan, J.; Zhu, Y.; Wang, J.; Liu, Z.; He, M.; Zhang, T.; Li, P.; Qiu, F. Facile Modification of Biochar Derived from Agricultural Straw Waste with Effective Adsorption and Removal of Phosphorus from Domestic Sewage. *J. Inorg. Organomet. Polyme* **2021**, *31*, 3867–3879. [CrossRef]
78. Kera, N.H.; Bhaumik, M.; Pillay, K.; Ray, S.S.; Maity, A. Selective Removal of Toxic Cr(VI) from Aqueous Solution by Adsorption Combined with Reduction at a Magnetic Nanocomposite Surface. *J. Colloid Interf. Sci.* **2017**, *503*, 214–228. [CrossRef]
79. Zhan, Y.; Zhao, S.; Wan, X.; He, S. Hierarchical Fe_3O_4-derived Organic/inorganic Hybrids Constructed by Modified Bio-inspired Functionalization: Efficient Adsorbents for Water-Soluble Methylene Blue and Mechanism. *J. Chem. Technol. Biot.* **2019**, *94*, 1638–1650. [CrossRef]
80. Wu, B.L.; Fang, L.P.; Fortner, J.D.; Guan, X.H.; Lo, I.M.C. Highly Efficient and Selective Phosphate Removal from Wastewater by Magnetically Recoverable $La(OH)_3$/Fe_3O_4 Nanocomposites. *Water Res.* **2017**, *126*, 179–188. [CrossRef] [PubMed]
81. Xie, J.; Wang, Z.; Lu, S.Y.; Wu, D.Y.; Zhang, Z.J.; Kong, H.N. Removal and Recovery of Phosphate from Water by Lanthanum Hydroxide Materials. *Chem. Eng. J.* **2014**, *254*, 163–170. [CrossRef]
82. Wu, Z.J.; Wang, M.J.; An, X.W.; Du, J.; Fan, H.L.; Lin, N.M.; Wang, Z.D.; Hao, X.G.; Guan, G.Q. Intelligent Nanospheres with Potential-triggered Undamaged Regeneration Ability and Superparamagnetism for Selective Separation of Cesium Ion. *Chem. Eng. J.* **2017**, *325*, 229–238. [CrossRef]
83. Kadam, A.A.; Jang, J.; Lee, D.S. Facile Synthesis of Pectin-stabilized Magnetic Graphene Oxide Prussian Blue Nanocomposites for Selective Cesium Removal from Aqueous Solution. *Bioresource Technol.* **2016**, *216*, 391–398. [CrossRef] [PubMed]
84. Cai, Y.W.; Wu, C.F.; Liu, Z.Y.; Zhang, L.J.; Chen, L.H.; Wang, J.Q.; Wang, X.K.; Yang, S.T.; Wang, S. Fabrication of a Phosphorylated Graphene Oxide-chitosan Composite for Highly Effective and Selective Capture of U(VI). *Environ. Sci. Nano* **2017**, *4*, 1876–1886. [CrossRef]
85. Zhou, Y.S.; He, Z.Y.; Tao, Y.; Xiao, Y.H.; Zhou, T.T.; Jing, T.; Zhou, Y.K.; Mei, S.R. Preparation of a Functional Silica Membrane Coated on Fe_3O_4 Nanoparticle for Rapid and Selective Removal of Perfluorinated Compounds from Surface Water Sample. *Chem. Eng. J.* **2016**, *303*, 156–166. [CrossRef]
86. Du, Z.W.; Deng, S.B.; Zhang, S.Y.; Wang, W.; Wang, B.; Huang, J.; Wang, Y.J.; Yu, G.; Xing, B.S. Selective and Fast Adsorption of Perfluorooctanesulfonate from Wastewater by Magnetic Fluorinated Vermiculite. *Environ. Sci. Technol.* **2017**, *51*, 8027–8035. [CrossRef] [PubMed]

Disclaimer/Publisher's Note: The statements, opinions and data contained in all publications are solely those of the individual author(s) and contributor(s) and not of MDPI and/or the editor(s). MDPI and/or the editor(s) disclaim responsibility for any injury to people or property resulting from any ideas, methods, instructions or products referred to in the content.

toxics

MDPI

Review

Recent Progress in Molecular Oxygen Activation by Iron-Based Materials: Prospects for Nano-Enabled In Situ Remediation of Organic-Contaminated Sites

Fangru He [1], Lianrui Xu [1], Hongyang Wang [2,*] and Chuanjia Jiang [1,*]

[1] College of Environmental Science and Engineering, Ministry of Education Key Laboratory of Pollution Processes and Environmental Criteria, Tianjin Key Laboratory of Environmental Remediation and Pollution Control, Nankai University, Tianjin 300350, China
[2] State Key Laboratory of Environmental Criteria and Risk Assessment, Chinese Research Academy of Environmental Sciences, Beijing 100012, China
* Correspondence: wanghongyang_why@126.com (H.W.); jiangcj@nankai.edu.cn (C.J.)

Abstract: In situ chemical oxidation (ISCO) is commonly used for the remediation of contaminated sites, and molecular oxygen (O_2) after activation by aquifer constituents and artificial remediation agents has displayed potential for efficient and selective removal of soil and groundwater contaminants via ISCO. In particular, Fe-based materials are actively investigated for O_2 activation due to their prominent catalytic performance, wide availability, and environmental compatibility. This review provides a timely overview on O_2 activation by Fe-based materials (including zero-valent iron-based materials, iron sulfides, iron (oxyhydr)oxides, and Fe-containing clay minerals) for degradation of organic pollutants. The mechanisms of O_2 activation are systematically summarized, including the electron transfer pathways, reactive oxygen species formation, and the transformation of the materials during O_2 activation, highlighting the effects of the coordination state of Fe atoms on the capability of the materials to activate O_2. In addition, the key factors influencing the O_2 activation process are analyzed, particularly the effects of organic ligands. This review deepens our understanding of the mechanisms of O_2 activation by Fe-based materials and provides further insights into the application of this process for in situ remediation of organic-contaminated sites.

Keywords: oxygen activation; Fe-based materials; reactive oxygen species; organic pollutants; groundwater contamination

1. Introduction

Groundwater is a vital resource for agricultural irrigation, drinking water supply, and industrial use. However, this valuable resource is becoming increasingly more scarce due to the excessive extraction and consumption, as well as the widely occurring groundwater pollution [1–4]. Among the various sources of groundwater contamination are closed landfills without proper maintenance and "brownfield" sites, which are abandoned lands left behind after the closure or relocation of industrial or commercial facilities [5–7]. Soil and groundwater in a majority of the sites are contaminated with various organic pollutants [8], including organic solvents (especially chlorinated solvents) [9], petroleum hydrocarbons and gasoline products (e.g., benzene, toluene, ethylbenzene, and xylenes, collectively known as BTEX) [10], polycyclic aromatic hydrocarbons (PAHs) [11,12], pesticides [13], polybrominated diphenyl ethers [14], and perfluoroalkyl and polyfluoroalkyl substances (PFASs) [15]. For example, in approximately 80% of the Superfund sites, the groundwater is contaminated with chlorinated aliphatic hydrocarbons [16]. Due to the chemical stability, low water solubility, and propensity to adsorb onto the soil medium, these organic pollutants can persist in the subsurface for many years and pose long-term environmental risks [17,18]. These organic pollutants not only cause harm to the soil and subsurface ecological environment [19,20] but also lead to adverse effects on human health [21–24]. Numerous studies have indicated that exposure to these pollutants can lead to cancer, diabetes, respiratory and neurological diseases, and reproductive disorders [25–27]. For example, exposure to PAHs accounted for a significant proportion of lung cancer cases, especially in e-waste processing areas [28]. In recent years, the toxicities and health risks of emerging organic pollutants have raised increasing attention. Notably, PFASs can cause immunotoxicity, cardiotoxicity, and pancreatic and liver damage, as well as endocrine-disrupting effects [29,30]. Therefore, it is urgent to formulate effective strategies for mitigating the ecological and health risks posed by these legacy and emerging organic pollutants.

Intensive studies have been conducted to develop remediation technologies for soil and groundwater with organic contamination [31–33]. Among the various remediation technologies, in situ chemical oxidation (ISCO) methods have received increasing attention due to their high efficiency and simple operation [34–36]. During ISCO processes, oxidants such as ozone, potassium permanganate, hydrogen peroxide (H_2O_2), and persulfate are injected into the contaminated source zone and then activated when needed, generating stronger oxidizing species, such as hydroxyl radical (•OH) and sulfate radical [34,36,37]. However, these commonly used oxidants still suffer from some drawbacks, such as low selectivity, rapid consumption by aquifer constituents, and the risk of secondary pollution. In particular, their efficiency and cost-effectiveness for removing residual non-aqueous phase liquids (NAPL) pollutants is low.

Recent studies have found that molecular oxygen (O_2) is an ideal green alternative to the traditional oxidants for ISCO remediation of contaminated soil/sediment and groundwater. O_2 is relatively stable due to unfavorable one-electron reduction chemistry and spin restriction [38], which is favorable for its delivery to the pollutants without extensive consumption by aquifer constituents during its transport in the subsurface porous media. Additionally, groundwater table fluctuation, which can be caused by evaporation and rainfall, tide, lateral recharge, and drainage, results in the trapping of O_2 in unsaturated soil and saturated aquifers [39]. Under certain conditions, O_2 could be activated to form reactive oxygen species (ROS), such as •OH, singlet oxygen ($^1O_2$) and superoxide radical ($O_2^{\bullet-}$), and H_2O_2 [40], mainly via electrochemical- [41], photochemical- [42], and chemical-activation approaches [38]. The electrochemical- and photochemical-activation methods require electrical power and light irradiation, as well as devices that may not be facilely emplaced underground, which greatly hinders the application of these methods for in situ remediation of contaminated sites. In contrast, the chemical activation of O_2 by earth-abundant elements holds great promise for in situ soil and groundwater remediation.

Iron is the fourth-most-abundant element in the Earth's crust, and Fe-containing minerals are ubiquitous in soils and aquifers [43–45]. Iron-based materials are extensively investigated for applications in environmental remediation, exhibiting high efficiency in degrading a range of organic pollutants (Table 1) due to the redox and catalytic activities of the Fe element and the versatility, availability, and environmental compatibility of Fe-based materials [46,47]. For example, zero-valent iron (ZVI)-based materials are the most widely used agents for in situ chemical-reduction remediation [48–53]. Meanwhile, Fe-based materials can promote ISCO remediation by the activation of H_2O_2/CaO_2 [54–58], persulfates [59–63], and other oxidants [64–66]. Moreover, Fe-based materials such as ZVI [67,68], iron sulfides [69,70], iron (oxyhydr)oxides [71], and Fe-containing clay minerals [72] can mediate O_2 activation to degrade organic pollutants, and the potential of Fe-mediated O_2 activation for ISCO remediation has been actively explored in recent years. However, there is a lack of a timely review of the mechanisms and key influencing factors of O_2 activation by iron-based materials for potential applications for in situ remediation of soil- and groundwater-suffering organic contamination.

Table 1. Degradation of organic pollutants by Fe-based materials via different mechanisms.

Pollutant	Material	Reaction Mechanism	Reference
Trichloroethylene	nZVI	Reductive dechlorination	[48]
Trichloroethylene	mZVI	Reductive dechlorination	[49]
Trichloroethylene	S-nZVI	Reductive dechlorination	[50]
Trichloroethylene	S-mZVI	Reductive dechlorination	[49]
Trichloroethylene	Fe_xN	Reductive dechlorination	[51]
Chloroform	S–N(C)–ZVI	Reductive dechlorination	[52]
Florfenicol	S-nZVI	Reductive dehalogenation	[53]
Fluorenone	Fe_3O_4	H_2O_2 activation	[54]
Trichloroethylene	Reduced nontronite	H_2O_2 activation	[55]
Diethyl phthalate	Reduced nontronite	H_2O_2 activation	[56]
Trichloroethylene	FeS	CaO_2 activation	[57]
Sulfanilamide	FeS_2	CaO_2 activation	[58]
Phenanthrene	FeCo-BDC	Peroxymonosulfate activation	[59]
Perfluorooctanic acid	Fe/AC	Persulfate activation	[60]
Trichloroethylene	nZVI	Persulfate activation	[61]
Ciprofloxacin	FeS_2	Persulfate activation	[62]
Bisphenol A	Fe_3S_4	Peroxymonosulfate activation	[63]
Bisphenol AF	FeS	Periodate activation	[64]
Tetracycline	ZVI	Peracetic acid activation	[65]
Sulfamethoxazole	FeS	Peracetic acid activation	[66]

Note: ZVI, zero-valent iron; nZVI, nanoscale ZVI; mZVI, microscale ZVI; S-nZVI, sulfidated nZVI; S-mZVI, sulfidated mZVI; Fe_xN, iron nitrides; S–N(C)–ZVI, ZVI treated by nitridation and sulfidation; BDC, bimetallic metal-organic frameworks; AC, activated carbon.

Herein, we comprehensively review the current status of research on the activation of O_2 by Fe-based materials, including ZVI-based materials, iron sulfides, iron (oxyhydr)oxides, and Fe-containing clay minerals for degrading contaminants commonly found in soil/sediment and groundwater. Unlike previous reviews on O_2 activation, which highlight more efficient techniques such as electrochemical and photochemical activation for rapid abatement of pollutants (e.g., in wastewater treatment) [73–76], this review focuses on O_2 activation by Fe-based materials without external energy input, which holds better promise for application in the remediation of organic-contaminated sites. Particularly, this review includes discussions on recent research in O_2 activation by reduced Fe-bearing minerals abundant in soils and sediments, which has important implications for slower but sustained remediation via natural attenuation processes. The major mechanisms involved in the activation of O_2 by the Fe-based materials are summarized, highlighting electron transfer and utilization, reaction intermediates and ROS chain reactions, and the oxidative transformation of the materials during the O_2-activation process are discussed. Additionally, we discussed the influences of environmental and operational factors, including O_2 concentration, organic ligands, inorganic anions, and microbial activity, on the O_2 activation and pollutant-degradation performance by iron-based materials. This review also identifies limitations of current studies and suggests future research directions to enhance understanding of O_2 activation by iron-based materials and its applications in soil and groundwater remediation.

2. Activation of O_2 by Fe-Based Materials

2.1. ZVI-Based Materials

Materials containing ZVI have been commonly employed as remediation agents for reductive degradation of organic pollutants under anaerobic conditions due to the high reducing capacity of elemental Fe (Table 2) [77–83]. However, studies have indicated that the degradation efficiency of organic pollutants by ZVI is significantly higher in O_2-containing aqueous solutions than under anaerobic conditions [67,84,85], and O_2 activation by ZVI-based materials has recently been extensively explored for degrading various organic pollutants (Table 3). This increased efficiency is primarily attributed to the reaction between ZVI and O_2, which leads to the generation of ROS [67,68]. The mechanism of ZVI-mediated O_2 activation for the generation of ROS involves two-electron transfers from Fe^0 to adsorbed O_2, producing Fe(II) and H_2O_2. And the release of Fe(II) further induces O_2 activation via a sequential single-electron transfer process, generating •OH, $O_2^{\bullet-}$, and other ROS (Equations (1)–(6)) [86]. The yield of •OH is primarily affected by the reaction between Fe^0 and O_2 via four-electron transfer without ROS generation (Equation (7)) [87]. Due to this reaction pathway, only less than 10% of the ZVI is utilized for contaminant transformation under oxic conditions [88]. Therefore, the yield of ROS decreases with increasing pH due to the inhibition of Fe(II) release via the four-electron-transfer reaction under high pH conditions [84]. Notably, the ROS generation is influenced by the oxide layer on the surface of ZVI, in a fashion dependent on the thickness of the layer [89]. The iron oxide layer can adsorb ferrous ions, which can activate O_2 through a single-electron-transfer pathway. When the iron oxide layer is thin, both the Fe^0 core-mediated two-electron transfer and the surface-bound/adsorbed Fe(II)-mediated single-electron transfer play significant roles in O_2 activation [73]. However, as the thickness of the oxide layer increases, it inhibits electron transfers from the Fe^0 core to adsorbed O_2. Meanwhile, more ferrous ions are adsorbed on the surface, and these surface-bound Fe(II) become the primary species responsible for O_2 activation (Figure 1a) [75,89,90].

$$Fe^0 + O_2 + 2H^+ \rightarrow Fe(II) + H_2O_2 \quad (1)$$

$$Fe(II) + O_2 \rightarrow Fe(III) + O_2^{\bullet-} \quad (2)$$

$$Fe(II) + O_2^{\bullet-} + 2H^+ \rightarrow Fe(III) + H_2O_2 \quad (3)$$

$$O_2^{\bullet-} + H^+ \rightarrow HO_2^{\bullet} \quad (4)$$

$$HO_2^{\bullet} + HO_2^{\bullet} \rightarrow H_2O_2 + O_2 \quad (5)$$

$$Fe(II) + H_2O_2 \rightarrow Fe(III) + \bullet OH + OH^- \text{ or } Fe(IV) + 2OH^- \quad (6)$$

$$2Fe^0 + O_2 + 4H^+ \rightarrow 2Fe(II) + 2H_2O \quad (7)$$

Table 2. Redox potential of different Fe species.

Species	Redox Potential (V)	Reference
Fe^0/Fe(II)	−0.44 (vs. SHE)	[80]
Structural Fe(II)/Fe(III) of pyrite	0.66 (vs. SHE)	[81]
Structural Fe(II)/Fe(III) of clay mineral	−0.6 to +0.6 (vs. SHE)	[82]
Fe(III)/Fe(II)-CA	0.37 (vs. NHE)	[83]
Fe(III)/Fe(II)-OA	0.002 (vs. NHE)	[83]
Fe(III)/Fe(II)-EDTA	0.12, 0.11, and 0.096 (vs. NHE)	[83]
Fe(III)/Fe(II)-EDDS	0.19 (vs. NHE)	[83]
Fe(III)/Fe(II)-NTA	0.10 and 0.39 (vs. NHE)	[83]
$Fe(H_2O)_6^{2+}/Fe(H_2O)_6^{3+}$	0.77 (vs. NHE)	[83]

Note: SHE, standard hydrogen electrode; NHE, normal hydrogen electrode; CA, citrate; OA, oxalate; EDTA, ethylenediaminetetraacetic acid; EDDS, N,N′-1,2-ethanediylbis-1-aspartic acid; NTA, nitilotriacetic acid.

Table 3. Degradation of organic pollutants via O_2 activation by ZVI-based materials.

Material	Pollutant	Removal Ratio (%)	Reaction Time (h)	pH	Rate Constant	Reference
ZVI	EDTA	100	2.5	6.0 ± 0.2	$1.02\ h^{-1}$	[68]
nZVI	2-Chlorobiphenyl	59.4	4	5.0	$0.0035\ min^{-1}$	[84]
S-nZVI	Bisphenol A	100	6	5.0	$59.2 \pm 2.29\ h^{-1}$	[85]
$Fe@Fe_2O_3$	4-Chlorophenol	77.8	7	6.0	$0.22\ h^{-1}$	[89]
Al–Fe	4-Chlorophenol	43.7	5	2.5	N/A	[91]
Fe/Cu	4-Chlorophenol	100	2	3.0	N/A	[92]
Fe/Cu	Diclofenac	96	2	6.0	N/A	[93]
Mg/Fe	4-Chlorophenol	100	0.75	3.0	N/A	[94]

Table 3. *Cont.*

Material	Pollutant	Removal Ratio (%)	Reaction Time (h)	pH	Rate Constant	Reference
Fe/Mn	Enrofloxacin	100	1	3.0	N/A	[95]
ZVI	Enrofloxacin	58.6	1	3.0	N/A	[95]
S-nZVI	p-Nitrophenol	99.3 *	2	7.6	0.769 min^{-1}	[96]
mZVI/NGB	Tetracycline	100	0.83	5.8	N/A	[97]
3D-GN@nZVI	Sulfadiazine	81.0	2	3.0	N/A	[98]
Cu/Fe-BC	Ciprofloxacin	93.2 *	1.5	5.0	0.052 min^{-1}	[99]
Cu/Fe-BC	Enrofloxacin	88.9 *	1.5	5.0	0.036 min^{-1}	[99]
Cu/Fe-BC	Norfloxacin	95.4 *	1.5	5.0	0.096 min^{-1}	[99]
Cu/Fe-BC	Tetracycline	82.3 *	1.5	5.0	0.037 min^{-1}	[99]
Cu/Fe-BC	Methylene blue	95.6 *	1.5	5.0	0.145 min^{-1}	[99]
ZVI-BC	Tetracycline	93.1	6	Unadjusted	N/A	[100]
Zn-Fe-CNTs	Sulfamethoxazole	95.3	0.33	1.5	N/A	[101]
Zn-Fe-CNTs	4-Chlorophenol	90.8	0.33	2.0	N/A	[102]
nZVI@MSN	Nitrobenzene	96.5	0.33	3.0	0.201 min^{-1}	[103]

Note: N/A, data not available. *, data obtained from the literature using the Getdata 2.26 software.

To improve the O_2-activation efficiency and ROS yield, various modified ZVI materials have been developed, notably by doping with metal and non-metal elements or immobilizing ZVI on porous materials. Inspired by the galvanic corrosion between connected dissimilar metals [91], a series of ZVI-based bimetallic materials were designed. In a Cu^0/ZVI material, Cu^0 can enhance the reaction potential by forming infinite galvanic cells with Fe^0, thereby significantly accelerating Fe-mediated O_2 activation [92]. Additionally, Cu accelerates the release of Fe(II) species during O_2 activation (Figure 1b). Furthermore, Cu species can also effectively facilitate the iron cycle and serve as new active sites (Equations (8)–(12)) [93]. However, it has also been suggested that Cu may inhibit •OH generation due to the formation of a passivation layer of Cu oxides [104]. This discrepancy may be attributed to the complex chain reactions and the dosage of Cu, which should earn more attention in further studies. Another metal element, Ni, has been shown to reduce the proportion of the four-electron-reaction pathway of ZVI by increasing Fe(II) release [88]. Alternatively, the Fe–Mg bimetallic material can increase the degradation of 4-chlorophenol by enhancing the generation of surface-bound •OH [94]. Recently, a Fe–Mn core-shell bimetallic material was reported for O_2 activation and on-site generation of H_2O_2, and it was proposed that the amorphous Mn shell can not only protect the Fe core from excessive oxidation, thereby increasing electron utilization, but also contain abundant structural defects, which serve as efficient catalytic sites [95]. Doping with non-metal elements can also affect the efficiency of O_2 activation by ZVI materials. For example, the incorporation of chloride ions into microscale zero-valent iron (mZVI) can create oxygen vacancies (OVs), resulting in abundant adsorbed ferrous ions and accelerated electron transfer [105]. Sulfidation is one of the most effective methods to improve the efficiency and selectivity of reductive degradation of pollutants by nanoscale ZVI (nZVI) [106–109], and it has been demonstrated that the presence of S could enhance electron transfers from the Fe^0 core to surface Fe(III) and O_2 via the single-electron pathway, thus promoting pollutant degradation under aerobic conditions [96,110].

$$Cu^0 + Fe(III) \rightarrow Cu(I) + Fe(II) \quad (8)$$

$$Cu(I) + Fe(III) \rightarrow Cu(II) + Fe(II) \quad (9)$$

$$Cu^0 + O_2 + 2H^+ \rightarrow Cu(II) + H_2O_2 \quad (10)$$

$$Cu(I) + O_2 \rightarrow Cu(II) + O_2^{\bullet -} \quad (11)$$

$$Cu(I) + O_2^{\bullet -} + 2H^+ \rightarrow Cu(II) + H_2O_2 \quad (12)$$

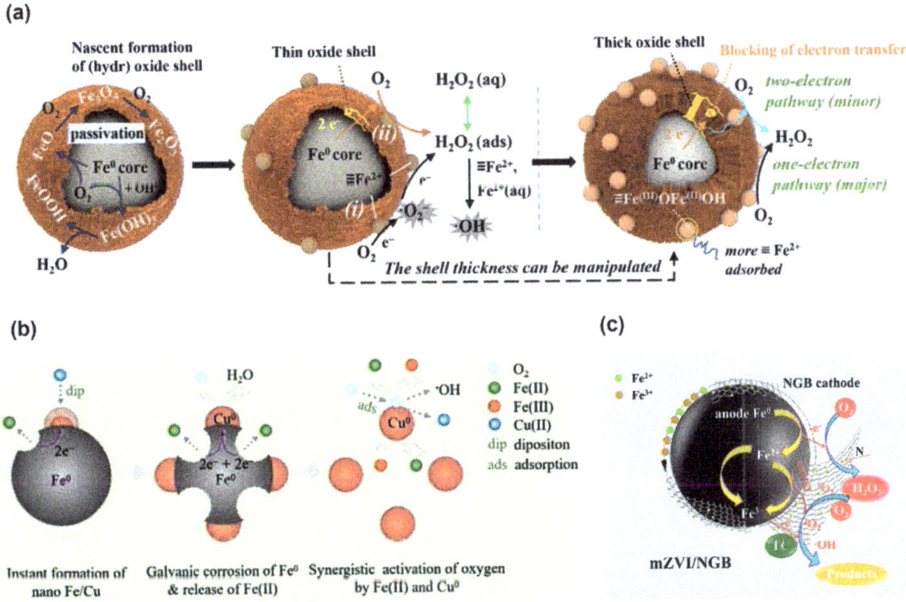

Figure 1. (**a**) Illustration of the mechanisms of O_2 activation by Fe^0 with an oxide shell of different thickness [73]. (**b**) Illustration of the mechanisms of Cu-enhanced •OH production in Cu^0/ZVI system under oxic condition [93]. (**c**) Illustration of the catalytic mechanism of O_2 by mZVI/N-doped graphene-like biochar (mZVI/NGB) [97].

Another approach to enhancing ZVI performance involves immobilizing ZVI on porous supports, particularly carbon materials. The carbon materials (e.g., graphene), usually with a conjugated network skeleton of sp^2 hybridized carbon atoms, could enhance electron transfer and modify the adsorption/dissociation energy of O_2, making O_2 activation thermodynamically and kinetically more favorable [111–113]. Meanwhile, the combination of ZVI with graphene (3D-GN@nZVI) can also greatly inhibit the reduction of O_2 to H_2O via a four-electron process, thereby increasing the yield of ROS [98]. Recent studies have indicated that a mZVI/N-doped graphene-like biochar composite (mZVI/NGB) could promote the contribution of non-radical pathways ($^1O_2$ and electron transfer) during O_2 activation for antibiotics degradation (Figure 1c) [97]. Similarly, for Fe/Cu-biochar (Cu/Fe-BC) materials, both radicals (e.g., $O_2^{\bullet -}$, •OH) and non-radicals (e.g., $^1O_2$) were detected as the dominant reactive species, enabling the efficient degradation of a wide spectrum of organic pollutants, even in the presence of various interfering substances [99]. In contrast, $O_2^{\bullet -}$ and H_2O_2 are the key ROS in the ZVI–biochar system, indicating that the secondary metal has a significant influence on the O_2-activation pathway [100]. Carbon nanotubes (CNTs) have also been demonstrated as an excellent support material to regulate ROS generation dynamics by bimetallic Fe-based materials. For example, in a Zn–Fe–CNT composite, CNTs effectively collect electrons from Zn^0 nanoparticles and reduce O_2 to H_2O_2, which was subsequently converted to •OH by Fe^0 nanoparticles [101,102]. Due to the synergistic effects, the composite achieved excellent performance for the degradation of 4-chlorophenol and sulfamethoxazole (Table 3). In addition to carbon materials, nZVI can also be successfully incorporated within the channels of monodisperse mesoporous silica nanospheres (nZVI@MSN) to increase its stability and durability [103].

2.2. Iron Sulfides

2.2.1. Pyrite

Pyrite (FeS_2) is the most widely distributed stable-phase iron sulfide mineral in Earth's crust [114]. The oxidation of natural pyrite, which can lead to the generation of H_2O_2, has been confirmed under anaerobic conditions [115]. The mechanism involved in this process is primarily attributed to the presence of surface defects, arising from the cleavage of the S–S bond [116]. As a result, transient S– and ≡Fe(III) dangling bonds are generated at these sulfur-deficient defect sites, which can induce the formation of •OH by extracting an electron from adsorbed H_2O, and the •OH radicals then combine to form H_2O_2 in the absence of O_2 [117]. In a recent study investigating oxidation of benzoic acid by sulfur vacancy (SV)-rich FeS_2 in isotopically labeled $H_2{}^{18}O$, the generation of ^{18}O-containing *p*-hydroxybenzoic acid was observed, which further provided direct evidence that the anaerobic oxidation of water by SV-rich FeS_2 is responsible for the generation of •OH at the pyrite–water interface [118]. However, it was noted that the •OH generated through this mechanism was insufficient to achieve an obvious degradation of organic pollutants. In contrast, efficient pollutant removal can be achieved in pyrite suspension with sufficient O_2 (Table 4), which highlights the crucial role of pyrite oxidation by O_2 [70,119].

Table 4. Degradation of organic pollutants via O_2 activation by iron sulfides.

Material	Pollutant	Removal Ratio (%)	Reaction Time (h)	pH	Rate Constant	Reference
Pyrite	Trichloroethylene	100	323	4.0	$0.013\ h^{-1}$	[70]
Pyrite	Acid orange 7	52.8	5	6.3	N/A	[120]
Pyrite	Carbamazepine	81.5	24	7.0	$0.103 \pm 0.001\ h^{-1}$	[121]
Pyrite	Phenol	76.8	24	7.0	$0.084 \pm 0.001\ h^{-1}$	[121]
Pyrite	Bisphenol A	100	24	7.0	$0.147 \pm 0.001\ h^{-1}$	[121]
SV-rich pyrite	Sulfamethoxazole	88.3	12	8.5	$29.2 \times 10^{-4}\ h^{-1}$	[122]
Pyrite	Sulfamethoxazole	70.0	12	4.0	$0.095\ h^{-1}$	[123]
Mackinawite	Flumequine	79.8	4	7.0	$51.6 \times 10^{-3}\ min^{-1}$	[124]
Mackinawite	Enrofloxacin	87.7	4	7.0	$34.0 \times 10^{-3}\ min^{-1}$	[124]
Mackinawite	Ciprofloxacin	81.5	4	7.0	$25.4 \times 10^{-3}\ min^{-1}$	[124]
Mackinawite	Trichloroethylene	23.4 *	3	7.0	$2.07 \times 10^{-3}\ min^{-1}$	[69]
Mackinawite	Phenol	34.1 *	3	7.0	$3.53 \times 10^{-3}\ min^{-1}$	[69]
Surface-oxidized mackinawite	Phenol	17.5 *	1.5	7.3	$3.8 \times 10^{-3}\ min^{-1}$	[125]

Note: N/A, data not available. SV, sulfur vacancy. *, data obtained from the literature using the Getdata 2.26 software.

The mechanism for O_2 activation by pyrite can be understood from an analysis of the Fe species in pyrite suspension (Equations (13)–(18)). Previous studies have proposed that structural Fe(II) and surface-bound Fe(II) can mediate either a two-electron-transfer pathway or two separate one-electron-transfer processes, with H_2O_2 or $O_2{}^{\bullet-}$ as intermediates, ultimately leading to the formation of •OH (Figure 2a) [119,126]. Another pathway involves the leaching of dissolved Fe(II) from bulk FeS_2, which mediates O_2 activation in aqueous solution via a single-electron-transfer pathway [127]. However, the contribution of this approach is generally minor due to the stable structure and extremely low dissolution rate of pyrite even under acidic conditions [120,128]. Moreover, the reactivity of dissolved Fe(II) is lower than the structural Fe(II) and surface-bound Fe(II) according to their redox potentials (Table 2) [72,129]. Note that, although $S_2{}^{2-}$ has the capability to reduce surface-bound Fe(III), it does not directly participate in O_2 reduction [130]. This inference is substantiated by the results of in situ horizontal attenuated total reflectance infrared spectroscopy and isotope analysis of reaction products (e.g., $SO_4{}^{2-}$ and iron oxyhydroxide), which demonstrated that the O atoms in $SO_4{}^{2-}$ primarily originate from H_2O, while the O atoms in the iron oxyhydroxide are derived from O_2 [131,132].

$$\equiv Fe(II) + O_2 + 2H^+ \rightarrow Fe(III) + H_2O_2 \qquad (13)$$

$$Fe(II)_{ad} + O_2 \rightarrow Fe(III) + O_2^{\bullet-} \quad (14)$$

$$Fe(II) + O_2^{\bullet-} + 2H^+ \rightarrow Fe(III) + H_2O_2 \quad (15)$$

$$Fe(II)_{(aq)} + O_2 \rightarrow Fe(III)_{(aq)} + O_2^{\bullet-} \quad (16)$$

$$Fe(II)_{(aq)} + O_2^{\bullet-} + 2H^+ \rightarrow Fe(III)_{(aq)} + H_2O_2 \quad (17)$$

$$H_2O_2 + Fe(II) \rightarrow Fe(III) + \bullet OH + OH^- \quad (18)$$

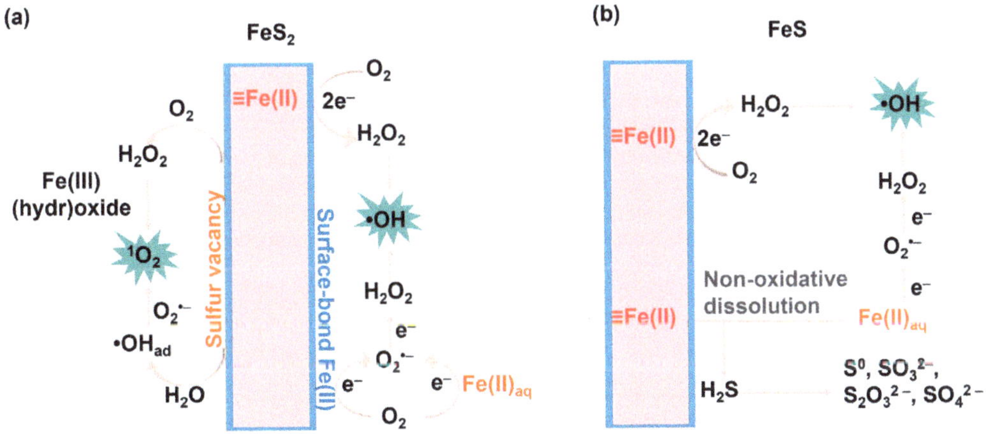

Figure 2. Illustration of the mechanisms of O_2 activation by (**a**) FeS_2 and (**b**) FeS.

As an interfacial reaction, the efficiency of O_2 activation by pyrite is dictated by the surface properties of pyrite, which in turn is dependent on the exposed facets, the presence of surface defects, and the formation of the iron (oxyhydr)oxide layer. The exposed facets of pyrite significantly influence the O_2-activation configurations and electron-transfer ability [133]. Notably, it has been recently revealed that pyrite crystals with more exposed {210} facets exhibit higher generation rates of •OH and other ROS (e.g., $O_2^{\bullet-}$ and H_2O_2) due to different surface electron-donating capacities and kinetics among different facets. Correspondingly, facet-dependent degradation of organic pollutants (e.g., carbamazepine, phenol, and bisphenol A) was achieved [121]. Another recent study highlighted that SV sites in pyrite can activate O_2 via a two-electron-transfer mechanism, generating $^1O_2$, which played a key role in the selective degradation of sulfamethoxazole [122]. The generation of $^1O_2$ arises from the breakage of the O–H bond in H_2O_2, facilitated by Fe(III) (oxyhydr)oxide on the pyrite surface, as presented by Equations (19) and (20) [122], whereas another recent study proposed that $^1O_2$ could also form through the interaction between surface-bound •OH and $O_2^{\bullet-}$ in the pyrite-oxidation process (Equations (21) and (22)) [123]. The iron (oxyhydr)oxides on the surface of pyrite, which quickly forms after pyrite is exposed to O_2, can also affect ROS generation during O_2 activation in the pyrite system in both positive and negative ways. The formation of iron (oxyhydr)oxides can provide fast electron-transfer channels by establishing a potential gradient between the two mineral phases [134]. Additionally, the iron (oxyhydr)oxides can adsorb more surface-bound Fe(II) by forming inner-sphere complexes with surface groups, thereby enhancing O_2-reduction efficiency [135]. However, other studies indicated that the (oxyhydr)oxide coating tends to lower the H_2O_2 utilization by catalyzing its transformation into H_2O, resulting in the decrease of ROS concentration in the system [136–138].

$$Fe(III)_{(SV\text{-}Pyrite)} + O_2 \rightarrow H_2O_2 \quad (19)$$

$$H_2O_2 \rightarrow {}^1O_2 + 2H^+ \quad (20)$$

$$Fe(III)_{(SV\text{-}Pyrite)} + H_2O \rightarrow Fe(II)_{(SV\text{-}Pyrite)} + \bullet OH_{ad} + H^+ \quad (21)$$

$$O_2^{\bullet -} + \bullet OH_{ad} \rightarrow {}^1O_2 + OH^- \quad (22)$$

2.2.2. Mackinawite

Mackinawite (FeS) is a metastable iron sulfide mineral, which can readily convert into more stable phases, such as FeS_2 and greigite (Fe_3S_4) in a natural environment [139]. Due to its structural instability and reducing power, FeS is prone to oxidation by O_2, and oxidative transformation of pollutants (e.g., As(III) and U(IV)) has been observed during FeS oxidation under aerobic conditions, and different oxidative species (e.g., •OH, Fe(IV), and transient surface Fe(III) species) have been proposed to initiate the pollutant degradation [140–142]. Yuan's group confirmed the production of •OH during the oxidation of FeS by O_2 and demonstrated that the produced •OH played a key role in the oxidation of As(III) [143]. The rate of •OH formation by mackinawite is one-to-two orders of magnitude higher than that observed for other forms of reduced iron minerals, such as nontronite, pyrite, and siderite ($FeCO_3$), under comparable conditions due to the metastable structure and higher Fe(II) content [143,144]. The ROS produced during FeS oxidation can also efficiently degrade various organic pollutants, such as phenol, trichloroethylene (TCE), and fluoroquinolone antibiotics [69,124].

The mechanism of FeS oxidation by O_2 is dependent on the pH conditions. When pH is lower than 3, FeS primarily undergoes non-oxidative dissolution, and most of the structural Fe(II) in FeS enters the aqueous solution before undergoing oxidation, which leads to an increase in dissolved ferrous ion concentration and the generation of H_2S (Equation (23)) (Figure 2b) [140,145]. The dissolved ferrous ions can mediate homogeneous Fenton processes [127]. Surface-mediated oxidative dissolution also occurs under acidic conditions, due to the formation of an iron (oxyhydr)oxide layer, which can adsorb Fe(II) on the surface, activating O_2 to form ROS [135]. Under neutral pH conditions, due to the low concentration of dissolved iron ions, surface-mediated oxidation mechanisms predominate, with surface species of FeS transforming from ≡Fe(II)–S through ≡Fe(III)–S to ≡Fe(III)–O in the presence of O_2 [140]. This structural Fe(II)-mediated heterogeneous reaction dominates the O_2 activation and ROS production reactions in the FeS system under neutral conditions [143], which involves a two-electron-transfer mechanism, leading to the generation of H_2O_2 intermediate and subsequent formation of •OH [143,146]. Notably, in addition to aqueous •OH, other active species such as high-valent iron, surface-bound •OH, or sulfur-based radicals may also be present in the FeS/O_2 system [147]. Apart from structural Fe(II), S(–II) can also act as the electron donator, mediating the iron cycle without directly participating in the O_2-activation process [143]. As with FeS_2, the Fe (oxyhydr)oxide coatings formed on the surface of FeS could affect O_2-activation efficiency by mediating electron transfers from FeS to O_2. Interestingly, the storage of partially oxidized FeS under anoxic conditions could change its mineralogical structure and surface Fe speciation, forming new Fe(II) species in the (oxyhydr)oxide layer, which leads to enhanced reactivity toward O_2 and the production of ROS [125].

$$FeS + 2H^+ \rightarrow Fe(II)_{(aq)} + H_2S_{(aq)} \quad (23)$$

2.3. Iron (Oxyhydr)Oxides

Magnetite (Fe_3O_4) is one of the most widely distributed reductive iron oxides in the subsurface environment, and it can activate O_2 to generate $O_2^{\bullet -}$, H_2O_2, and •OH and has been used for organic pollutant degradation (Table 5) [71,148,149]. Structural Fe(II) is considered the dominant species participating in O_2 activation in magnetite via a single-electron transfer under alkaline conditions, whereas dissolved iron ions originating from the dissolution of magnetite can participate in O_2 activation at $pH < 6.5$ [71]. Interior structural iron could facilitate O_2 activation by transferring electrons to surface iron and accelerating the iron cycle [150]. Nevertheless, studies have indicated that only half of the total Fe(II) in magnetite could be effectively utilized due to the low inner-electron-transfer ability, leading to a low O_2-activation efficiency [71]. Notably, it has been proposed that the presence of OVs can change the Gibbs free energy for the generation of adsorbed O_2 intermediate, making the reduction of O_2 thermodynamically more favorable and facilitating the electron transfer [148].

Table 5. Degradation of organic pollutants via O_2 activation by iron (oxyhydr)oxides.

Material	Pollutant	Removal Ratio (%)	Reaction Time (h)	pH	Rate Constant	Reference
Magnetite	2-Chlorobiphenyl	80	4	3.0	N/A	[149]
Cu^0/Fe_3O_4	4-Chlorophenol	99.5	1	7.0	0.073 min^{-1}	[151]
Zn^0-CNTs-Fe_3O_4	4-Chlorophenol	99	0.33	1.5	N/A	[152]
CNTs-Fe_3O_4	4-Chlorophenol	25	0.33	1.5	N/A	[152]
b-CoS_2/Fe_3O_4	4-Nitrophenol	62.3 *	0.25	5.0	N/A	[153]
b-CoS_2/Fe_3O_4	Methyl orange	85.7	0.25	5.0	N/A	[153]
b-CoS_2/Fe_3O_4	Sulfadiazine	67.1	0.25	5.0	N/A	[153]
b-CoS_2/Fe_3O_4	Tetracycline	96.0	0.25	5.0	N/A	[153]
b-CoS_2/Fe_3O_4	Rhodamine b	98.6 *	0.25	5.0	N/A	[153]
b-CoS_2/Fe_3O_4	Malachite green	91.5 *	0.25	5.0	N/A	[153]
Ferrihydrite	Phenol	29.8 *	10	7.0	N/A	[154]

Note: N/A, data not available. *, data obtained from the literature using the Getdata 2.26 software.

To improve the efficiency of O_2 activation, many synthetic Fe_3O_4 based composite materials have also been designed. For example, in a carboxylated Cu^0/Fe_3O_4 system, Cu^0 can act as the reducing agent and accelerate the regeneration of surface iron. In addition, Cu^0 serves as a new O_2 activation site to further increase the generation of H_2O_2 in the system via two-electron transfer, leading to the efficient removal of chlorophenol [151]. Similarly, Fe_3O_4 can mediate the ROS-generation dynamics of carbon-supported zero-valent metal, thus increasing the overall O_2-activation efficiency. For example, in a Zn^0–CNT–Fe_3O_4 composite, with Zn^0 and Fe_3O_4 nanoparticles well dispersed on the surface of CNTs, the self-decomposition of H_2O_2 (generated from O_2 reduction by Zn^0 on CNTs surface) into H_2O is significantly inhibited. Meanwhile, the conversion of H_2O_2 into •OH rapidly occurs with a high yield [152]. As a result, the Zn^0–CNT–Fe_3O_4 composite exhibited approximately four times higher removal efficiency for 4-chlorophenol than that by CNT-Fe_3O_4 and Zn^0–CNT materials [152]. Moreover, sulfidation can lead to higher H_2O_2 and •OH production during Fe_3O_4 oxidation because surface sulfur species can decrease electron-transfer resistances of Fe_3O_4, thereby accelerating the electron transfer from interior structural iron to the surface Fe(III) and facilitating the reaction between surface iron and O_2 [155]. Moreover, for a vacancy-rich iron-cobalt bimetallic composite prepared by ball milling CoS_2 and Fe_3O_4 (b-CoS_2/Fe_3O_4), the interfacial interaction between CoS_2 and Fe_3O_4 can change the Fe–O bond energy of Fe_3O_4, thereby accelerating the formation of surface-bound Fe(II), which in turn promotes O_2 activation by the composite [153].

Fe(III) (oxyhydr)oxides such as goethite (α-FeOOH) and hematite (α-Fe_2O_3) generally lack the ability to reduce O_2 due to the +3 valence state of iron in these compounds. However, studies have indicated that Fe(III) (oxyhydr)oxides can adsorb Fe(II) on their surface, and these surface-bound Fe(II) can activate O_2 through a single electron-transfer process [156,157]. Furthermore, the incorporation of Cu into goethite can increase the O_2-activation ability of surface-bound Fe(II) by modifying the adsorption energy of O_2, lowering its oxidation potential, and increasing the interfacial electron-transfer process on Fe(III) (oxyhydr)oxides [156]. The incorporation of secondary metal atoms can also increase the content of OVs in hematite, further enhancing the electron-transfer efficiency [150]. Notably, the incorporation of secondary metals does not necessarily have a positive effect on the O_2-activation efficiency. For example, the incorporation of Zn results in a higher oxidation potential for Fe(II) oxidation, which is unfavorable for the O_2-activation process [150]. Moreover, it was recently proposed that Fe(III) (oxyhydr)oxides can serve as an electron-transfer mediator for O_2 reduction by reducing organic compounds to generate ROS. Specifically, Fe(III) on their surface receives electrons from the reducing organic compounds, such as thiols, to form surface Fe(II), which then mediate the activation of O_2 efficiently [154].

2.4. Fe(II)-Containing Clay Minerals

Fe(II)-containing clay minerals are widely present in subsurface environments, such as sediments and soils [158]. Recent studies have shown that the oxidation of Fe(II)-containing clay minerals, such as smectites [72] (particularly reduced nontronite [159,160]) and illite [161], is one of the important sources of environmental radicals, which deeply affects the attenuation behavior of pollutants, including 1,4-dioxane, TCE, phenol, and PAHs (Table 6). The iron contents in these clay minerals, which range from about 2 wt.% in montmorillonite to about 30 wt.% in nontronite [162], is a crucial factor influencing the ROS yield [72]. In addition to total Fe content, the different forms of Fe species in Fe-containing clay minerals, including structural Fe(II), surface-bound Fe(II), and exchangeable Fe(II), also have a significant impact on the generation of ROS. Structural Fe(II) is generally the dominant species for O_2 activation by Fe-containing clay minerals [163,164]. The reactivity of structural Fe(II) is highly affected by its coordination environment. Specifically, structural Fe(II) at the edge ($Fe(II)_{edge}$) is coordinated by electron-rich ligands (e.g., ≡O–, ≡HO–, and ≡Fe(II)–O–), which exhibit high activity for O_2 activation and preferentially lead to the generation of Fe(IV), along with a low •OH yield (Figure 3a). In contrast, interior Fe(II) ($Fe(II)_{int}$) tends to be coordinated by electron-poor ligands (e.g., ≡Al(III)–O– and ≡Fe(III)–O–). Although $Fe(II)_{int}$ is much less active than $Fe(II)_{edge}$, it can selectively activate O_2 to •OH [165]. Additionally, the $Fe(III)_{edge}$ can be regenerated through various pathways, such as the continuous supply of electrons from interior adjacent Fe(II) via single-electron transfer until the electrons in the interior Fe(II) are eventually depleted [72,166].

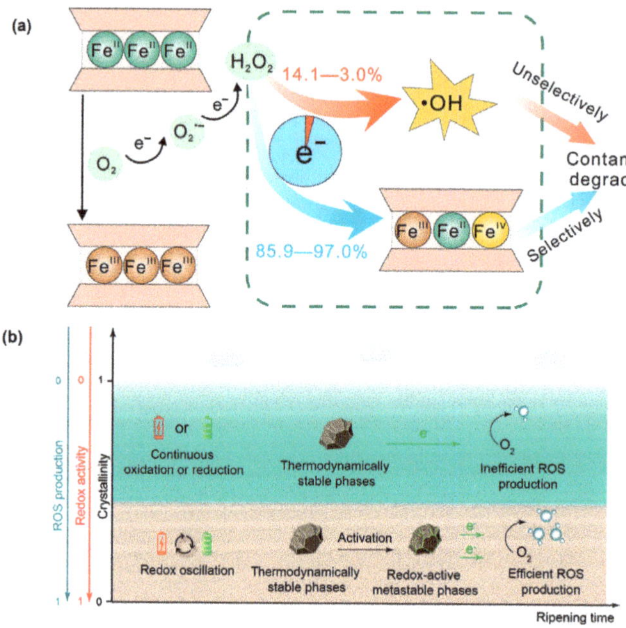

Figure 3. (**a**) Illustration of edge surface Fe(II) in clay mineral favoring the Fe(IV) generation over •OH generation [165]. (**b**) Illustration of redox oscillations activating thermodynamically stable iron minerals for enhanced ROS production [167].

Table 6. Degradation of organic pollutants via O_2 activation by Fe(II)-containing clay minerals and soils/sediments.

Material	Pollutant	Removal Ratio (%)	Reaction Time (h)	pH	Rate Constant	Reference
Nontronite	1,4-Dioxane	78.8 *	120	7.0	N/A	[161]
Illite	1,4-Dioxane	34.3 *	120	7.0	N/A	[161]
Montmorillonite	1,4-Dioxane	27.4 *	120	7.0	N/A	[161]
Reduced nontronite	Trichloroethylene	50.0	0.5	7.5	N/A	[160]
Riparian sediment	Trichloroethylene	27.6	6	7.0	N/A	[164]
Lakeshore sediment	Trichloroethylene	19.1	6	7.0	N/A	[164]
Pond sediment	Trichloroethylene	15.4	6	7.0	N/A	[164]
Nontronite	Phenol	43.1	6	7.0	N/A	[165]
Montmorillonite	Phenol	59.8	6	7.3	N/A	[165]
Sandbeach sediment	Phenol	9.95 *	10	6.96	N/A	[168]
Lakeshore sediment	Phenol	39.3 *	10	7.15	N/A	[168]
Farmland sediment	Phenol	48.5 *	10	Unadjusted	N/A	[168]
Paddy soils	Naphthalene	76.0 *	252	Unadjusted	N/A	[169]
Paddy soils	Phenanthrene	49.6 *	252	Unadjusted	N/A	[169]
Paddy soils	Pyrene	28.6 *	252	Unadjusted	N/A	[169]

Note: N/A, data not available. *, data obtained from the literature using the Getdata 2.26 software.

In addition to structural Fe(II), surface-bound Fe(II) also plays an important role in ROS generation and pollutant degradation by Fe(II)-containing clay minerals [168]. Similar to structural Fe(II), the reactivity of surface bound Fe(II) is also highly dependent on its coordination environment. Specifically, the sequence of surface-bound Fe(II) reactivity probably follows the order of ≡Si(IV)–O–Fe(II) < ≡Al(III)–O–Fe(II) < ≡Fe(III)–O–Fe(II) < ≡Fe(II)–O–Fe(II) < ≡HO–Fe(II) [169]. Additionally, the coordination environment of surface-bound Fe(II) also affects the O_2-activation mechanism. When surface-bound Fe(II) is coordinated with electron-rich ligands, it can efficiently activate O_2. Nonetheless, due to the inner-sphere complexation with O_2, more non-•OH species (e.g., Fe(IV)) are generated. Conversely, when coordinated with electron-poor ligands, surface-bound Fe(II) exhibits a low reactivity toward O_2, but more •OH is generated, via an outer-sphere interaction mechanism with O_2 [168,169]. Compared to structural Fe(II) and surface-bound Fe(II), exchangeable Fe(II) contributes minimally to •OH formation and can even exhibit a scavenging effect against •OH [169].

In addition to engineered Fe-based materials and isolated Fe-bearing minerals, recent studies have increasingly focused on the generation of ROS in actual soils and sediments under the redox-fluctuation condition, and Fe(II)-bearing compounds play key roles in this process [164,170–175]. The soil–water interface was generally considered as the active zone for intense H_2O_2 and •OH production due to the limited oxygen penetration and the rapid turnover of the reducing and oxidizing substances at the redox interfaces [170,171]. Although the yield of ROS varies across different sediments due to their unique physicochemical properties, for specific sediment, surface-bound Fe(II) and structural Fe(II) in poorly crystalline iron minerals are the primary contributors to ROS production [168,176]. Model studies indicate that the relative contributions of surface-adsorbed Fe(II) and structural Fe(II) in •OH production are 16.4–33.9% and 66.1–83.6% in sediment, respectively [164]. It has recently been proposed that tidal hydrology-triggered redox fluctuation could promote ROS generation by accelerating the production of reactive ferrous ions and amorphous ferric oxyhydroxides, thereby promoting surface electrochemical activities and O_2-activation capability (Figure 3b) [167]. These results further confirm the vital role of natural iron minerals in ROS generation under dark conditions.

3. Influencing Factors

The efficiency of O_2 activation by iron-based materials is influenced by several key environmental or operational factors, including O_2 concentration, organic ligands, inorganic anions, and microbial activity. Moreover, some factors can change the O_2-activation mechanism and pathways. This section summarizes the effects of these factors on the efficiency and mechanisms of O_2 activation by iron-based materials.

3.1. O_2 Concentration

With O_2 being the precursor to the generated ROS, increasing the O_2 concentration commonly results in a higher ROS concentration during O_2 activation by Fe-based materials [69,70,72]. However, it has been reported that, for Fe-containing clay minerals, excessively high levels of O_2 would lead to adverse effects on the generation of ROS due to the ineffective oxidation of structural Fe(II) [177]. Additionally, in systems where oxidation and reduction transformation processes of pollutants occur simultaneously, the concentration of O_2 also influences the reaction pathway and products of organic pollutants. For example, in the aerobic degradation of TCE by ferrous minerals in natural sediments, more low-molecular-weight acids were generated when O_2 concentration exceeded 120 μM, while only acetylene and/or ethene were observed when O_2 concentration was lower than 26 μM [178]. Furthermore, O_2 concentration also affects the transformation behaviors of the iron-bearing minerals. For example, high O_2 concentration promotes the dissolution of FeS and facilitates the formation of reactive iron hydroxides/oxides, such as lepidocrocite, while Fe_3S_4 is generated in the absence of O_2 [144]. These O_2-concentration-dependent transformation products exhibit different capabilities in generating ROS such as •OH [144].

3.2. Organic Ligands

Both natural and synthetic organic ligands exist in the subsurface environment and significantly affect the efficiency of Fe-based materials in mediating O_2 activation. In general, synthetic organic ligands have greater influences on O_2 activation than natural ligands. Synthetic organic ligands, such as ethylenediaminetetraacetic acid (EDTA) and N,N′-1,2-ethanediylbis-1-aspartic acid (NTA), are well known for their outstanding ability to form stable complexes with iron ions across a wide pH range (Figure 4a) [179], thereby influencing Fe(II) oxidation kinetics and the iron cycle via modifying the redox potential [180,181]. Furthermore, these ligands have been found to promote the release of active iron species from the bulk materials through surface-polarization reactions [182] or the proton-coupled electron-transfer process (Figure 4b) [183–185]. The release of Fe into the aqueous solution depends on the concentration and complexing ability of ligands (Table 7) [83,186,187]. Specifically, EDTA, which has a particularly strong complexation ability, can form monodentate inner-sphere Fe(II)-EDTA complexes on the surface ZVI. When the concentration of EDTA is significantly lower than that of ZVI, only low levels of dissolved Fe(II) are detected in the solution. In this case, a heterogeneous reaction dominated by $Fe(II)_{ad}$–EDTA complexes is responsible for O_2 activation and ROS generation [188]. However, when the dosages of EDTA and ZVI are on the same order of magnitude, the $Fe(II)_{aq}$–EDTA complex in the solution predominantly drives the reactions, and free •OH in the solution is primarily responsible for removing organic pollutants [184]. Additionally, the configurations of the Fe–ligand complexes also significantly affect the efficiency of O_2 activation. For example, the Fe(II)–NTA complex is more efficient in activating O_2 than the Fe(II)–EDTA complex due to the relatively open structure, leading to more ROS generation under the same condition [188]. Moreover, some organic ligands with reducing capabilities, such as hydroxylamine [189], can even directly reduce Fe(III) to Fe(II), resulting in a more efficient Fe cycle and higher ROS yield. Notably, the quenching effect of the ligands on •OH (Table 7) should be also considered for a more accurate analysis of the contributions of ROS to pollutant degradation kinetics.

Table 7. Stability constants of Fe(II)/Fe(III)-complexes with common organic ligands and rate constants for reaction of these ligands with •OH.

Species	Log β of Fe(II)-Complex	Log β of Fe(III)-Complex	Reaction Rate Constants with •OH (M^{-1} s^{-1})	Reference
EDTA	14.3	25.1	2.0×10^9	[83]
EDDS	N/A	20.6	2.5×10^9	[83,187]
NTA	8.05	15.90	5.5×10^8	[83,186]
CA	3.2	8.36–12.38 and 11.5	3.2×10^8	[83]
OA	> 4.70	9.4	1.0×10^7	[83]
HA	N/A	6.65–7.59	N/A	[83]

Note: N/A, data not available.

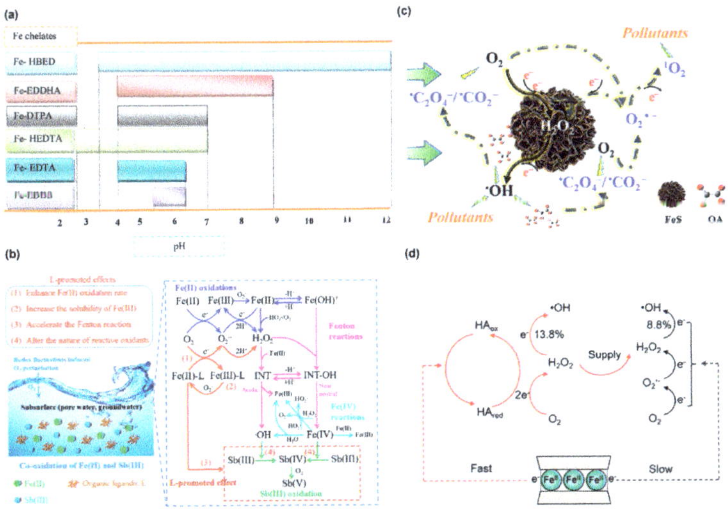

Figure 4. (**a**) Summary of pH ranges over which Fe chelates are stable [179]. (**b**) Reaction mechanisms of ligand-enhanced Fe(II) oxidation [183]. (**c**) Illustration of OA-enhanced FeS oxygenation mechanism [190]. (**d**) Illustration of the mechanisms for HA-enhanced oxygenation of Fe-containing clay mineral [191].

Compared to synthetic ligands, natural organic acids, including citric acid (CA), oxalate (OA), and humic acid (HA), generally exhibit a relatively weak ability to form complexes with iron. However, they can still modify the redox potential of iron ions and accelerate the iron cycle and O_2 activation effectively [192]. Moreover, reductive natural organic acids, such as glutathione [193], ascorbic acid [194], and protocatechuic acid [195], have the potential to reduce Fe(III) to Fe(II) directly. Studies have indicated that the dissolved Fe(II)–ligand complexes can mediate one electron-transfer process and play a dominant role in O_2 activation by FeS in the presence of CA and OA [192]. However, the capability of these organic acids to promote the dissolution of FeS_2 is lower than for FeS due to the more stable structure of FeS_2. Correspondingly, dissolved Fe(II)-mediated O_2 activation is less important in FeS_2 suspension [182,192]. These results indicated that the effect of organic acids on the contribution of homogeneous reactions to overall O_2 activation efficiency is closely related to the properties of the materials. Notably, it was recently revealed that organic ligands can promote ROS generation during the oxygenation of FeS minerals by producing abundant carbon-centered radicals. For example, OA could be preferentially oxidized by •OH, leading to the generation of carbon-centered radicals (e.g., $\bullet C_2O_4^-$ and $\bullet CO_2^-$), which further supply electrons to O_2 and contribute to at least 93.6% of the total •OH production in the FeS/OA/O_2 system (Figure 4c) [190].

As an important constituent of natural organic matter, HA has been confirmed to promote the generation of •OH by forming an aqueous Fe–HA complex and promoting the regeneration of Fe(II) via its reduced functional groups [196–199]. Meanwhile, microbially or chemically reduced HA has the potential to directly activate O_2 via the active quinone groups (−137 to −225 mV vs. NHE) to generate ROS [200]. Moreover, HA can mediate heterogeneous O_2 activation by Fe-containing minerals. It has been recently proposed that the presence of HA could change the reaction mechanism of nontronite oxygenation, where HA accepts electrons from the structural Fe and then delivers the elections to O_2 through two-electron-transfer pathways (Figure 4d) [191]. Compared to the direct electron transfer from structural Fe to O_2, reduced HA exhibits a faster O_2-reduction rate and higher selectivity for •OH. The HA-mediated pathways contributed to 70% of H_2O_2 and 62.1% of •OH generation in the HA/nontronite system. However, other studies have shown that the presence of HA could slow down the oxidation of reduced Fe-bearing clay minerals due to the competitive adsorption with O_2 [201].

Organic ligands can also affect the generation of non-hydroxyl radical species during the Fe-based material-mediated O_2-activation process. For example, the presence of CA could promote the generation of $^1O_2$ in the S–nZVI/O_2 systems because the Fe(II)–CA complex promotes the generation of more $O_2^{\bullet -}$, which could further react with H_2O/H^+ to generate $^1O_2$ and H_2O_2 [202].

3.3. Inorganic Anions

Inorganic anions are prevalent in the subsurface environment and play a significant role in influencing the oxidation behavior of Fe-based materials, mainly by inducing aggregation, increasing the hydrodynamic diameter, and competing with ROS [110]. Studies have shown that the inhibitory effects of common inorganic anions on O_2 activation by Fe-based materials, such as S–nZVI and nZVI, can be ranked in the order of $Cl^- < NO_3^- < SO_4^{2-} < HCO_3^- < HPO_4^{2-}$. This discrepancy could be primarily attributed to the varying degrees of competition that these anions exhibit with ROS [203]. Moreover, for Cl^-, its reaction with •OH to generate secondary chlorine radicals, such as $Cl^\bullet$, $Cl_2^{\bullet -}$, and $ClOH^{\bullet -}$ (Table 8), can partially mitigate the negative effects caused by •OH consumption. The impact of these inorganic anions on the degradation of organic pollutants also depends on the concentrations of the anions and ROS. For a FeS suspension exposed to air, 5 mg/L of Cl^- significantly hinders •OH generation. However, with sufficient O_2 (e.g., through O_2 purging of the suspension), even 500 mg/L of Cl^- has a limited effect on •OH concentration in the suspension due to the abundance of ROS involved in the •OH generation process. Interestingly, the addition of 50,000 mg/L Cl^- has been reported to enhance •OH generation, likely due to the production of secondary chlorine radicals [144]. Particularly noteworthy is the effect of the orthophosphate ion (PO_4^{3-}), which could adsorb on the surface of Fe minerals (e.g., green rust) and form the $[FeII(OH)_2\text{-}PO_4]^{3-}$ complex (Figure 5a), facilitating O_2 activation and the generation of $O_2^{\bullet -}$ [204]. When PO_4^{3-} is introduced to an aerated suspension of surface-oxidized ZVI (Fe@Fe_2O_3), the PO_4^{3-} ions can change the O_2-reduction pathway from a four-electron to a one-electron process (Figure 5b) [205]. Moreover, the surface phosphate layer induces the in situ generation of atomic hydrogen (•H) on the Fe@Fe_2O_3 surface, which can further promote the sequential one-electron O_2-reduction pathway (Figure 5b) [205].

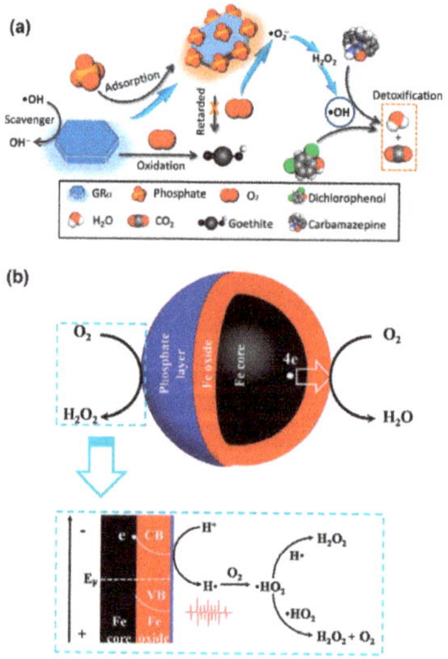

Figure 5. (a) Illustration of phosphate-enhanced O_2 activation by green rust [204]. (b) Illustration of phosphate-enhanced O_2 activation in Fe@Fe_2O_3 system [205].

Table 8. Reaction rate constants between •OH and inorganic anions.

Reaction	Rate Constant (M^{-1} s^{-1})	Reference
$Cl^- + \bullet OH \rightarrow ClOH^{\bullet-}$	4.3×10^9	[206]
$ClOH^{\bullet-} \rightarrow Cl^- + \bullet OH$	6.1×10^9	[206]
$ClOH^{\bullet-} + H^+ \rightarrow Cl^\bullet + H_2O$	4.3×10^{10}	[206]
$Cl^\bullet + Cl^- \rightarrow Cl_2^{\bullet-}$	1.0×10^5	[206]
$\bullet OH + HCO_3^- \rightarrow CO_3^{\bullet-} + H_2O$	8.5×10^6	[207]
$\bullet OH + CO_3^{2-} \rightarrow CO_3^{\bullet-} + OH^-$	3.9×10^8	[206]
$HPO_4^{2-} + \bullet OH \rightarrow HPO_4^{\bullet-} + OH^-$	8.0×10^5	[206]

3.4. Microbial Activity

Microbially mediated iron redox reactions are crucial geochemical processes in the environment [208]. Although the presence of O_2 can lead to a 38–64% decrease in the abundance of iron-reducing bacteria, these microorganisms can recover to 121–793% of their original levels after the restoration of anoxic conditions [209]. Therefore, the role of microorganisms in redox dynamics should garner more attention. Iron-reducing microorganisms could enhance the cycle of Fe(II)/Fe(III) via direct electron transfer [210], serving as extracellular electron shuttles [211] and releasing reduced species (e.g., flavins) [212,213], thus leading to a continuous supply of Fe(II) and a higher ROS yield. Notably, the extent of Fe(III) reduction is related to the mineral composition. For example, goethite with lower crystallinity is preferentially reduced compared to illite by *Shewanella putrefaciens* CN32, a metal-reducing bacterium [213]. Sulfate-reducing microorganisms also have important effects on the iron cycle [214]. Additionally, the presence of certain bacteria, such as neutrophilic iron-oxidizing bacteria [215], can alter the surface properties of Fe minerals and result in the renewal of mineral surfaces through continuous oxidative dissolution [216]. This microbially mediated transformation of iron minerals has significant impacts on O_2 activation by these minerals.

4. Conclusions and Perspectives

Iron-based materials have demonstrated significant potential in activating O_2 for in situ remediation of sites with organic contaminants. This review systematically examines the current research on O_2 activation by ZVI, iron sulfide, iron oxide, and Fe-bearing clay minerals. Notably, we have summarized recent findings about the roles of Fe-bearing components in natural soils and sediments for O_2 activation and ROS generation. The mechanisms for O_2 activation by these Fe-based materials are thoroughly discussed, including the active sites/species, electron-transfer pathways, and transformation of the materials, and the environmental and operational factors influencing O_2 activation and ROS generation are analyzed. Despite these significant advances, O_2 activation by Fe-based materials has not yet been applied for in situ remediation of organic-contaminated sites. Further investigations are needed to address the potential limitations of this promising remediation technology and overcome the barriers to its real-world application:

1. The ROS-generation dynamics under environmental conditions need thorough elucidation and characterization to achieve more accurate prediction and precise control of pollutant removal performance in practical applications. Current studies have demonstrated that the major reactive species generated in O_2 activation by Fe-based materials are H_2O_2, •OH, and $O_2^{\bullet-}$. However, the potential contribution of other reactive species, especially $^1O_2$, should be further explored, which has shown tremendous potential in selective oxidation of various contaminants [217,218]. Meanwhile, high-resolution monitoring of ROS-generation dynamics in actual subsurface environments is indispensable, which requires further exploration of novel tools suitable for in situ analysis of trace-level ROS. A notable example of such analytical tools is flow-injection chemiluminescence analysis, which can be performed with a portable device, achieving on-site quantification of •OH in environmental matrices [219].

2. While O_2 activation by Fe-based materials can degrade a variety of organic pollutants (e.g., TCE, PAHs, phenols, organic dyes, and antibiotics), future efforts are needed to explore its potential for degrading recalcitrant emerging pollutants (e.g., PFASs). The configuration of surface iron sites and interfacial microenvironment significantly affect the efficiency of O_2 activation. Further research is needed to elucidate the relationship between the functional groups of pollutants and the electron-shuttle mechanism for the tailored development of efficient Fe-based materials for the removal of emerging pollutants. For example, •OH is ineffective in degrading PFASs, whereas $O_2^{\bullet-}$ has demonstrated the capability to degrade perfluorocarboxylic acids with varying chain lengths [220]. Although $O_2^{\bullet-}$ is an easily formed intermediate during O_2 activation by Fe-based materials, it is quickly converted to other ROS. Therefore, nanotechnology-enabled rational material design is needed to manipulate the generation and consumption pathways of $O_2^{\bullet-}$ during O_2 activation and improve its selectivity toward reaction with PFASs. This can benefit from theoretical simulations of the interaction between the material surface and O_2/pollutant molecules under environmentally realistic conditions. Furthermore, the rational design of Fe-based materials for controllable O_2 activation and ROS generation can be substantially expedited by incorporating machine-learning analysis of large datasets on the structure–reactivity relationships [221,222].
3. Attention should be directed toward conducting pilot-scale applications of this technology to validate its effectiveness in real-world scenarios. In particular, for remediating contaminated sites lacking reactive Fe minerals, it is necessary to introduce Fe-based materials capable of efficient O_2 activation. Iron-based materials that show excellent performance in laboratory studies may not work when applied in real aquifers, and it is vital to ensure that mass-produced remediation agents exhibit activity comparable to those tested in the initial research and development (R&D) stage. Moreover, the effective delivery of these Fe-based remediation agents can be a bottleneck for ISCO remediation via Fe-mediated O_2 activation. This calls for a simultaneous evaluation of the transport properties of the Fe-based materials while optimizing their O_2 activation efficiency.
4. In addition to the above technical challenges, other barriers to the real-world applications of O_2 activation for site remediation need to be overcome. To be economically viable and competitive, costs associated with the materials, equipment, and power need to be lowered. While Fe is an earth-abundant element, the R&D and scale-up production of sophisticated Fe-based materials still may be costly. Moreover, despite the abundance and availability of O_2 in the air, the energy required to deliver it into deep aquifers adds to the total cost of this technology. Finally, since Fe-based materials (e.g., nZVI) have shown toxicity to a variety of soil organisms [223–225], it is critical to evaluate the potential environmental impact of these materials before they can be safely applied in the subsurface environment. A comprehensive consideration of these factors is warranted to ensure the successful utilization of O_2, a green and inexhaustible oxidant, for sustainable in situ remediation of organic-contaminated sites.

Author Contributions: F.H.: Investigation, Writing—original draft, Writing—review and editing. L.X.: Investigation, Validation. H.W.: Writing—review and editing, Funding acquisition, Project administration. C.J.: Writing—review and editing, Funding acquisition, Supervision, Conceptualization. All authors have read and agreed to the published version of the manuscript.

Funding: This research was funded by the National Natural Science Foundation of China (22276101, 52370053), Tianjin Municipal Science and Technology Bureau (23JCZDJC00740), and the Fundamental Research Funds for the Central Universities (63241632).

Institutional Review Board Statement: Not applicable.

Informed Consent Statement: Not applicable.

Data Availability Statement: No new data were created or analyzed in this study. Data sharing is not applicable to this article.

Conflicts of Interest: The authors declare no conflicts of interest.

References

1. Bunting, S.Y.; Lapworth, D.J.; Crane, E.J.; Grima-Olmedo, J.; Koroša, A.; Kuczyńska, A.; Mali, N.; Rosenqvist, L.; Van Vliet, M.E.; Togola, A.; et al. Emerging Organic Compounds in European Groundwater. *Environ. Pollut.* **2021**, *269*, 115945. [CrossRef] [PubMed]
2. Hua, T.; Propp, V.R.; Power, C.; Brown, S.J.; Collins, P.; Smith, J.E.; Roy, J.W. Multizone Aquatic Ecological Exposures to Landfill Contaminants from a Groundwater Plume Discharging to a Pond. *Environ. Toxicol. Chem.* **2023**, *42*, 1667–1684. [CrossRef] [PubMed]
3. Jia, X.; O'Connor, D.; Hou, D.; Jin, Y.; Li, G.; Zheng, C.; Ok, Y.S.; Tsang, D.C.W.; Luo, J. Groundwater Depletion and Contamination: Spatial Distribution of Groundwater Resources Sustainability in China. *Sci. Total Environ.* **2019**, *672*, 551–562. [CrossRef] [PubMed]
4. Kumar, M.; Goswami, R.; Patel, A.K.; Srivastava, M.; Das, N. Scenario, Perspectives and Mechanism of Arsenic and Fluoride Co-Occurrence in the Groundwater: A Review. *Chemosphere* **2020**, *249*, 126126. [CrossRef] [PubMed]
5. Abiriga, D.; Vestgarden, L.S.; Klempe, H. Groundwater Contamination from a Municipal Landfill: Effect of Age, Landfill Closure, and Season on Groundwater Chemistry. *Sci. Total Environ.* **2020**, *737*, 140307. [CrossRef] [PubMed]
6. Propp, V.R.; De Silva, A.O.; Spencer, C.; Brown, S.J.; Catingan, S.D.; Smith, J.E.; Roy, J.W. Organic Contaminants of Emerging Concern in Leachate of Historic Municipal Landfills. *Environ. Pollut.* **2021**, *276*, 116474. [CrossRef] [PubMed]
7. Sun, Y.; Li, H.; Lei, S.; Semple, K.T.; Coulon, F.; Hu, Q.; Gao, J.; Guo, G.; Gu, Q.; Jones, K.C. Redevelopment of Urban Brownfield Sites in China: Motivation, History, Policies and Improved Management. *Eco-Environ. Health* **2022**, *1*, 63–72. [CrossRef] [PubMed]
8. Wu, S.; Xiang, Z.; Lin, D.; Zhu, L. Multimedia Distribution and Health Risk Assessment of Typical Organic Pollutants in a Retired Industrial Park. *Front. Environ. Sci. Eng.* **2023**, *17*, 142. [CrossRef]
9. Moran, M.J.; Zogorski, J.S.; Squillace, P.J. Chlorinated Solvents in Groundwater of the United States. *Environ. Sci. Technol.* **2007**, *41*, 74–81. [CrossRef]
10. Bulatović, S.; Ilić, M.; Šolević Knudsen, T.; Milić, J.; Pucarević, M.; Jovančićević, B.; Vrvić, M.M. Evaluation of Potential Human Health Risks from Exposure to Volatile Organic Compounds in Contaminated Urban Groundwater in the Sava River Aquifer, Belgrade, Serbia. *Environ. Geochem. Health* **2022**, *44*, 3451–3472. [CrossRef]
11. Montuori, P.; De Rosa, E.; Cerino, P.; Pizzolante, A.; Nicodemo, F.; Gallo, A.; Rofrano, G.; De Vita, S.; Limone, A.; Triassi, M. Estimation of Polycyclic Aromatic Hydrocarbons in Groundwater from Campania Plain: Spatial Distribution, Source Attribution and Health Cancer Risk Evaluation. *Toxics* **2023**, *11*, 435. [CrossRef] [PubMed]
12. Peng, B.; Dong, Q.; Li, F.; Wang, T.; Qiu, X.; Zhu, T. A Systematic Review of Polycyclic Aromatic Hydrocarbon Derivatives: Occurrences, Levels, Biotransformation, Exposure Biomarkers, and Toxicity. *Environ. Sci. Technol.* **2023**, *57*, 15314–15335. [CrossRef] [PubMed]
13. Li, M.; Chen, Q.; Yang, L.; Zhang, Y.; Jiang, J.; Deng, S.; Wan, J.; Fan, T.; Long, T.; Zhang, S.; et al. Contaminant Characterization at Pesticide Production Sites in the Yangtze River Delta: Residue, Distribution, and Environmental Risk. *Sci. Total Environ.* **2023**, *860*, 160156. [CrossRef] [PubMed]
14. Huo, Z.; Xi, M.; Xu, L.; Jiang, C.; Chen, W. Colloid-Facilitated Release of Polybrominated Diphenyl Ethers at an e-Waste Recycling Site: Evidence from Undisturbed Soil Core Leaching Experiments. *Front. Environ. Sci. Eng.* **2024**, *18*, 21. [CrossRef]
15. Schaefer, C.E.; Lavorgna, G.M.; Lippincott, D.R.; Nguyen, D.; Schaum, A.; Higgins, C.P.; Field, J. Leaching of Perfluoroalkyl Acids during Unsaturated Zone Flushing at a Field Site Impacted with Aqueous Film Forming Foam. *Environ. Sci. Technol.* **2023**, *57*, 1940–1948. [CrossRef] [PubMed]
16. Stroo, H.F.; Unger, M.; Ward, C.H.; Kavanaugh, M.C.; Vogel, C.; Leeson, A.; Marqusee, J.A.; Smith, B.P. Remediating Chlorinated Solvent Source Zones. *Environ. Sci. Technol.* **2003**, *37*, 224A–230A. [CrossRef]
17. Cheng, Y.; Zhu, J. Significance of Mass–Concentration Relation on the Contaminant Source Depletion in the Nonaqueous Phase Liquid (NAPL) Contaminated Zone. *Transp. Porous Med.* **2021**, *137*, 399–416. [CrossRef]
18. Tatti, F.; Papini, M.P.; Sappa, G.; Raboni, M.; Arjmand, F.; Viotti, P. Contaminant Back-Diffusion from Low Permeability Layers as Affected by Groundwater Velocity: A Laboratory Investigation by Box Model and Image Analysis. *Sci. Total Environ.* **2018**, *622–623*, 164–171. [CrossRef]
19. Ma, T.; Pan, X.; Wang, T.; Li, X.; Luo, Y. Toxicity of Per- and Polyfluoroalkyl Substances to Nematodes. *Toxics* **2023**, *11*, 593. [CrossRef]
20. Ge, Y.; Wang, Z.; Chen, X.; Wang, W.; Liu, Z.; Sun, H.; Zhang, L. Comparative Toxicological Effects of Perfluorooctane Sulfonate and Its Alternative 6:2 Chlorinated Polyfluorinated Ether Sulfonate on Earthworms. *Environ. Toxicol. Chem.* **2023**, *43*, 170–181. [CrossRef]
21. Ordaz, J.D.; Damayanti, N.P.; Irudayaraj, J.M.K. Toxicological Effects of Trichloroethylene Exposure on Immune Disorders. *Immunopharmacol. Immunotoxicol.* **2017**, *39*, 305–317. [CrossRef] [PubMed]

22. Grazuleviciene, R.; Nieuwenhuijsen, M.J.; Vencloviene, J.; Kostopoulou-Karadanelli, M.; Krasner, S.W.; Danileviciute, A.; Balcius, G.; Kapustinskiene, V. Individual Exposures to Drinking Water Trihalomethanes, Low Birth Weight and Small for Gestational Age Risk: A Prospective Kaunas Cohort Study. *Environ. Health* **2011**, *10*, 32. [CrossRef] [PubMed]
23. Han, Y.; Cheng, J.; He, L.; Zhang, M.; Ren, S.; Sun, J.; Xing, X.; Tang, Z. Polybrominated Diphenyl Ethers in Soils from Tianjin, North China: Distribution, Health Risk, and Temporal Trends. *Environ. Geochem. Health* **2021**, *43*, 1177–1191. [CrossRef] [PubMed]
24. Sankar, T.K.; Kumar, A.; Mahto, D.K.; Das, K.C.; Narayan, P.; Fukate, M.; Awachat, P.; Padghan, D.; Mohammad, F.; Al-Lohedan, H.A.; et al. The Health Risk and Source Assessment of Polycyclic Aromatic Hydrocarbons (PAHs) in the Soil of Industrial Cities in India. *Toxics* **2023**, *11*, 515. [CrossRef] [PubMed]
25. Daraban, G.M.; Hlihor, R.-M.; Suteu, D. Pesticides vs. Biopesticides: From Pest Management to Toxicity and Impacts on the Environment and Human Health. *Toxics* **2023**, *11*, 983. [CrossRef]
26. Horzmann, K.A.; Portales, A.M.; Batcho, K.G.; Freeman, J.L. Developmental Toxicity of Trichloroethylene in Zebrafish (*Danio Rerio*). *Environ. Sci. Processes Impacts* **2020**, *22*, 728–739. [CrossRef]
27. Yan, Z.G.; Li, Z.G.; Du, J.Z. Effects of Environmental Organic Pollutants on Environment and Human Health: The Latest Updates. *Toxics* **2024**, *12*, 231. [CrossRef]
28. Wang, J.; Chen, S.; Tian, M.; Zheng, X.; Gonzales, L.; Ohura, T.; Mai, B.; Simonich, S.L.M. Inhalation Cancer Risk Associated with Exposure to Complex Polycyclic Aromatic Hydrocarbon Mixtures in an Electronic Waste and Urban Area in South China. *Environ. Sci. Technol.* **2012**, *46*, 9745–9752. [CrossRef]
29. Mahoney, H.; Xie, Y.; Brinkmann, M.; Giesy, J.P. Next Generation Per- and Poly-Fluoroalkyl Substances: Status and Trends, Aquatic Toxicity, and Risk Assessment. *Eco-Environ. Health* **2022**, *1*, 117–131. [CrossRef]
30. Palazzolo, S.; Caligiuri, I.; Sfriso, A.A.; Mauceri, M.; Rotondo, R.; Campagnol, D.; Canzonieri, V.; Rizzolio, F. Early Warnings by Liver Organoids on Short- and Long-Chain PFAS Toxicity. *Toxics* **2022**, *10*, 91. [CrossRef]
31. Stroo, H.F.; Leeson, A.; Marqusee, J.A.; Johnson, P.C.; Ward, C.H.; Kavanaugh, M.C.; Sale, T.C.; Newell, C.J.; Pennell, K.D.; Lebrón, C.A.; et al. Chlorinated Ethene Source Remediation: Lessons Learned. *Environ. Sci. Technol.* **2012**, *46*, 6438–6447. [CrossRef] [PubMed]
32. Cheng, M.; Zeng, G.; Huang, D.; Lai, C.; Xu, P.; Zhang, C.; Liu, Y. Hydroxyl Radicals Based Advanced Oxidation Processes (AOPs) for Remediation of Soils Contaminated with Organic Compounds: A Review. *Chem. Eng. J.* **2016**, *284*, 582–598. [CrossRef]
33. Liu, J.W.; Wei, K.H.; Xu, S.W.; Cui, J.; Ma, J.; Xiao, X.L.; Xi, B.D.; He, X.S. Surfactant-Enhanced Remediation of Oil-Contaminated Soil and Groundwater: A Review. *Sci. Total Environ.* **2021**, *756*, 144142. [CrossRef] [PubMed]
34. Tsitonaki, A.; Petri, B.; Crimi, M.; Mosbæk, H.; Siegrist, R.L.; Bjerg, P.L. In Situ Chemical Oxidation of Contaminated Soil and Groundwater Using Persulfate: A Review. *Crit. Rev. Environ. Sci. Technol.* **2010**, *40*, 55–91. [CrossRef]
35. Ranc, B.; Faure, P.; Croze, V.; Simonnot, M.O. Selection of Oxidant Doses for in Situ Chemical Oxidation of Soils Contaminated by Polycyclic Aromatic Hydrocarbons (PAHs): A Review. *J. Hazard. Mater.* **2016**, *312*, 280–297. [CrossRef]
36. Wei, K.H.; Ma, J.; Xi, B.-D.; Yu, M.D.; Cui, J.; Chen, B.L.; Li, Y.; Gu, Q.B.; He, X.S. Recent Progress on In-Situ Chemical Oxidation for the Remediation of Petroleum Contaminated Soil and Groundwater. *J. Hazard. Mater.* **2022**, *432*, 128738. [CrossRef]
37. Liu, H.; Bruton, T.A.; Doyle, F.M.; Sedlak, D.L. In Situ Chemical Oxidation of Contaminated Groundwater by Persulfate: Decomposition by Fe(III)- and Mn(IV)-Containing Oxides and Aquifer Materials. *Environ. Sci. Technol.* **2014**, *48*, 10330–10336. [CrossRef]
38. Pan, C.; Wang, C.; Zhao, X.; Xu, P.; Mao, F.; Yang, J.; Zhu, Y.; Yu, R.; Xiao, S.; Fang, Y.; et al. Neighboring sp-Hybridized Carbon Participated Molecular Oxygen Activation on the Interface of Sub-Nanocluster CuO/Graphdiyne. *J. Am. Chem. Soc.* **2022**, *144*, 4942–4951. [CrossRef]
39. Wei, Y.; Chen, Y.; Cao, X.; Xiang, M.; Huang, Y.; Li, H. A Critical Review of Groundwater Table Fluctuation: Formation, Effects on Multifields, and Contaminant Behaviors in a Soil and Aquifer System. *Environ. Sci. Technol.* **2024**, *58*, 2185–2203. [CrossRef]
40. Ji, J.; Wang, Z.; Xu, Q.; Zhu, Q.; Xing, M. In Situ H_2O_2 Generation and Corresponding Pollutant Removal Applications: A Review. *Chem. Eur. J.* **2023**, *29*, e202203921. [CrossRef]
41. Guo, X.; Lin, S.; Gu, J.; Zhang, S.; Chen, Z.; Huang, S. Simultaneously Achieving High Activity and Selectivity toward Two-Electron O_2 Electroreduction: The Power of Single-Atom Catalysts. *ACS Catal.* **2019**, *9*, 11042–11054. [CrossRef]
42. Zheng, Y.; Yu, Z.; Ou, H.; Asiri, A.M.; Chen, Y.; Wang, X. Black Phosphorus and Polymeric Carbon Nitride Heterostructure for Photoinduced Molecular Oxygen Activation. *Adv. Funct. Mater.* **2018**, *28*, 1705407. [CrossRef]
43. Colombo, C.; Palumbo, G.; He, J.-Z.; Pinton, R.; Cesco, S. Review on Iron Availability in Soil: Interaction of Fe Minerals, Plants, and Microbes. *J. Soils Sediments* **2014**, *14*, 538–548. [CrossRef]
44. Guo, H.; Barnard, A.S. Naturally Occurring Iron Oxide Nanoparticles: Morphology, Surface Chemistry and Environmental Stability. *J. Mater. Chem. A* **2013**, *1*, 27–42. [CrossRef]
45. Wang, Y.; Jin, X.; Peng, A.; Gu, C. Transformation and Toxicity of Environmental Contaminants as Influenced by Fe Containing Clay Minerals: A Review. *Bull. Environ. Contam. Toxicol.* **2020**, *104*, 8–14. [CrossRef]
46. Gong, Y.; Tang, J.; Zhao, D. Application of Iron Sulfide Particles for Groundwater and Soil Remediation: A Review. *Water Res.* **2016**, *89*, 309–320. [CrossRef]
47. Zeng, G.; Wang, J.; Dai, M.; Meng, Y.; Luo, H.; Zhou, Q.; Lin, L.; Zang, K.; Meng, Z.; Pan, X. Natural Iron Minerals in an Electrocatalytic Oxidation System and in Situ Pollutant Removal in Groundwater: Applications, Mechanisms, and Challenges. *Sci. Total Environ.* **2023**, *871*, 161826. [CrossRef]

48. Quinn, J.; Geiger, C.; Clausen, C.; Brooks, K.; Coon, C.; O'Hara, S.; Krug, T.; Major, D.; Yoon, W.-S.; Gavaskar, A.; et al. Field Demonstration of DNAPL Dehalogenation Using Emulsified Zero-Valent Iron. *Environ. Sci. Technol.* **2005**, *39*, 1309–1318. [CrossRef]
49. Gu, Y.; Wang, B.; He, F.; Bradley, M.J.; Tratnyek, P.G. Mechanochemically Sulfidated Microscale Zero Valent Iron: Pathways, Kinetics, Mechanism, and Efficiency of Trichloroethylene Dechlorination. *Environ. Sci. Technol.* **2017**, *51*, 12653–12662. [CrossRef]
50. Xu, J.; Wang, Y.; Weng, C.; Bai, W.; Jiao, Y.; Kaegi, R.; Lowry, G.V. Reactivity, Selectivity, and Long-Term Performance of Sulfidized Nanoscale Zerovalent Iron with Different Properties. *Environ. Sci. Technol.* **2019**, *53*, 5936–5945. [CrossRef]
51. Brumovský, M.; Oborná, J.; Micić, V.; Malina, O.; Kašlík, J.; Tunega, D.; Kolos, M.; Hofmann, T.; Karlický, F.; Filip, J. Iron Nitride Nanoparticles for Enhanced Reductive Dechlorination of Trichloroethylene. *Environ. Sci. Technol.* **2022**, *56*, 4425–4436. [CrossRef] [PubMed]
52. Gong, L.; Chen, J.; Hu, Y.; He, K.; Bylaska, E.J.; Tratnyek, P.G.; He, F. Degradation of Chloroform by Zerovalent Iron: Effects of Mechanochemical Sulfidation and Nitridation on the Kinetics and Mechanism. *Environ. Sci. Technol.* **2023**, *57*, 9811–9821. [CrossRef]
53. Chen, Z.; Chen, J.; Tan, S.; Yang, Z.; Zhang, Y. Dechlorination Helps Defluorination: Insights into the Defluorination Mechanism of Florfenicol by S-nZVI and DFT Calculations on the Reaction Pathways. *Environ. Sci. Technol.* **2024**, *58*, 2542–2553. [CrossRef] [PubMed]
54. Usman, M.; Faure, P.; Ruby, C.; Hanna, K. Remediation of PAH-Contaminated Soils by Magnetite Catalyzed Fenton-like Oxidation. *Appl. Catal. B* **2012**, *117–118*, 10–17. [CrossRef]
55. Liu, X.; Yuan, S.; Zhang, P.; Zhu, J.; Tong, M. Reduced Nontronite-Activated H_2O_2 for Contaminants Degradation: The Beneficial Role of Clayed Fractions in ISCO Treatments. *J. Hazard. Mater.* **2020**, *386*, 121945. [CrossRef] [PubMed]
56. Chen, N.; Fang, G.; Liu, G.; Zhou, D.; Gao, J.; Gu, C. The Effects of Fe-Bearing Smectite Clays on •OH Formation and Diethyl Phthalate Degradation with Polyphenols and H_2O_2. *J. Hazard. Mater.* **2018**, *357*, 483–490. [CrossRef]
57. Sun, Y.; Danish, M.; Ali, M.; Shan, A.; Li, M.; Lyu, Y.; Qiu, Z.; Sui, Q.; Zang, X.; Lyu, S. Trichloroethene Degradation by Nanoscale CaO_2 Activated with Fe(II)/FeS: The Role of FeS and the Synergistic Activation Mechanism of Fe(II)/FeS. *Chem. Eng. J.* **2020**, *394*, 124830. [CrossRef]
58. Kim, J.-G.; Kim, H.-B.; Jeong, W.-G.; Baek, K. Enhanced-Oxidation of Sulfanilamide in Groundwater Using Combination of Calcium Peroxide and Pyrite. *J. Hazard. Mater.* **2021**, *419*, 126514. [CrossRef]
59. Li, H.; Yao, Y.; Zhang, J.; Du, J.; Xu, S.; Wang, C.; Zhang, D.; Tang, J.; Zhao, H.; Zhou, J. Degradation of Phenanthrene by Peroxymonosulfate Activated with Bimetallic Metal-Organic Frameworks: Kinetics, Mechanisms, and Degradation Products. *Chem. Eng. J.* **2020**, *397*, 125401. [CrossRef]
60. Lee, Y.C.; Li, Y.; Chen, M.J.; Chen, Y.-C.; Kuo, J.; Lo, S.L. Efficient Decomposition of Perfluorooctanic Acid by Persulfate with Iron-Modified Activated Carbon. *Water Res.* **2020**, *174*, 115618. [CrossRef]
61. Dong, H.; He, Q.; Zeng, G.; Tang, L.; Zhang, L.; Xie, Y.; Zeng, Y.; Zhao, F. Degradation of Trichloroethene by Nanoscale Zero-Valent Iron (nZVI) and nZVI Activated Persulfate in the Absence and Presence of EDTA. *Chem. Eng. J.* **2017**, *316*, 410–418. [CrossRef]
62. Liu, H.; Fu, P.; Liu, F.; Hou, Q.; Tong, Z.; Bi, W. Degradation of Ciprofloxacin by Persulfate Activated with Pyrite: Mechanism, Acidification and Tailwater Reuse. *RSC Adv.* **2022**, *12*, 29991–30000. [CrossRef] [PubMed]
63. Liu, X.; Shao, P.; Gao, S.; Bai, Z.; Tian, J. Benzoquinone-Assisted Heterogeneous Activation of PMS on Fe_3S_4 via Formation of Active Complexes to Mediate Electron Transfer towards Enhanced Bisphenol A Degradation. *Water Res.* **2022**, *226*, 119218. [CrossRef] [PubMed]
64. Wang, Q.; Zeng, H.; Liang, Y.; Cao, Y.; Xiao, Y.; Ma, J. Degradation of Bisphenol AF in Water by Periodate Activation with FeS (Mackinawite) and the Role of Sulfur Species in the Generation of Sulfate Radicals. *Chem. Eng. J.* **2021**, *407*, 126738. [CrossRef]
65. Zhang, P.; Zhang, X.; Zhao, X.; Jing, G.; Zhou, Z. Activation of Peracetic Acid with Zero-Valent Iron for Tetracycline Abatement: The Role of Fe(II) Complexation with Tetracycline. *J. Hazard. Mater.* **2022**, *424*, 127653. [CrossRef]
66. Li, L.; Wu, Y.; Dong, W. Enhancement in Sulfamethoxazole Degradation via Efficient Heterogeneous Activation of Peracetic Acid by FeS. *Water* **2024**, *16*, 2405. [CrossRef]
67. Joo, S.H.; Feitz, A.J.; Waite, T.D. Oxidative Degradation of the Carbothioate Herbicide, Molinate, Using Nanoscale Zero-Valent Iron. *Environ. Sci. Technol.* **2004**, *38*, 2242–2247. [CrossRef]
68. Noradoun, C.E.; Cheng, I.F. EDTA Degradation Induced by Oxygen Activation in a Zerovalent Iron/Air/Water System. *Environ. Sci. Technol.* **2005**, *39*, 7158–7163. [CrossRef]
69. Cheng, D.; Neumann, A.; Yuan, S.; Liao, W.; Qian, A. Oxidative Degradation of Organic Contaminants by FeS in the Presence of O_2. *Environ. Sci. Technol.* **2020**, *54*, 4091–4101. [CrossRef]
70. Pham, H.T.; Kitsuneduka, M.; Hara, J.; Suto, K.; Inoue, C. Trichloroethylene Transformation by Natural Mineral Pyrite. The Deciding Role of Oxygen. *Environ. Sci. Technol.* **2008**, *42*, 7470–7475. [CrossRef]
71. Ardo, S.G.; Nélieu, S.; Ona-Nguema, G.; Delarue, G.; Brest, J.; Pironin, E.; Morin, G. Oxidative Degradation of Nalidixic Acid by Nano-Magnetite via Fe^{2+}/O_2-Mediated Reactions. *Environ. Sci. Technol.* **2015**, *49*, 4506–4514. [CrossRef] [PubMed]
72. Chen, N.; Fang, G.; Liu, G.; Zhou, D.; Gao, J.; Gu, C. The Degradation of Diethyl Phthalate by Reduced Smectite Clays and Dissolved Oxygen. *Chem. Eng. J.* **2019**, *355*, 247–254. [CrossRef]
73. Pi, L.; Cai, J.; Xiong, L.; Cui, J.; Hua, H.; Tang, D.; Mao, X. Generation of H_2O_2 by On-Site Activation of Molecular Dioxygen for Environmental Remediation Applications: A Review. *Chem. Eng. J.* **2020**, *389*, 123420. [CrossRef]

74. Liu, Y.; Zhao, Y.; Wang, J. Fenton/Fenton-like Processes with in-Situ Production of Hydrogen Peroxide/Hydroxyl Radical for Degradation of Emerging Contaminants: Advances and Prospects. *J. Hazard. Mater.* **2021**, *404*, 124191. [CrossRef] [PubMed]
75. Li, Q.; Li, F. Recent Advances in Molecular Oxygen Activation via Photocatalysis and Its Application in Oxidation Reactions. *Chem. Eng. J.* **2021**, *421*, 129915. [CrossRef]
76. Ni, Y.; Zhou, C.; Xing, M.; Zhou, Y. Oxidation of Emerging Organic Contaminants by In-Situ H_2O_2 Fenton System. *Green Energy Environ.* **2024**, *9*, 417–434. [CrossRef]
77. Wilson, G.S. Determination of Oxidation-Reduction Potentials. *Methods Enzymol.* **1978**, *54*, 396–410.
78. Petersen, E.J.; Pinto, R.A.; Shi, X.; Huang, Q. Impact of Size and Sorption on Degradation of Trichloroethylene and Polychlorinated Biphenyls by Nano-Scale Zerovalent Iron. *J. Hazard. Mater.* **2012**, *243*, 73–79. [CrossRef]
79. Li, Q.; Yin, J.; Wu, L.; Li, S.; Chen, L. Effects of Biochar and Zero Valent Iron on the Bioavailability and Potential Toxicity of Heavy Metals in Contaminated Soil at the Field Scale. *Sci. Total Environ.* **2023**, *897*, 165386. [CrossRef]
80. Wei, K.; Li, H.; Gu, H.; Liu, X.; Ling, C.; Cao, S.; Li, M.; Liao, M.; Peng, X.; Shi, Y.; et al. Strained Zero-Valent Iron for Highly Efficient Heavy Metal Removal. *Adv. Funct. Mater.* **2022**, *32*, 2200498. [CrossRef]
81. Li, Y.; Kawashima, N.; Li, J.; Chandra, A.P.; Gerson, A.R. A Review of the Structure, and Fundamental Mechanisms and Kinetics of the Leaching of Chalcopyrite. *Adv. Colloid Interface Sci.* **2013**, *197–198*, 1–32. [CrossRef] [PubMed]
82. Zeng, Q.; Dong, H.; Wang, X. Effect of Ligands on the Production of Oxidants from Oxygenation of Reduced Fe-Bearing Clay Mineral Nontronite. *Geochim. Cosmochim. Acta* **2019**, *251*, 136–156. [CrossRef]
83. Zhang, Y.; Zhou, M. A Critical Review of the Application of Chelating Agents to Enable Fenton and Fenton-like Reactions at High pH Values. *J. Hazard. Mater.* **2019**, *362*, 436–450. [CrossRef] [PubMed]
84. Wang, Y.; Liu, L.; Fang, G.; Wang, L.; Kengara, F.O.; Zhu, C. The Mechanism of 2-Chlorobiphenyl Oxidative Degradation by Nanoscale Zero-Valent Iron in the Presence of Dissolved Oxygen. *Environ. Sci. Pollut. Res.* **2018**, *25*, 2265–2272. [CrossRef]
85. Wu, J.; Zhao, J.; Hou, J.; Zeng, R.J.; Xing, B. Degradation of Tetrabromobisphenol A by Sulfidated Nanoscale Zerovalent Iron in a Dynamic Two-Step Anoxic/Oxic Process. *Environ. Sci. Technol.* **2019**, *53*, 8105–8114. [CrossRef]
86. Keenan, C.R.; Sedlak, D.L. Factors Affecting the Yield of Oxidants from the Reaction of Nanoparticulate Zero-Valent Iron and Oxygen. *Environ. Sci. Technol.* **2008**, *42*, 1262–1267. [CrossRef]
87. Harada, T.; Yatagai, T.; Kawase, Y. Hydroxyl Radical Generation Linked with Iron Dissolution and Dissolved Oxygen Consumption in Zero-Valent Iron Wastewater Treatment Process. *Chem. Eng. J.* **2016**, *303*, 611–620. [CrossRef]
88. Lee, C.; Sedlak, D.L. Enhanced Formation of Oxidants from Bimetallic Nickel−Iron Nanoparticles in the Presence of Oxygen. *Environ. Sci. Technol.* **2008**, *42*, 8528–8533. [CrossRef]
89. Ai, Z.; Gao, Z.; Zhang, L.; He, W.; Yin, J.J. Core–Shell Structure Dependent Reactivity of $Fe@Fe_2O_3$ Nanowires on Aerobic Degradation of 4-Chlorophenol. *Environ. Sci. Technol.* **2013**, *47*, 5344–5352. [CrossRef]
90. He, D.; Ma, J.; Collins, R.N.; Waite, T.D. Effect of Structural Transformation of Nanoparticulate Zero-Valent Iron on Generation of Reactive Oxygen Species. *Environ. Sci. Technol.* **2016**, *50*, 3820–3828. [CrossRef]
91. Liu, X.; Fan, J.H.; Ma, L.M. Elimination of 4-Chlorophenol in Aqueous Solution by the Bimetallic Al–Fe/O_2 at Normal Temperature and Pressure. *Chem. Eng. J.* **2014**, *236*, 274–284. [CrossRef]
92. Fan, J.; Qin, H.; Zhang, Y.; Jiang, S. Degradation of 4-chlorophenol by BM Fe/Cu-O_2 System: The Symbiosis of and •OH Radicals. *Water Environ. Res.* **2019**, *91*, 770–779. [CrossRef] [PubMed]
93. Xiang, W.; Huang, M.; Wang, Y.; Wu, X.; Zhang, F.; Li, D.; Zhou, T. New Insight in the O_2 Activation by Nano Fe/Cu Bimetals: The Synergistic Role of Cu(0) and Fe(II). *Chin. Chem. Lett.* **2020**, *31*, 2831–2834. [CrossRef]
94. Yang, Z.; Zhang, X.; Pu, S.; Ni, R.; Lin, Y.; Liu, Y. Novel Fenton-like System (Mg/Fe-O_2) for Degradation of 4-Chlorophenol. *Environ. Pollut.* **2019**, *250*, 906–913. [CrossRef]
95. Wu, J.; Lin, M.; Liu, M.; Chen, Z. Novel Crystalline/Amorphous Heterophase Fe-Mn Core–Shell Chains on-Site Generate Hydrogen Peroxide in Aqueous Solution. *J. Colloid Interface Sci.* **2024**, *676*, 227–237. [CrossRef] [PubMed]
96. Tang, J.; Tang, L.; Feng, H.; Zeng, G.; Dong, H.; Zhang, C.; Huang, B.; Deng, Y.; Wang, J.; Zhou, Y. pH-Dependent Degradation of p-Nitrophenol by Sulfidated Nanoscale Zerovalent Iron under Aerobic or Anoxic Conditions. *J. Hazard. Mater.* **2016**, *320*, 581–590. [CrossRef] [PubMed]
97. Zhang, X.; Sun, W.; Wang, Y.; Li, Z.; Huang, X.; Li, T.; Wang, H. Mechanochemical Synthesis of Microscale Zero-Valent Iron/N-Doped Graphene-like Biochar Composite for Degradation of Tetracycline via Molecular O_2 Activation. *J. Colloid Interface Sci.* **2024**, *659*, 1015–1028. [CrossRef] [PubMed]
98. Yang, Y.; Xu, L.; Li, W.; Fan, W.; Song, S.; Yang, J. Adsorption and Degradation of Sulfadiazine over Nanoscale Zero-Valent Iron Encapsulated in Three-Dimensional Graphene Network through Oxygen-Driven Heterogeneous Fenton-like Reactions. *Appl. Catal. B* **2019**, *259*, 118057. [CrossRef]
99. Luo, Y.; Li, H.; Yang, H.; Yang, Z.; Li, C.; Liu, S.; Chen, Q.; Xu, W.; Zhang, W.; Tan, X. Critical Role of Dissolved Oxygen and Iron–Copper Synergy in Dual-Metal/Char Catalyst Systems. *Environ. Sci. Nano* **2024**, *11*, 2091–2102. [CrossRef]
100. Duan, L.; Liu, X.; Zhang, H.; Liu, F.; Liu, X.; Zhang, X.; Dong, L. A Novel Way for Hydroxyl Radicals Generation: Biochar-Supported Zero-Valent Iron Composite Activates Oxygen to Generate Hydroxyl Radicals. *J. Environ. Chem. Eng.* **2022**, *10*, 108132. [CrossRef]
101. Liu, Y.; Fan, Q.; Wang, J. Zn-Fe-CNTs Catalytic in Situ Generation of H_2O_2 for Fenton-like Degradation of Sulfamethoxazole. *J. Hazard. Mater.* **2018**, *342*, 166–176. [CrossRef] [PubMed]

102. Liu, Y.; Fan, Q.; Liu, Y.; Wang, J. Fenton-like Oxidation of 4-Chlorophenol Using H_2O_2 in Situ Generated by Zn-Fe-CNTs Composite. *J. Environ. Manag.* **2018**, *214*, 252–260. [CrossRef] [PubMed]
103. Lu, H.; Wen, C.; Gao, S.; Dong, Y.; Zhang, M.; Li, B.; Hu, W.; Dong, J. Incorporation of Nanoscale Zero-Valent Iron Particles in Monodisperse Mesoporous Silica Nanospheres: Characterization, Reactivity, Transport in Porous Media. *Colloids Surf. A* **2018**, *553*, 28–34. [CrossRef]
104. Yamaguchi, R.; Kurosu, S.; Suzuki, M.; Kawase, Y. Hydroxyl Radical Generation by Zero-Valent Iron/Cu (ZVI/Cu) Bimetallic Catalyst in Wastewater Treatment: Heterogeneous Fenton/Fenton-like Reactions by Fenton Reagents Formed in-Situ under Oxic Conditions. *Chem. Eng. J.* **2018**, *334*, 1537–1549. [CrossRef]
105. Chen, X.; Su, J.; Meng, Y.; Yu, M.; Zheng, M.; Sun, Y.; Xi, B. Oxygen Vacancy Promoted Heterogeneous Fenton-like Degradation of Sulfamethazine by Chlorine-Incorporated Micro Zero-Valent Iron. *Chem. Eng. J.* **2023**, *463*, 142360. [CrossRef]
106. Rajajayavel, S.R.C.; Ghoshal, S. Enhanced Reductive Dechlorination of Trichloroethylene by Sulfidated Nanoscale Zerovalent Iron. *Water Res.* **2015**, *78*, 144–153. [CrossRef]
107. Li, J.; Zhang, X.; Sun, Y.; Liang, L.; Pan, B.; Zhang, W.; Guan, X. Advances in Sulfidation of Zerovalent Iron for Water Decontamination. *Environ. Sci. Technol.* **2017**, *51*, 13533–13544. [CrossRef]
108. Xu, J.; Li, H.; Lowry, G.V. Sulfidized Nanoscale Zero-Valent Iron: Tuning the Properties of This Complex Material for Efficient Groundwater Remediation. *Acc. Mater. Res.* **2021**, *2*, 420–431. [CrossRef]
109. Garcia, A.N.; Zhang, Y.; Ghoshal, S.; He, F.; O'Carroll, D.M. Recent Advances in Sulfidated Zerovalent Iron for Contaminant Transformation. *Environ. Sci. Technol.* **2021**, *55*, 8464–8483. [CrossRef]
110. Song, S.; Su, Y.; Adeleye, A.S.; Zhang, Y.; Zhou, X. Optimal Design and Characterization of Sulfide-Modified Nanoscale Zerovalent Iron for Diclofenac Removal. *Appl. Catal. B* **2017**, *201*, 211–220. [CrossRef]
111. Ma, D.; Jia, S.; Zhao, D.; Lu, Z.; Yang, Z. O_2 Activation on the Outer Surface of Carbon Nanotubes Modified by Encapsulated Iron Clusters. *Appl. Surf. Sci.* **2014**, *300*, 91–97. [CrossRef]
112. Wang, B.; Zhou, X.; Wang, D.; Yin, J.-J.; Chen, H.; Gao, X.; Zhang, J.; Ibrahim, K.; Chai, Z.; Feng, W.; et al. Structure and Catalytic Activities of Ferrous Centers Confined on the Interface between Carbon Nanotubes and Humic Acid. *Nanoscale* **2015**, *7*, 2651–2658. [CrossRef] [PubMed]
113. Zhang, C.; Wang, D.; Liu, Q.; Tang, J. Ligand-Citric Acid Enhanced in-Situ ROS Generation by GBC@nZVI to Promote the Aerobic Degradation of Adsorbed 2,4-Dichlorophenol. *Chem. Eng. J.* **2023**, *477*, 147126. [CrossRef]
114. Rickard, D.; Luther, G.W. Chemistry of Iron Sulfides. *Chem. Rev.* **2007**, *107*, 514–562. [CrossRef] [PubMed]
115. Borda, M.J.; Elsetinow, A.R.; Schoonen, M.A.; Strongin, D.R. Pyrite-Induced Hydrogen Peroxide Formation as a Driving Force in the Evolution of Photosynthetic Organisms on an Early Earth. *Astrobiology* **2001**, *1*, 283–288. [CrossRef]
116. Nesbitt, H.W.; Bancroft, G.M.; Pratt, A.R.; Scaini, M.J. Sulfur and Iron Surface States on Fractured Pyrite Surfaces. *Am. Mineral.* **1998**, *83*, 1067–1076. [CrossRef]
117. Borda, M.J.; Elsetinow, A.R.; Strongin, D.R.; Schoonen, M.A. A Mechanism for the Production of Hydroxyl Radical at Surface Defect Sites on Pyrite. *Geochim. Cosmochim. Acta* **2003**, *67*, 935–939. [CrossRef]
118. Ling, C.; Liu, X.; Li, M.; Wang, X.; Shi, Y.; Qi, J.; Zhao, J.; Zhang, L. Sulphur Vacancy Derived Anaerobic Hydroxyl Radical Generation at the Pyrite-Water Interface: Pollutants Removal and Pyrite Self-Oxidation Behavior. *Appl. Catal. B* **2021**, *290*, 120051. [CrossRef]
119. Zhang, P.; Yuan, S.; Liao, P. Mechanisms of Hydroxyl Radical Production from Abiotic Oxidation of Pyrite under Acidic Conditions. *Geochim. Cosmochim. Acta* **2016**, *172*, 444–457. [CrossRef]
120. Hao, F.; Guo, W.; Lin, X.; Leng, Y.; Wang, A.; Yue, X.; Yan, L. Degradation of Acid Orange 7 in Aqueous Solution by Dioxygen Activation in a Pyrite/H_2O/O_2 System. *Environ. Sci. Pollut. Res.* **2014**, *21*, 6723–6728. [CrossRef]
121. Tan, M.; Zheng, X.; Yu, W.; Chen, B.; Chu, C. Facet-Dependent Productions of Reactive Oxygen Species from Pyrite Oxidation. *Environ. Sci. Technol.* **2024**, *58*, 432–439. [CrossRef] [PubMed]
122. Zhu, L.; Wang, H.; Sun, J.; Lu, L.; Li, S. Sulfur Vacancies in Pyrite Trigger the Path to Nonradical Singlet Oxygen and Spontaneous Sulfamethoxazole Degradation: Unveiling the Hidden Potential in Sediments. *Environ. Sci. Technol.* **2024**, *58*, 6753–6762. [CrossRef] [PubMed]
123. Gao, X.; Dai, C.; Tian, X.; Nie, Y.; Shi, J. Self-Acclimation Mechanism of Pyrite to Sulfamethoxazole Concentration in Terms of Degradation Behavior and Toxicity Effects Caused by Reactive Oxygen Species. *J. Hazard. Mater.* **2024**, *464*, 132962. [CrossRef] [PubMed]
124. Chen, X.; Xian, Z.; Gao, S.; Bai, L.; Liang, S.; Tian, H.; Wang, C.; Gu, C. Mechanistic Insights into Surface Catalytic Oxidation of Fluoroquinolone Antibiotics on Sediment Mackinawite. *Water Res.* **2023**, *232*, 119651. [CrossRef]
125. Cheng, D.; Liao, W.; Yuan, S. Effect of in Situ Generated Iron Oxyhydroxide Coatings on FeS Oxygenation and Resultant Hydroxyl Radical Production for Contaminant Degradation. *Chem. Eng. J.* **2020**, *394*, 124961. [CrossRef]
126. Cohn, C.A.; Mueller, S.; Wimmer, E.; Leifer, N.; Greenbaum, S.; Strongin, D.R.; Schoonen, M.A. Pyrite-Induced Hydroxyl Radical Formation and Its Effect on Nucleic Acids. *Geochem. Trans.* **2006**, *7*, 3. [CrossRef]
127. Kong, L.; Hu, X.; He, M. Mechanisms of Sb(III) Oxidation by Pyrite-Induced Hydroxyl Radicals and Hydrogen Peroxide. *Environ. Sci. Technol.* **2015**, *49*, 3499–3505. [CrossRef]
128. Song, B.; Zeng, Z.; Almatrafi, E.; Shen, M.; Xiong, W.; Zhou, C.; Wang, W.; Zeng, G.; Gong, J. Pyrite-Mediated Advanced Oxidation Processes: Applications, Mechanisms, and Enhancing Strategies. *Water Res.* **2022**, *211*, 118048. [CrossRef]

129. Schoonen, M.A.A.; Cohn, C.A.; Roemer, E.; Laffers, R.; Simon, S.R.; O'Riordan, T. Mineral-Induced Formation of Reactive Oxygen Species. *Rev. Mineral. Geochem.* **2006**, *64*, 179–221. [CrossRef]
130. Fan, D.; Lan, Y.; Tratnyek, P.G.; Johnson, R.L.; Filip, J.; O'Carroll, D.M.; Nunez Garcia, A.; Agrawal, A. Sulfidation of Iron-Based Materials: A Review of Processes and Implications for Water Treatment and Remediation. *Environ. Sci. Technol.* **2017**, *51*, 13070–13085. [CrossRef]
131. Dos Santos, E.C.; de Mendonça Silva, J.C.; Duarte, H.A. Pyrite Oxidation Mechanism by Oxygen in Aqueous Medium. *J. Phys. Chem. C* **2016**, *120*, 2760–2768. [CrossRef]
132. Usher, C.R.; Cleveland, C.A.; Strongin, D.R.; Schoonen, M.A. Origin of Oxygen in Sulfate during Pyrite Oxidation with Water and Dissolved Oxygen: An In Situ Horizontal Attenuated Total Reflectance Infrared Spectroscopy Isotope Study. *Environ. Sci. Technol.* **2004**, *38*, 5604–5606. [CrossRef] [PubMed]
133. Tang, B.; Liang, J.; Wen, Z.; Zhou, Y.; Yan, Z.; Zhou, Y.; He, P.; Gu, C.; Gan, M.; Zhu, J. Insight into the Crystal Facet-Dependent Cr(VI) Reduction: A Comparative Study of Pyrite {100} and {111} Facets. *J. Environ. Sci.* **2025**, *150*, 78–79. [CrossRef] [PubMed]
134. Zhou, S.; Gan, M.; Wang, X.; Zhang, Y.; Fang, Y.; Gu, G.; Wang, Y.; Qiu, G. ROS Formation Driven by Pyrite-Mediated Arsenopyrite Oxidation and Its Potential Role on Arsenic Transformation. *J. Hazard. Mater.* **2023**, *443*, 130151. [CrossRef] [PubMed]
135. Wehrli, B.; Sulzberger, B.; Stumm, W. Redox Processes Catalyzed by Hydrous Oxide Surfaces. *Chem. Geol.* **1989**, *78*, 167–179. [CrossRef]
136. Schoonen, M.A.A.; Harrington, A.D.; Laffers, R.; Strongin, D.R. Role of Hydrogen Peroxide and Hydroxyl Radical in Pyrite Oxidation by Molecular Oxygen. *Geochim. Cosmochim. Acta* **2010**, *74*, 4971–4987. [CrossRef]
137. Nicholson, R.V.; Gillham, R.W.; Reardon, E.J. Pyrite Oxidation in Carbonate-Buffered Solution: 2. Rate Control by Oxide Coatings. *Geochim. Cosmochim. Acta* **1990**, *54*, 395–402. [CrossRef]
138. Liu, R.; Dai, Y.; Feng, Y.; Sun, S.; Zhang, X.; An, C.; Zhao, S. Hydroxyl Radical Production by Abiotic Oxidation of Pyrite under Estuarine Conditions: The Effects of Aging, Seawater Anions and Illumination. *J. Environ. Sci.* **2024**, *135*, 715–727. [CrossRef]
139. Morgan, B.; Rate, A.W.; Burton, E.D. Water Chemistry and Nutrient Release during the Resuspension of FeS-Rich Sediments in a Eutrophic Estuarine System. *Sci. Total Environ.* **2012**, *432*, 47–56. [CrossRef]
140. Jeong, H.Y.; Han, Y.-S.; Park, S.W.; Hayes, K.F. Aerobic Oxidation of Mackinawite (FeS) and Its Environmental Implication for Arsenic Mobilization. *Geochim. Cosmochim. Acta* **2010**, *74*, 3182–3198. [CrossRef]
141. Jeong, H.Y.; Han, Y.-S.; Hayes, K.F. X-Ray Absorption and X-Ray Photoelectron Spectroscopic Study of Arsenic Mobilization during Mackinawite (FeS) Oxidation. *Environ. Sci. Technol.* **2010**, *44*, 955–961. [CrossRef] [PubMed]
142. Bi, Y.; Stylo, M.; Bernier-Latmani, R.; Hayes, K.F. Rapid Mobilization of Noncrystalline U(IV) Coupled with FeS Oxidation. *Environ. Sci. Technol.* **2016**, *50*, 1403–1411. [CrossRef] [PubMed]
143. Cheng, D.; Yuan, S.; Liao, P.; Zhang, P. Oxidizing Impact Induced by Mackinawite (FeS) Nanoparticles at Oxic Conditions Due to Production of Hydroxyl Radicals. *Environ. Sci. Technol.* **2016**, *50*, 11646–11653. [CrossRef] [PubMed]
144. Zhu, A.; Guo, Y.; Liu, G.; Song, M.; Liang, Y.; Cai, Y.; Yin, Y. Hydroxyl Radical Formation upon Dark Oxidation of Reduced Iron Minerals: Effects of Iron Species and Environmental Factors. *Chin. Chem. Lett.* **2019**, *30*, 2241–2244. [CrossRef]
145. Chiriţă, P.; Schlegel, M.L. Oxidative Dissolution of Iron Monosulfide (FeS) in Acidic Conditions: The Effect of Solid Pretreatment. *Int. J. Miner. Process.* **2015**, *135*, 57–64. [CrossRef]
146. Xiao, H.; Wang, Y.; Peng, H.; Zhu, Y.; Fang, D.; Wu, G.; Li, L.; Zeng, Z. Highly Efficient Degradation of Tetracycline Hydrochloride in Water by Oxygenation of Carboxymethyl Cellulose-Stabilized FeS Nanofluids. *Int. J. Environ. Res. Public Health* **2022**, *19*, 11447. [CrossRef]
147. He, J.; Miller, C.J.; Collins, R.; Wang, D.; Waite, T.D. Production of a Surface-Localized Oxidant during Oxygenation of Mackinawite (FeS). *Environ. Sci. Technol.* **2020**, *54*, 1167–1176. [CrossRef]
148. Meng, F.; Tong, H.; Feng, C.; Huang, Z.; Wu, P.; Zhou, J.; Hua, J.; Wu, F.; Liu, C. Structural Fe(II)-Induced Generation of Reactive Oxygen Species on Magnetite Surface for Aqueous As(III) Oxidation during Oxygen Activation. *Water Res.* **2024**, *252*, 121232. [CrossRef]
149. Fang, G.D.; Zhou, D.M.; Dionysiou, D.D. Superoxide Mediated Production of Hydroxyl Radicals by Magnetite Nanoparticles: Demonstration in the Degradation of 2-Chlorobiphenyl. *J. Hazard. Mater.* **2013**, *250–251*, 68–75. [CrossRef]
150. Fang, L.; Gao, B.; Li, F.; Liu, K.; Chi, J. The Nature of Metal Atoms Incorporated in Hematite Determines Oxygen Activation by Surface-Bound Fe(II) for As(III) Oxidation. *Water Res.* **2022**, *227*, 119351. [CrossRef]
151. Ding, Y.; Ruan, Y.; Zhu, L.; Tang, H. Efficient Oxidative Degradation of Chlorophenols by Using Magnetic Surface Carboxylated Cu^0/Fe_3O_4 Nanocomposites in a Wide pH Range. *J. Environ. Chem. Eng.* **2017**, *5*, 2681–2690. [CrossRef]
152. Yang, Z.; Gong, X.; Peng, L.; Yang, D.; Liu, Y. Zn^0-CNTs-Fe_3O_4 Catalytic in Situ Generation of H_2O_2 for Heterogeneous Fenton Degradation of 4-Chlorophenol. *Chemosphere* **2018**, *208*, 665–673. [CrossRef] [PubMed]
153. Wu, Y.; Tan, X.; Wang, E.; Zhao, J.; Ma, J. A Facile Strategy to Convert Low-Activity Commercial Iron Oxides and Cobalt Disulfide into High-Activity Fenton-like Catalysts: Spontaneous Generation of $^1O_2$ for Aqueous Decontamination. *Mater. Today Chem.* **2022**, *26*, 101106. [CrossRef]
154. Zhang, P.; Zhang, W.; Yu, H.; Chen, R.; Liu, Y.; Tian, Y.; Yuan, S. Fe(III) Oxyhydroxides Mediated Electron Transfer from Thiols to O_2 for Hydroxyl Radical Production. *Chem. Geol.* **2024**, *648*, 121962. [CrossRef]

155. Li, D.; Sun, J.; Fu, Y.; Hong, W.; Wang, H.; Yang, Q.; Wu, J.; Yang, S.; Xu, J.; Zhang, Y.; et al. Fluctuating Redox Conditions Accelerate the Electron Storage and Transfer in Magnetite and Production of Dark Hydroxyl Radicals. *Water Res.* **2024**, *248*, 120884. [CrossRef]
156. Hong, Z.; Li, F.; Borch, T.; Shi, Q.; Fang, L. Incorporation of Cu into Goethite Stimulates Oxygen Activation by Surface-Bound Fe(II) for Enhanced As(III) Oxidative Transformation. *Environ. Sci. Technol.* **2023**, *57*, 2162–2174. [CrossRef]
157. Amstaetter, K.; Borch, T.; Larese-Casanova, P.; Kappler, A. Redox Transformation of Arsenic by Fe(II)-Activated Goethite (α-FeOOH). *Environ. Sci. Technol.* **2010**, *44*, 102–108. [CrossRef]
158. Warr, L.N. Earth's Clay Mineral Inventory and Its Climate Interaction: A Quantitative Assessment. *Earth-Sci. Rev.* **2022**, *234*, 104198. [CrossRef]
159. Yuan, S.; Liu, X.; Liao, W.; Zhang, P.; Wang, X.; Tong, M. Mechanisms of Electron Transfer from Structrual Fe(II) in Reduced Nontronite to Oxygen for Production of Hydroxyl Radicals. *Geochim. Cosmochim. Acta* **2018**, *223*, 422–436. [CrossRef]
160. Liu, X.; Yuan, S.; Tong, M.; Liu, D. Oxidation of Trichloroethylene by the Hydroxyl Radicals Produced from Oxygenation of Reduced Nontronite. *Water Res.* **2017**, *113*, 72–79. [CrossRef]
161. Zeng, Q.; Dong, H.; Wang, X.; Yu, T.; Cui, W. Degradation of 1,4-Dioxane by Hydroxyl Radicals Produced from Clay Minerals. *J. Hazard. Mater.* **2017**, *331*, 88–98. [CrossRef] [PubMed]
162. Guo, J.; Zhang, X. The Impact of Electron Transfer between Structural Fe(II) and Aqueous Fe(III) on the Redox Transformation of Arsenic. *Environ. Earth Sci.* **2023**, *82*, 304. [CrossRef]
163. Yu, C.; Qian, A.; Lu, Y.; Liao, W.; Zhang, P.; Tong, M.; Dong, H.; Zeng, Q.; Yuan, S. Electron Transfer Processes Associated with Structural Fe in Clay Minerals. *Crit. Rev. Environ. Sci. Technol.* **2024**, *54*, 13–38. [CrossRef]
164. Zhang, P.; Liu, J.; Yu, H.; Cheng, D.; Liu, H.; Yuan, S. Kinetic Models for Hydroxyl Radical Production and Contaminant Removal during Soil/Sediment Oxygenation. *Water Res.* **2023**, *240*, 120071. [CrossRef]
165. Yu, C.; Ji, W.; Li, X.; Yuan, S.; Zhang, P.; Pu, S. Critical Role of Mineral Fe(IV) Formation in Low Hydroxyl Radical Yields during Fe(II)-Bearing Clay Mineral Oxygenation. *Environ. Sci. Technol.* **2024**, *58*, 9669–9678. [CrossRef]
166. Rothwell, K.A.; Pentrak, M.P.; Pentrak, L.A.; Stucki, J.W.; Neumann, A. Reduction Pathway-Dependent Formation of Reactive Fe(II) Sites in Clay Minerals. *Environ. Sci. Technol.* **2023**, *57*, 10231–10241. [CrossRef]
167. Zhao, G.; Tan, M.; Wu, B.; Zheng, X.; Xiong, R.; Chen, B.; Kappler, A.; Chu, C. Redox Oscillations Activate Thermodynamically Stable Iron Minerals for Enhanced Reactive Oxygen Species Production. *Environ. Sci. Technol.* **2023**, *57*, 8628–8637. [CrossRef]
168. Xie, W.; Yuan, S.; Tong, M.; Ma, S.; Liao, W.; Zhang, N.; Chen, C. Contaminant Degradation by •OH during Sediment Oxygenation: Dependence on Fe(II) Species. *Environ. Sci. Technol.* **2020**, *54*, 2975–2984. [CrossRef]
169. Chen, N.; Huang, D.; Liu, G.; Chu, L.; Fang, G.; Zhu, C.; Zhou, D.; Gao, J. Active Iron Species Driven Hydroxyl Radicals Formation in Oxygenation of Different Paddy Soils: Implications to Polycyclic Aromatic Hydrocarbons Degradation. *Water Res.* **2021**, *203*, 117484. [CrossRef]
170. Liu, J.; Zhu, C.; Zhu, F.; Sun, H.; Wang, J.; Fang, G.; Zhou, D. Strong Substance Exchange at Paddy Soil-Water Interface Promotes Nonphotochemical Formation of Reactive Oxygen Species in Overlying Water. *Environ. Sci. Technol.* **2024**, *58*, 7403–7414. [CrossRef]
171. Huang, D.; Chen, N.; Zhu, C.; Sun, H.; Fang, G.; Zhou, D. Dynamic Production of Hydroxyl Radicals during the Flooding–Drainage Process of Paddy Soil: An In Situ Column Study. *Environ. Sci. Technol.* **2023**, *57*, 16340–16347. [CrossRef] [PubMed]
172. Feng, Y.; Dai, Y.; Liu, R.; Zhao, D.; Sun, S.; Xu, X.; Chen, Y.; Yuan, X.; Zhang, B.; Zhao, S. Production and Prediction of Hydroxyl Radicals in Distinct Redox-Fluctuation Zones of the Yellow River Estuary. *J. Hazard. Mater.* **2024**, *469*, 133980. [CrossRef] [PubMed]
173. Liu, F.; Wang, Z.; Liu, J.; Latif, J.; Qin, J.; Yang, H.; Jiang, W.; Deng, Y.; Yang, K.; Ni, Z.; et al. Seasonal and Spatial Fluctuations of Reactive Oxygen Species in Riparian Soils and Their Contributions on Organic Carbon Mineralization. *Environ. Sci. Technol.* **2024**, *58*, 7066–7077. [CrossRef] [PubMed]
174. Zhao, G.; Wu, B.; Zheng, X.; Chen, B.; Kappler, A.; Chu, C. Tide-Triggered Production of Reactive Oxygen Species in Coastal Soils. *Environ. Sci. Technol.* **2022**, *56*, 11888–11896. [CrossRef]
175. Chen, N.; Fu, Q.; Wu, T.; Cui, P.; Fang, G.; Liu, C.; Chen, C.; Liu, G.; Wang, W.; Wang, D.; et al. Active Iron Phases Regulate the Abiotic Transformation of Organic Carbon during Redox Fluctuation Cycles of Paddy Soil. *Environ. Sci. Technol.* **2021**, *55*, 14281–14293. [CrossRef]
176. Lu, J.; Yu, P.; Zhang, J.; Guo, Z.; Li, Y.; Wang, S.; Hu, Z. Biotic/Abiotic Transformation Mechanisms of Phenanthrene in Iron-Rich Constructed Wetland under Redox Fluctuation. *Water Res.* **2024**, *261*, 122033. [CrossRef]
177. Sun, Z.; Huang, M.; Liu, C.; Fang, G.; Chen, N.; Zhou, D.; Gao, J. The Formation of •OH with Fe-Bearing Smectite Clays and Low-Molecular-Weight Thiols: Implication of As(III) Removal. *Water Res.* **2020**, *174*, 115631. [CrossRef]
178. Schaefer, C.E.; Ho, P.; Berns, E.; Werth, C. Mechanisms for Abiotic Dechlorination of Trichloroethene by Ferrous Minerals under Oxic and Anoxic Conditions in Natural Sediments. *Environ. Sci. Technol.* **2018**, *52*, 13747–13755. [CrossRef]
179. Kasozi, N.; Tandlich, R.; Fick, M.; Kaiser, H.; Wilhelmi, B. Iron Supplementation and Management in Aquaponic Systems: A Review. *Aquac. Rep.* **2019**, *15*, 100221. [CrossRef]
180. Ren, H.; He, F.; Liu, S.; Li, T.; Zhou, R. Enhancing Fenton–like Process at Neutral pH by Fe(III)–GLDA Complexation for the Oxidation Removal of Organic Pollutants. *J. Hazard. Mater.* **2021**, *416*, 126077. [CrossRef]

181. Jones, A.M.; Griffin, P.J.; Waite, T.D. Ferrous Iron Oxidation by Molecular Oxygen under Acidic Conditions: The Effect of Citrate, EDTA and Fulvic Acid. *Geochim. Cosmochim. Acta* **2015**, *160*, 117–131. [CrossRef]
182. Zhang, P.; Yuan, S. Production of Hydroxyl Radicals from Abiotic Oxidation of Pyrite by Oxygen under Circumneutral Conditions in the Presence of Low-Molecular-Weight Organic Acids. *Geochim. Cosmochim. Acta* **2017**, *218*, 153–166. [CrossRef]
183. Wang, Y.; Kong, L.; He, M.; Lin, C.; Ouyang, W.; Liu, X.; Peng, X. Mechanistic Insights into Sb(III) and Fe(II) Co-Oxidation by Oxygen and Hydrogen Peroxide: Dominant Reactive Oxygen Species and Roles of Organic Ligands. *Water Res.* **2023**, *242*, 120296. [CrossRef] [PubMed]
184. Wang, Q.; Pan, Y.; Fu, W.; Wu, H.; Zhou, M.; Zhang, Y. Aminopolycarboxylic Acids Modified Oxygen Reduction by Zero Valent Iron: Proton-Coupled Electron Transfer, Role of Iron Ion and Reactive Oxidant Generation. *J. Hazard. Mater.* **2022**, *430*, 128402. [CrossRef] [PubMed]
185. Checa-Fernandez, A.; Santos, A.; Romero, A.; Dominguez, C.M. Application of Chelating Agents to Enhance Fenton Process in Soil Remediation: A Review. *Catalysts* **2021**, *11*, 722. [CrossRef]
186. De Laat, J.; Dao, Y.H.; Hamdi El Najjar, N.; Daou, C. Effect of Some Parameters on the Rate of the Catalysed Decomposition of Hydrogen Peroxide by Iron(III)-Nitrilotriacetate in Water. *Water Res.* **2011**, *45*, 5654–5664. [CrossRef]
187. Zhang, Y.; Klamerth, N.; Messele, S.A.; Chelme-Ayala, P.; Gamal El-Din, M. Kinetics Study on the Degradation of a Model Naphthenic Acid by Ethylenediamine-N,N′-Disuccinic Acid-Modified Fenton Process. *J. Hazard. Mater.* **2016**, *318*, 371–378. [CrossRef]
188. Zhou, H.; Sun, Q.; Wang, X.; Wang, L.; Chen, J.; Zhang, J.; Lu, X. Removal of 2,4-Dichlorophenol from Contaminated Soil by a Heterogeneous ZVI/EDTA/Air Fenton-like System. *Sep. Purif. Technol.* **2014**, *132*, 346–353. [CrossRef]
189. Huang, M.; Fang, G.; Chen, N.; Zhou, D. Hydroxylamine Promoted Hydroxyl Radical Production and Organic Contaminants Degradation in Oxygenation of Pyrite. *J. Hazard. Mater.* **2022**, *429*, 128380. [CrossRef]
190. Li, K.; Ma, S.; Zou, C.; Latif, J.; Jiang, Y.; Ni, Z.; Shen, S.; Feng, J.; Jia, H. Unrecognized Role of Organic Acid in Natural Attenuation of Pollutants by Mackinawite (FeS): The Significance of Carbon-Center Free Radicals. *Environ. Sci. Technol.* **2023**, *57*, 20871–20880. [CrossRef]
191. Yu, C.; Zhang, Y.; Lu, Y.; Qian, A.; Zhang, P.; Cui, Y.; Yuan, S. Mechanistic Insight into Humic Acid-Enhanced Hydroxyl Radical Production from Fe(II)-Bearing Clay Mineral Oxygenation. *Environ. Sci. Technol.* **2021**, *55*, 13366–13375. [PubMed]
192. Cheng, D.; Ding, H.; Tan, Y.; Yang, D.; Pan, Y.; Liao, W.; He, F. Dramatically Enhanced Phenol Degradation upon FeS Oxygenation by Low-Molecular-Weight Organic Acids. *J. Hazard. Mater.* **2023**, *459*, 132260. [CrossRef] [PubMed]
193. Mohamad, N.D.; Zaki, Z.M.; Amir, A. Mechanisms of Enhanced Oxidative Degradation of Tetrachloroethene by Nano-Magnetite Catalysed with Glutathione. *Chem. Eng. J.* **2020**, *393*, 124760. [CrossRef]
194. Zhao, S.; Liu, Z.; Zhang, R.; Liu, J.; Liu, J.; Dai, Y.; Zhang, C.; Jia, H. Interfacial Reaction between Organic Acids and Iron-Containing Clay Minerals: Hydroxyl Radical Generation and Phenolic Compounds Degradation. *Sci. Total Environ.* **2021**, *783*, 147025. [CrossRef]
195. Qin, Y.; Song, F.; Ai, Z.; Zhang, P.; Zhang, L. Protocatechuic Acid Promoted Alachlor Degradation in Fe(III)/H_2O_2 Fenton System. *Environ. Sci. Technol.* **2015**, *49*, 7948–7956. [CrossRef]
196. Xie, L.; Shang, C. Role of Humic Acid and Quinone Model Compounds in Bromate Reduction by Zerovalent Iron. *Environ. Sci. Technol.* **2005**, *39*, 1092–1100. [CrossRef]
197. Liu, Z.; Fu, J.; Liu, A.; Zhang, W.-X. Influence of Natural Organic Matter on Nanoscale Zero-Valent Iron for Contaminants Removal in Water: A Critical Review. *Chem. Eng. J.* **2024**, *488*, 150836. [CrossRef]
198. Liu, Y.; Zhao, W.; Zhang, P.; Fu, Q.; Yu, C.; Yuan, S. Asymmetrical Changes of Electron-Donating and Electron-Accepting Capacities of Natural Organic Matter during Its Interaction with Fe Oxyhydroxides. *Chem. Geol.* **2024**, *661*, 122189. [CrossRef]
199. Dong, H.; Zeng, Q.; Sheng, Y.; Chen, C.; Yu, G.; Kappler, A. Coupled Iron Cycling and Organic Matter Transformation across Redox Interfaces. *Nat. Rev. Earth Environ.* **2023**, *4*, 659–673. [CrossRef]
200. Wolf, M.; Kappler, A.; Jiang, J.; Meckenstock, R.U. Effects of Humic Substances and Quinones at Low Concentrations on Ferrihydrite Reduction by *Geobacter Metallireducens*. *Environ. Sci. Technol.* **2009**, *43*, 5679–5685. [CrossRef]
201. Zeng, Q.; Wang, X.; Liu, X.; Huang, L.; Hu, J.; Chu, R.; Tolic, N.; Dong, H. Mutual Interactions between Reduced Fe-Bearing Clay Minerals and Humic Acids under Dark, Oxygenated Conditions: Hydroxyl Radical Generation and Humic Acid Transformation. *Environ. Sci. Technol.* **2020**, *54*, 15013–15023. [CrossRef] [PubMed]
202. Yuan, Y.; Wei, X.; Zhu, M.; Cai, Y.; Wang, Y.; Dang, Z.; Yin, H. Unravelling the Removal Mechanisms of Trivalent Arsenic by Sulfidated Nanoscale Zero-Valent Iron: The Crucial Role of Reactive Oxygen Species and the Multiple Effects of Citric Acid. *Sci. Total Environ.* **2024**, *916*, 170275. [CrossRef]
203. Cheng, X.; Liang, L.; Ye, J.; Li, N.; Yan, B.; Chen, G. Influence and Mechanism of Water Matrices on H_2O_2-Based Fenton-like Oxidation Processes: A Review. *Sci. Total Environ.* **2023**, *888*, 164086. [CrossRef] [PubMed]
204. Fang, L.; Xu, L.; Deng, J.; Gao, S.; Huang, L.Z. Induced Generation of Hydroxyl Radicals from Green Rust under Oxic Conditions by Iron-Phosphate Complexes. *Chem. Eng. J.* **2021**, *414*, 128780. [CrossRef]
205. Mu, Y.; Ai, Z.; Zhang, L. Phosphate Shifted Oxygen Reduction Pathway on Fe@Fe_2O_3 Core–Shell Nanowires for Enhanced Reactive Oxygen Species Generation and Aerobic 4-Chlorophenol Degradation. *Environ. Sci. Technol.* **2017**, *51*, 8101–8109. [CrossRef]

206. Wang, J.; Wang, S. Effect of Inorganic Anions on the Performance of Advanced Oxidation Processes for Degradation of Organic Contaminants. *Chem. Eng. J.* **2021**, *411*, 128392. [CrossRef]
207. Hug, S.J.; Leupin, O. Iron-Catalyzed Oxidation of Arsenic(III) by Oxygen and by Hydrogen Peroxide: pH-Dependent Formation of Oxidants in the Fenton Reaction. *Environ. Sci. Technol.* **2003**, *37*, 2734–2742. [CrossRef]
208. Ernstsen, V.; Gates, W.P.; Stucki, J.W. Microbial Reduction of Structural Iron in Clays—A Renewable Source of Reduction Capacity. *J. Environ. Qual.* **1998**, *27*, 761–766. [CrossRef]
209. Chen, R.; Liu, H.; Zhang, P.; Ma, J.; Jin, M. Co-Response of Fe-Reducing/Oxidizing Bacteria and Fe Species to the Dynamic Redox Cycles of Natural Sediment. *Sci. Total Environ.* **2022**, *815*, 152953. [CrossRef]
210. Han, R.; Lv, J.; Huang, Z.; Zhang, S.; Zhang, S. Pathway for the Production of Hydroxyl Radicals during the Microbially Mediated Redox Transformation of Iron (Oxyhydr)oxides. *Environ. Sci. Technol.* **2020**, *54*, 902–910. [CrossRef]
211. Dong, H.; Coffin, E.S.; Sheng, Y.; Duley, M.L.; Khalifa, Y.M. Microbial Reduction of Fe(III) in Nontronite: Role of Biochar as a Redox Mediator. *Geochim. Cosmochim. Acta* **2023**, *345*, 102–116. [CrossRef]
212. Pi, K.; Markelova, E.; Zhang, P.; Van Cappellen, P. Arsenic Oxidation by Flavin-Derived Reactive Species under Oxic and Anoxic Conditions: Oxidant Formation and pH Dependence. *Environ. Sci. Technol.* **2019**, *53*, 10897–10905. [CrossRef]
213. Dong, H.; Kukkadapu, R.K.; Fredrickson, J.K.; Zachara, J.M.; Kennedy, D.W.; Kostandarithes, H.M. Microbial Reduction of Structural Fe(III) in Illite and Goethite. *Environ. Sci. Technol.* **2003**, *37*, 1268–1276. [CrossRef]
214. You, X.; Liu, S.; Berns-Herrboldt, E.C.; Dai, C.; Werth, C.J. Kinetics of Hydroxyl Radical Production from Oxygenation of Reduced Iron Minerals and Their Reactivity with Trichloroethene: Effects of Iron Amounts, Iron Species, and Sulfate Reducing Bacteria. *Environ. Sci. Technol.* **2023**, *57*, 4892–4904. [CrossRef] [PubMed]
215. Edwards, K.J.; Bach, W.; McCollom, T.M.; Rogers, D.R. Neutrophilic Iron-Oxidizing Bacteria in the Ocean: Their Habitats, Diversity, and Roles in Mineral Deposition, Rock Alteration, and Biomass Production in the Deep-Sea. *Geomicrobiol. J.* **2004**, *21*, 393–404. [CrossRef]
216. Jones, S.; Santini, J.M. Mechanisms of Bioleaching: Iron and Sulfur Oxidation by Acidophilic Microorganisms. *Essays Biochem.* **2023**, *67*, 685–699.
217. Xie, Z.H.; He, C.S.; Pei, D.N.; Dong, Y.; Yang, S.R.; Xiong, Z.; Zhou, P.; Pan, Z.C.; Yao, G.; Lai, B. Review of Characteristics, Generation Pathways and Detection Methods of Singlet Oxygen Generated in Advanced Oxidation Processes (AOPs). *Chem. Eng. J.* **2023**, *468*, 143778. [CrossRef]
218. Sun, Y.; Han, W.; Zhang, F.; Li, H.; Zhang, Z.; Zhang, X.; Shen, B.; Guo, S.Q.; Ma, T. Dual Defect Regulation of BiOCl Halogen Layer Enables Photocatalytic O_2 Activation into Singlet Oxygen for Refractory Aromatic Pollutant Removal. *Appl. Catal. B* **2024**, *345*, 123689. [CrossRef]
219. Yu, W.; Zheng, X.; Tan, M.; Wang, J.; Wu, B.; Ma, J.; Pan, Y.; Chen, B.; Chu, C. Field Quantification of Hydroxyl Radicals by Flow-Injection Chemiluminescence Analysis with a Portable Device. *Environ. Sci. Technol.* **2024**, *58*, 2808–2816. [CrossRef]
220. Bai, L.; Jiang, Y.; Xia, D.; Wei, Z.; Spinney, R.; Dionysiou, D.D.; Minakata, D.; Xiao, R.; Xie, H.-B.; Chai, L. Mechanistic Understanding of Superoxide Radical-Mediated Degradation of Perfluorocarboxylic Acids. *Environ. Sci. Technol.* **2022**, *56*, 624–633. [CrossRef]
221. Wang, R.; Chen, H.; He, Z.; Zhang, S.; Wang, K.; Ren, N.; Ho, S.-H. Discovery of an End-to-End Pattern for Contaminant-Oriented Advanced Oxidation Processes Catalyzed by Biochar with Explainable Machine Learning. *Environ. Sci. Technol.* **2024**, *58*, 16867–16876. [CrossRef] [PubMed]
222. Cao, X.; Huang, J.; Tian, Y.; Hu, Z.; Luo, Z.; Wang, J.; Guo, Y. Machine-Learning-Assisted Descriptors Identification for Indoor Formaldehyde Oxidation Catalysts. *Environ. Sci. Technol.* **2024**, *58*, 8372–8379. [CrossRef] [PubMed]
223. Hjorth, R.; Coutris, C.; Nguyen, N.H.A.; Sevcu, A.; Gallego-Urrea, J.A.; Baun, A.; Joner, E.J. Ecotoxicity Testing and Environmental Risk Assessment of Iron Nanomaterials for Sub-Surface Remediation—Recommendations from the FP7 Project NanoRem. *Chemosphere* **2017**, *182*, 525–531. [CrossRef] [PubMed]
224. Zhang, S.; Yi, K.; Chen, A.; Shao, J.; Peng, L.; Luo, S. Toxicity of Zero-Valent Iron Nanoparticles to Soil Organisms and the Associated Defense Mechanisms: A Review. *Ecotoxicology* **2022**, *31*, 873–883. [CrossRef]
225. Zeng, G.; He, Y.; Wang, F.; Luo, H.; Liang, D.; Wang, J.; Huang, J.; Yu, C.; Jin, L.; Sun, D. Toxicity of Nanoscale Zero-Valent Iron to Soil Microorganisms and Related Defense Mechanisms: A Review. *Toxics* **2023**, *11*, 514. [CrossRef]

Disclaimer/Publisher's Note: The statements, opinions and data contained in all publications are solely those of the individual author(s) and contributor(s) and not of MDPI and/or the editor(s). MDPI and/or the editor(s) disclaim responsibility for any injury to people or property resulting from any ideas, methods, instructions or products referred to in the content.

toxics

MDPI

Article

Density Functional Theory Insight in Photocatalytic Degradation of Dichlorvos Using Covalent Triazine Frameworks Modified by Various Oxygen-Containing Acid Groups

Shouxi Yu and Zhongliao Wang *

Anhui Province Industrial Generic Technology Research Center for Alumics Materials, Anhui Province Key Laboratory of Pollutant Sensitive Materials and Environmental Remediation, School of Physics and Electronic Information, Huaibei Normal University, Huaibei 235000, China; chnu19810901912@163.com
* Correspondence: wangzl@chnu.edu.cn

Abstract: Dichlorvos (2,2-dichlorovinyl dimethyl phosphate, DDVP) is a highly toxic organophosphorus insecticide, and its persistence in air, water, and soil poses potential threats to human health and ecosystems. Covalent triazine frameworks (CTFs), with their sufficient visible-light harvesting capacity, ameliorated charge separation, and exceptional redox ability, have emerged as promising candidates for the photocatalytic degradation of DDVP. Nevertheless, pure CTFs lack effective oxidative active sites, resulting in elevated reaction energy barriers during the photodegradation of DDVP. In this work, density functional theory (DFT) calculations were employed to investigate the impact of various oxygen-containing acid groups (-COOH, -HSO_3, -H_2PO_3) on DDVP photodegradation performance. First, simulations of the structure and optical properties of modified CTFs reveal that oxygen-containing acid groups induce surface distortion and result in a redshift in the absorption edge. Subsequently, analysis of the density of states, frontier molecular orbitals, surface electrostatic potential, work function, and dipole moment demonstrates that oxygen-containing acid groups enhance CTF polarization, facilitate charge separation, and ameliorate their oxidative capability. Additionally, the free-energy diagram of DDVP degradation uncovers that oxygen-containing acid groups lower the energy barrier by elevating the adsorption and activation capability of DDVP. Notably, -H_2PO_3 presents optimal potential for the photodegradation of DDVP by unique electronic structure and activation capability. This work offers a valuable reference for the development of oxygen-containing acid CTF-based photocatalysts applied in degrading toxic organophosphate pesticides.

Keywords: organophosphate pesticide; photocatalytic degradation; covalent triazine frameworks; oxygen-containing acid groups; DFT calculation

1. Introduction

Dichlorvos (2,2-dichlorovinyl dimethyl phosphate, DDVP) is a highly effective organophosphorus insecticide extensively applied in agriculture to control various pests. However, due to its high toxicity and potential risks to ecosystems and human health, it has emerged as a critical environmental pollutant of concern. Current remediation strategies for DDVP include chemical degradation, microbial degradation [1], and environmental degradation. However, high costs, inefficiency, or dependency on environmental conditions limit their practical application [2]. In contrast, photocatalysis offers an eco-friendly, efficient, and economical channel for the sustainable degradation of DDVP. Traditional catalysts, such as ZnO [3], $BiVO_4$ [4], g-C_3N_4 [5], WO_3 [6], MnO_2 [7], et al., face challenges in effectively degrading DDVP due to severe electron-hole recombination, inefficient light harvesting, complex surface structure, and ambiguous active sites. Covalent triazine frameworks (CTFs) comprise donor-acceptor groups that facilitate fast charge separation, rendering them promising candidates for degrading DDVP [8–10]. Additionally, CTFs possess a relatively simple and well-defined structure with clearly identifiable active sites, providing an ideal platform for theoretical studies and mechanistic exploration [11,12]. However, the insufficient active sites and weak affinity capability lead to difficulty in the adsorption and activation of DDVP and further trigger a high degradation energy barrier [13]. Therefore, endowing CTFs with abundant active sites by tailored modification remains imperative.

Photodegradation reaction primarily relies on oxidation processes, where the oxidation capacity of active sites directly influences the degradation efficiency of the target pollutants [14]. Therefore, the development of active sites with high oxidation capabilities is crucial for enhancing photodegradation performance. However, oxidation sites are susceptible to photo-corrosion during photodegradation reactions, which can result in structural damage and performance deterioration. To achieve more efficient and durable photodegradation, it is imperative to enhance the oxidation capacity and the chemical stability of these active sites [15]. Due to their robust antioxidant properties, oxygen-containing acid functional groups exhibit distinct advantages for DDVP photodegradation, particularly by mitigating photo-corrosion and self-degradation under light irradiation. During the process of photodegradation, when a catalyst absorbs light energy, it triggers the excitation of electrons, causing them to transition from the valence band to the conduction band. Simultaneously, this process leads to the accumulation of photogenerated holes within the valence band. Owing to their greater electronegativity, photoexcited holes tend to accumulate in the oxygen-containing acid groups. Therefore, oxygen-containing acid groups can function as active sites for DDVP photodegradation. In addition, the presence of oxygen-rich acid groups induces an asymmetrical charge distribution and a robust dipole moment, thereby expediting the separation of electron-hole pairs. Consequently, the incorporation of appropriate oxygen-containing acid functional groups into CTFs is crucial for enhancing the photocatalytic degradation efficiency of DDVP [16].

Therefore, several common and stable oxygen-containing acid functional groups, such as COOH (carboxyl), HSO_3 (sulfonic), and H_2PO_3 (phosphoric) groups, were selected for the modification of CTFs. The X-ray diffraction (XRD), Fourier-transform infrared (FTIR) spectroscopy, and Raman spectra were simulated by theoretical calculations to examine the changes in structural and optical properties induced by oxygen-containing acid functional groups. Additionally, the influence of these functional groups on the light absorption properties and electronic structure of CTF was examined through ultraviolet-visible (UV–Vis) spectroscopy and density of states (DOS) calculations. Calculations of the dipole moment and surface electrostatic potential were conducted to further investigate charge separation efficiency and the distribution of active sites. Ultimately, the rate-determining step and reaction barriers for DDVP degradation were analyzed by calculating the reaction-free energy profile. These theoretical studies elucidated the reaction mechanism and underscored the potential advantages of oxygen-containing acid functional CTFs in DDVP photodegradation.

2. Computational Details

The analysis of spectroscopic properties, including vibrational modes and high-energy states, was performed using CP2K (version 2024.2), applying the PBE functional to ensure a consistent description of the systems. Electronic structure calculations were conducted within the unrestricted Kohn-Sham formalism, utilizing the Gaussian and plane waves (GPW) approach, with Goedecker-Teter-Hutter (GTH) pseudopotentials and a TZV2P-MOLOPT-GTH basis set for all elements. Computational parameters included a plane-wave cutoff of 410 Ry, an SCF convergence criterion of 5×10^{-6} hartree, and force convergence set to 5×10^{-4} Bohr per hartree. A vacuum layer of 20 Å was introduced to isolate the periodic slabs and minimize surface interactions. For capturing Van der Waals (vdW) forces, the DFT-D3 method with Grimme's zero-damping function was applied. The Gibbs free energy changes were computed using the equation $\Delta G = \Delta E + \Delta E_{ZPE} - T\Delta S$, where ΔE represents the electronic energy variation, ΔE_{ZPE} is the change in zero-point energy, and ΔS accounts for the temperature-dependent entropy variation. Zero-point energy corrections and vibrational frequencies, alongside entropy calculations for gaseous molecules, were evaluated using the Shermo software (version 2.6) [17]. Excited-state analysis and spectral studies were performed using Multiwfn software (version 3.8(dev)) [18].

3. Results and Discussion

Covalent triazine frameworks, characterized by their distinctive alternating arrangement of triazine (TR) and pyridine rings (Figure 1a), have emerged as a hotspot due to their versatile structural and functional properties. Therefore, single-layer CTFs incorporating various acidic functional groups (-COOH, -HSO_3, and -H_2PO_3) were engineered, which were named CTF-COOH, CTF-HSO_3, and CTF-H_2PO_3. Notably, the acidic functional groups are directly linked to the TR. The optimized structure illustrates that attaching different acidic functional groups to CTF markedly decreases its pore size. As depicted in Figure 1a–h, pristine CTF possesses the largest pore size of 7.64 Å. Upon functionalization, the pore sizes of CTF-COOH, CTF-HSO_3, and CTF-H_2PO_3 are reduced to 7.22 Å, 5.87 Å, and 5.77 Å, respectively, indicating that the incorporation of acidic groups effectively modifies the pore dimensions of CTF. Simulated X-ray diffraction (XRD) results indicate that the characteristic diffraction peaks of CTFs primarily fall within the 5–15° range (Figure 1i). Specifically, pristine CTF shows five distinct peaks at 5.89°, 7.00°, 9.15°, 11.79°, and 13.73°, which are attributed to the (001), (010), (101), (110), and (111) crystal planes, respectively. Notably, the XRD patterns of the functionalized derivatives, CTF-COOH, CTF-HSO_3, and CTF-H_2PO_3, exhibit the same diffraction peaks as the pristine CTF, suggesting that the introduction of acidic functional groups does not significantly affect the lattice parameters. This observation strongly suggests that the ordered structure of the CTF framework is well preserved after functionalization. Further analysis of the simulated structures for CTF-COOH, CTF-HSO_3, and CTF-H_2PO_3 reveals that their lattice parameters remain largely unchanged compared to pristine CTF. The lattice parameters for CTF-COOH were determined to be a = 14.56 Å, b = 14.55 Å, c = 15.00 Å; for CTF-HSO_3, a = b = 14.58 Å, c = 15.00 Å; and for CTF-H_2PO_3, a = 14.60 Å, b = 14.62 Å, c = 15.00 Å. These values are nearly identical to the lattice parameters of pristine CTF (a = b = 14.57 Å, c = 15.00 Å). The results suggest that the incorporation of functional groups effectively preserves the structural integrity of the CTF framework. This consistent crystalline arrangement highlights the robust nature of the CTF material, demonstrating its capacity to accommodate diverse functional groups while maintaining its ordered crystal structure and overall stability.

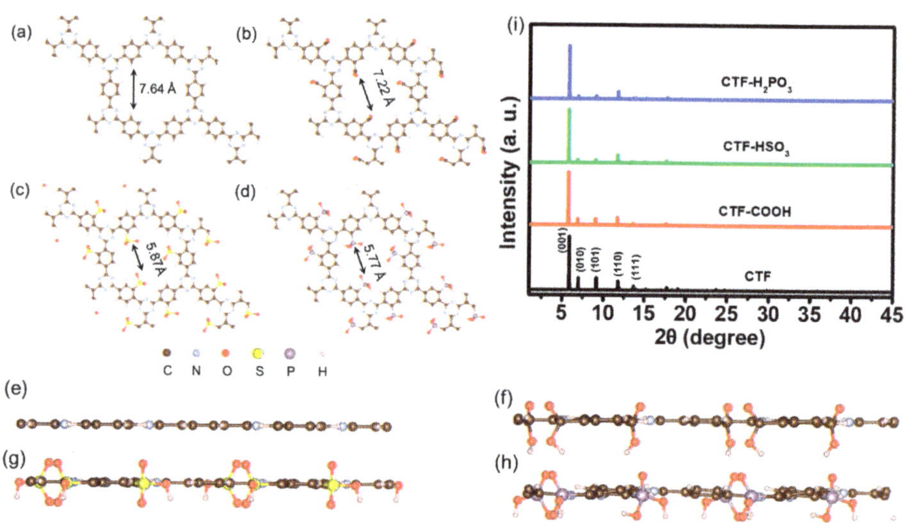

Figure 1. Top and side perspective of crystal structures: (**a**,**e**) CTF, (**b**,**f**) CTF-COOH, (**c**,**g**) CTF-HSO_3, and (**d**,**h**) CTF-H_2PO_3. Simulated (**i**) XRD pattern.

Infrared spectroscopy identifies vibrational frequencies and related parameters, offering insights into molecular interactions with their functional groups. As illustrated in Figure 2a, the infrared spectra of CTF, CTF-COOH, CTF-HSO_3, and CTF-H_2PO_3 were simulated within the 500–4000 cm^{-1} range. The CTF exhibits prominent infrared absorption peaks at 1412 and 1504 cm^{-1}. For the functionalized variants, CTF-COOH shows peaks at 1407 and 1495 cm^{-1}, CTF-HSO_3 at 1416 and 1496 cm^{-1}, and CTF-H_2PO_3 at 1415 and 1487 cm^{-1}, corresponding to the characteristic infrared absorption of the TR units [11]. Interestingly, further analysis of the spectra reveals additional peaks that are specific to the functional groups introduced. For instance, CTF-COOH displays a distinct absorption peak at 1272 cm^{-1}, which is absent in the pure CTF structure. This peak corresponds to the C-O stretch in the -COOH group, confirming the successful incorporation of the -COOH functional group into the CTF framework [19]. Similarly, CTF-HSO_3 presents a peak at 1229 cm^{-1}, attributable to the S=O stretching vibration in the -HSO_3 group [20], indicating its presence within the CTF structure. In addition, CTF-H_2PO_3 shows a characteristic peak at 944 cm^{-1}, corresponding to the P-O stretching vibration in the group [21], further corroborating the functionalization of CTF with this group. To further elucidate the crystal structure of CTF, Raman spectra (Figure 2b) simulations were conducted. The results revealed that the benzene rings within CTF exhibit four characteristic absorption peaks at 1453, 1556, 1623, and 1656 cm^{-1}, corresponding to the stretching vibrations of the benzene ring [22]. These peaks were similarly observed in the CTF-COOH, CTF-HSO_3, and CTF-H_2PO_3 crystals, though with a redshift. This redshift can be attributed to the influence of the functional groups. Furthermore, the C-H stretching peaks for CTF-COOH, CTF-HSO_3, and CTF-H_2PO_3 were observed at 1002, 991, and 991 cm^{-1}, respectively, indicating a blue shift relative to the peak at 988 cm^{-1} in the CTF sample. This blueshift is likely due to the interaction between the attached functional groups and the CTF framework, suggesting that the incorporation of functional groups induces changes in the electronic environment and alters the C-H bond vibrations within the material. This observation further emphasizes the impact of functionalization on the vibrational properties of the CTF structure.

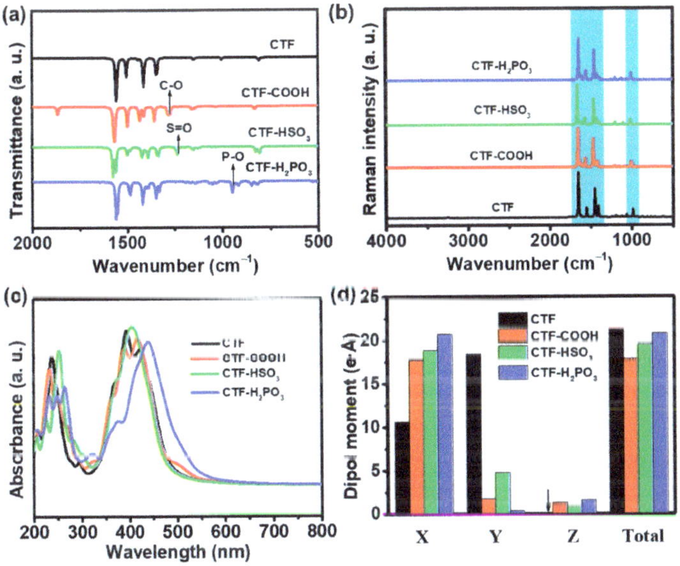

Figure 2. (**a**) IR spectra, (**b**) Raman spectra, and (**c**) UV–Vis spectra of CTF, CTF-COOH, CTF-HSO_3, and CTF-H_2PO_3. (**d**) Dipole moments on different components of CTF, CTF-COOH, CTF-HSO_3, and CTF-H_2PO_3.

To explore how the attachment of functional groups influences the optical absorption properties of CTFs, the UV–Vis absorption spectra simulated by electronic transition calculations were performed (Figure 2c). In its unmodified form, CTF displayed two absorption peaks: one in the 220–260 nm range, attributed to intramolecular electron transfer [11], and another in the 300–450 nm region, associated with electronic excitation processes [16]. The absorption edge of CTF is observed to be approximately 480 nm, indicating a limited capacity for visible light absorption. The absorption spectrum of CTF was expanded from 480 nm to 550 nm through functionalization with various groups. Among the functionalized CTFs, CTF-H_2PO_3 showed the largest shift, with an absorption edge at 550 nm. This enhanced absorption is likely due to the interactions between the functional groups and the CTF structure, which can modify the electronic energy levels of CTF. These interactions lower the energy required for electronic transitions, leading to an expanded absorption range. The extension of the absorption range into the visible region suggests that the functional groups modify the electronic structure of CTF, enhancing its photodegradation potential.

To further investigate the charge distribution after introducing oxygen-containing groups, dipole moment calculations for CTF, CTF-COOH, CTF-HSO_3, and CTF-H_2PO_3 (Figure 2d) were performed. The dipole moment is a crucial indicator of molecule interaction with light, particularly in the visible and near-infrared regions, which is important for photocatalytic performance. Notably, while the dipole moments of CTF-COOH, CTF-HSO_3, and CTF-H_2PO_3 in the X-direction are significantly larger than that of CTF, the overall dipole moments across all four materials are similar. This indicates that the functionalization does not significantly alter the absorption edge width, which is consistent with the trends observed in the UV–Vis absorption spectra. Furthermore, the dipole moment is also indicative of the internal charge distribution, with CTF-H_2PO_3 exhibiting the highest dipole moment in the X-direction. This suggests the presence of the strongest built-in electric field in CTF-H_2PO_3, which is likely to facilitate the efficient separation of photogenerated charge carriers, thus improving photocatalytic performance.

As depicted in Figure 3a–d, the calculated band gaps for CTF, CTF-COOH, CTF-HSO_3, and CTF-H_2PO_3 are 2.63, 2.42, 2.39, and 2.38 eV, respectively. Notably, CTF-H_2PO_3 exhibits the smallest band gap among these materials. A comparison of the absorption edges of CTF highlights that the theoretically determined band gaps align well with the optical absorption characteristics of these materials. Like pristine CTF, the functionalized derivatives CTF-COOH, CTF-HSO_3, and CTF-H_2PO_3 all exhibit direct band gaps. This indicates that the addition of functional groups does not alter the semiconducting nature of the band gap, preserving the high optical absorption efficiency and robust photogenerated carrier separation. In a direct band gap semiconductor, electrons transition from the valence band (VB) to the conduction band (CB) without a change in momentum, significantly enhancing photon absorption efficiency, particularly in the UV–Vis light regions. Further analysis of the DOS reveals distinct orbital contributions (Figure 3e–h). In pristine CTF, the VB is primarily composed of the N 2p orbitals, while the CB is largely derived from the C and N 2p orbitals. This disparity is attributed to the higher electronegativity of N, which favors electron acceptance, making the N 2p orbitals dominant in the VB. This phenomenon can be linked to the conjugated π-bond structure of TR, where π-electron clouds are delocalized between N and C atoms. Interestingly, when additional functional groups such as -H_2PO_3 are introduced to form CTF-H_2PO_3, the electronic structure undergoes significant modification. The VB shifts from being dominated by the N 2p orbitals to the O 2p orbitals.

To analyze the potential of CTF-based materials for photocatalytic degradation of pesticide contaminants such as dichlorvos (DDVP), the work function (W_f) of the CTF (001) crystal face was calculated (Figure 4a–c). The W_f for pristine CTF was found to be 6.13 eV (Figure 4a). Introducing electron-withdrawing functional groups like -COOH, -HSO_3, and -H_2PO_3 significantly alters the electronic properties of CTFs. These groups act as electron acceptors, lowering the Fermi levels of the modified CTFs (Figure 4a–d). Consequently, the W_f increased to 6.70, 6.60, and 7.10 eV for CTF-COOH, CTF-HSO_3, and CTF-H_2PO_3, respectively. A larger work function indicates stronger electron confinement by the catalyst. The introduction of H_2PO_3 results in the greatest increase in electronegativity, suggesting that H_2PO_3 has the strongest electron-binding ability. For donor-acceptor CTFs, the electron donor typically forms the major component of the conduction band, while the electron acceptor is generally part of the valence band [23]. Therefore, the trend of an increasing work function also suggests that these oxygen-containing acid functional groups primarily contribute to the valence band. This is consistent with the analysis of the band structure and density of states.

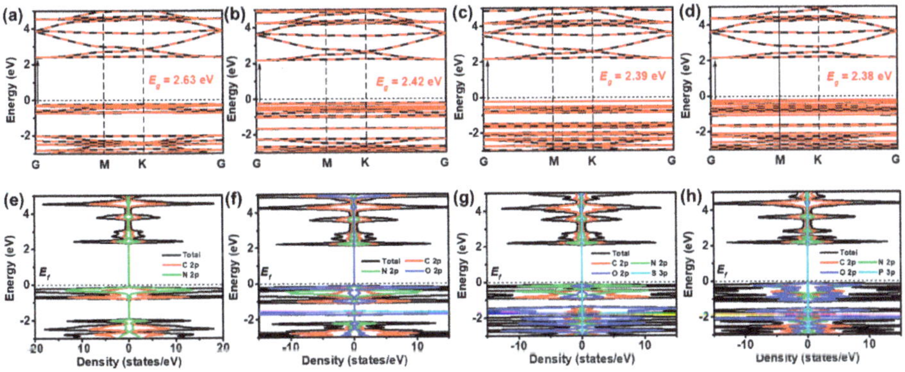

Figure 3. Band structure of (**a**) CTF, (**b**) CTF-COOH, (**c**) CTF-HSO_3, and (**d**) CTF-H_2PO_3. The density of states (DOS) of (**e**) CTF, (**f**) CTF-COOH, (**g**) CTF-HSO_3, and (**h**) CTF-H_2PO_3.

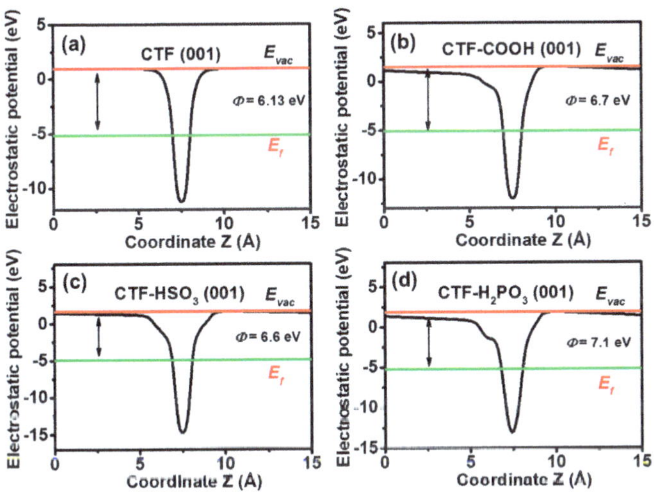

Figure 4. Electrostatic potentials curves of the (**a**) CTF, (**b**) CTF-COOH, (**c**) CTF-HSO_3, and (**d**) CTF-H_2PO_3.

To determine the effect of oxygen-containing acid groups on the surface distribution of potential and active sites, the surface electrostatic potential of the CTFs before and after modification was simulated. Figure 5a–d presents the top view distribution of surface electrostatic potentials for various CTF variants, offering a visual depiction of how distinct electron-accepting functional groups influence the electrostatic potential. The results highlight the varying degrees of electron redistribution caused by these functional groups, which directly affect the surface properties of the materials. The red regions have a lower electrostatic potential, while the blue regions have a higher electrostatic potential. For CTF, the electrostatic potential at the N sites is lower, while the potential at the H and C sites is higher. This indicates that the surface electrons of CTF are primarily concentrated at the N sites of the triazine rings, which is consistent with the electronic structure and work function analysis. Under light irradiation, photo-induced electrons in the red regions will transfer to the blue regions, causing the N sites to lose electrons and thereby retain more photo-induced holes. Consequently, the N sites act as oxidative active sites, participating in the oxidation of DDVP. In contrast to pure CTF, after the introduction of oxygen-containing acid groups, these groups become the regions with the lowest electrostatic potential and the highest electron density. Furthermore, the electron density and spatial distribution of these electron-rich regions are much larger than those at the N sites of pure CTF. This suggests that these functional groups not only serve as oxidation sites but also exhibit significantly stronger oxidative capabilities compared to pure CTF. Notably, the H_2PO_3 sites have the lowest potential and the largest effective range, indicating that H_2PO_3 can exert a stronger oxidative effect and degradation efficiency on DDVP during photodegradation processes.

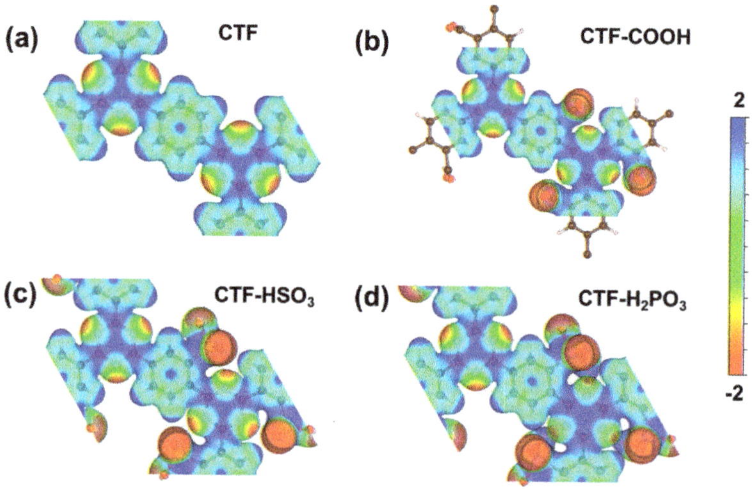

Figure 5. Surface electrostatic potential of (**a**) CTF, (**b**) CTF-COOH, (**c**) CTF-HSO_3, and (**d**) CTF-H_2PO_3.

Through a detailed analysis of molecular configuration, charge distribution, and electronic transfer properties, the modification mechanism of oxygen-containing acid functional groups on the performance of CTF was elucidated. As shown in Figure 6a, the adsorption distance between unmodified CTF and the DDVP molecule is 3.39 Å. After introducing functional groups, the adsorption distances were reduced to 2.37, 2.18, and 2.10 Å, respectively (Figure 6b–d). This reduction in distance can be attributed to the enhanced interaction between the CTF surface and the DDVP molecule due to the incorporation of oxygen-containing acid functional groups, thereby improving the adsorption capacity. The adsorption energy (ΔE_{ads}) of unmodified CTF was calculated to be −0.40 eV, while the introduction of oxygen-containing acid functional groups increased the adsorption energy to −0.46, −0.52, and −0.55 eV, respectively. These results indicate that the inclusion of functional groups significantly enhances the adsorption ability of CTF toward DDVP, primarily due to stronger electrostatic interactions, particularly van der Waals forces. Figure 6e–h present the differential charge density distributions after DDVP adsorption, where yellow regions represent electron accumulation and blue regions indicate electron depletion. These plots reveal minimal charge fluctuations between the two molecules, indicating that there is mutual electron transfer between DDVP and CTFs, which is one reason for the relatively low total interfacial electron transfer. To better determine the direction and electron transfer quantity, the electron gain and loss between DDVP and CTFs were analyzed using Bader charge (Figure 6i–l). Specifically, unmodified CTF exhibits a charge transfer of 0.036 e from CTFs to DDVP. After electron transfer, an electrostatic field is generated from CTFs towards DDVP. Under light irradiation, this electric field drives the photo-induced holes to transfer to DDVP, thereby facilitating its rapid degradation. The inclusion of COOH, HSO_3, and H_2PO_3 functional groups adjusts the charge transfer to 0.033, 0.013, and 0.016 e, respectively. This discrepancy between the electron transfer amount and adsorption energy may be because the adsorption energy is more influenced by the interaction distance, while the electron transfer is less dependent on it. Overall, these findings provide deeper insights into the role of oxygen-containing acid functional groups in enhancing the adsorption performance and oxidation behavior of CTFs.

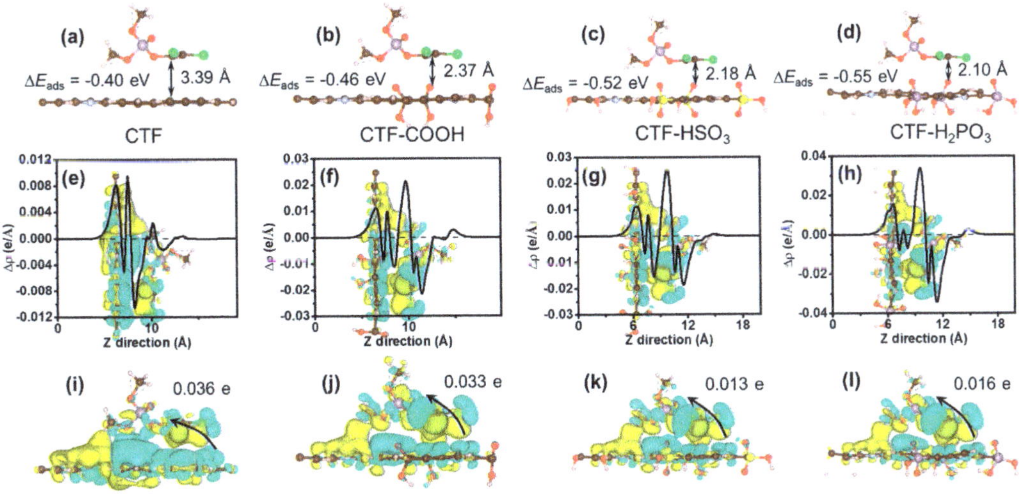

Figure 6. Interaction between DDVP and (**a**) CTF, (**b**) CTF-COOH, (**c**) CTF-HSO_3, and (**d**) CTF-H_2PO_3. Charge density difference and charge transfer of (**e**,**i**) CTF, (**f**,**j**) CTF-COOH, (**g**,**k**) CTF-HSO_3, (**h**,**l**) CTF-H_2PO_3.

To further explore the intermediate transformation during DDVP degradation on CTF, DFT was employed to assess the free energy changes associated with the oxidation of DDVP ($C_2H_7Cl_2O_4P$) to dichloroacetic acid (DCA, $C_2Cl_2O_2H$) on CTF (Figure 7a–d). Initially, DDVP adsorbs onto the TR active site of CTF, forming an adsorbed species (Equation (1)). This step is spontaneous and requires no energy input. Similarly, for CTF-COOH, CTF-HSO_3, and CTF-H_2PO_3, the initial adsorption process remains spontaneous, with modified CTFs showing more significant reductions in free energy, suggesting the process becomes more favorable. In the subsequent step, the adsorbed DDVP transforms into adsorbed 1,2-dichloroethane (DCE, C_2H_2Cl) and dimethyl phosphoric acid (DPA, $C_2H_7Cl_2P$) (Equation (2)), which is also a spontaneous reaction. This transformation remains spontaneous on CTF-COOH, CTF-HSO_3, and CTF-H_2PO_3, with notable free energy reductions, indicating that the reaction is more kinetically favorable. The next step involves the conversion of *DCE into *DCA (Equation (3)), which is also spontaneous. In the final step (Equation (4)), *DCA desorbs from the TR site of CTF. Unlike the earlier steps, this transformation is non-spontaneous and requires additional energy input. The rate-determining step (RDS) of the entire degradation process is the desorption of *DCA from the TR site on CTF (Equation (4)), with the energy barrier representing the critical bottleneck of the reaction. The reaction barriers for CTF, CTF-COOH, CTF-HSO_3, and CTF-H_2PO_3 are 0.75, 0.55, 0.45, and 0.33 eV, respectively. Compared with unmodified CTF, the reaction barrier of the modified CTFs is significantly reduced, primarily due to the electrostatic repulsion between the oxygen atoms in the DCA molecule and those in the modified functional groups. Additionally, CTF-H_2PO_3, with the lowest energy barrier (0.33 eV) among the four catalysts, demonstrates excellent DDVP degradation performance.

$$* + C_4H_7Cl_2O_4P \rightarrow {}^*C_4H_7Cl_2O_4P\,. \tag{1}$$

$$^*C_2H_7Cl_2O_4P \rightarrow {}^*C_2H_2Cl_2 + C_2H_7O_2P. \tag{2}$$

$$^*C_2H_2Cl_2 \rightarrow {}^*C_2Cl_2O_2H. \tag{3}$$

$$^*C_2Cl_2O_2H \rightarrow * \ C_2Cl_2O_2H \tag{4}$$

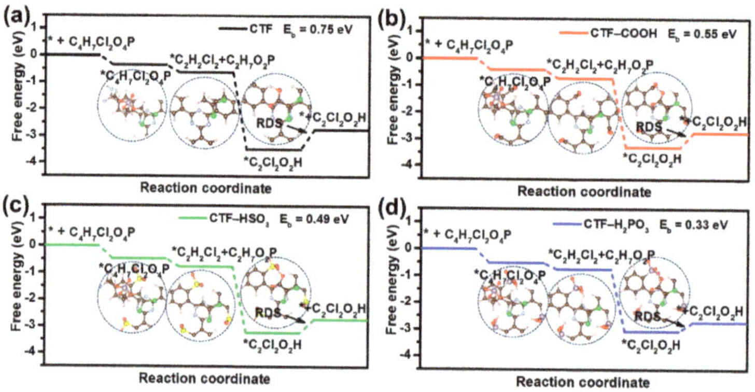

Figure 7. Free-energy diagrams for the photodegradation and reaction path of DDVP of (**a**) CTF, (**b**) CTF-COOH, (**c**) CTF-HSO_3, and (**d**) CTF-H_2PO_3.

4. Conclusions

This study systematically investigated the mechanism of photodegradation of DDVP using oxygen-containing acid functional CTFs. The results demonstrate that the introduction of -COOH, -HSO_3, and -H_2PO_3 functional groups offers several key advantages: (1) narrowing the band gap of CTFs, thereby promoting visible light absorption; (2) enhancing the dipole moment, which facilitates charge separation; (3) serving as the main component of the valence band and acting as oxidative active sites; (4) improving antioxidant capacity, thereby enhancing the stability of the photocatalytic degradation process; (5) strengthening the adsorption and activation of DDVP, promoting the desorption of acidic oxygenated products, and lowering the reaction energy barrier. Among them, the -H_2PO_3 group markedly reduces the reaction energy barrier, enhancing the activation capability and oxidation durability of CTFs, thereby demonstrating significant potential in the photodegradation of DDVP. This research presents valuable insights into the development of efficient, stable, and environmentally friendly photocatalysts for pesticide degradation.

Author Contributions: Writing—original draft, data curation, methodology, investigation, formal analysis, conceptualization, visualization, S.Y.; validation, writing—review and editing, S.Y. and Z.W.; supervision, resources, software, project administration, funding acquisition, Z.W. All authors have read and agreed to the published version of the manuscript.

Funding: This work was supported by the National Natural Science Foundation of China (NSFC) (No. 52402116) and the Key Research Project in Natural Sciences for Higher Education Institutions by the Ministry of Education (2022AH050396).

Institutional Review Board Statement: Not applicable.

Informed Consent Statement: Not applicable.

Data Availability Statement: Data are contained within the article.

Conflicts of Interest: The authors declare no conflicts of interest.

References

1. Mei, L.; Xia, X.; Cao, J.; Zhao, Y.; Huang, H.; Li, Y.; Zhang, Z. Degradation of Three Herbicides and Effect on Bacterial Communities under Combined Pollution. *Toxics* **2024**, *12*, 562. [CrossRef] [PubMed]
2. Do, K.H.; Kumar, D.P.; Rangappa, A.P.; Lee, J.; Yun, S.; Kim, T.K. Design and synthesis of a covalent organic framework bridging CdS nanoparticles and a homogeneous cobalt–bipyridine cocatalyst for a highly efficient photocatalytic CO_2 reduction. *J. Mater. Chem. A* **2023**, *11*, 8392–8403. [CrossRef]
3. Yi, Y.; Guan, Q.; Wang, W.; Jian, S.; Li, H.; Wu, L.; Zhang, H.; Jiang, C. Recyclable Carbon Cloth-Supported ZnO@Ag_3PO_4 Core-Shell Structure for Photocatalytic Degradation of Organic Dye. *Toxics* **2023**, *11*, 70. [CrossRef] [PubMed]
4. Lotfi, S.; Ouardi, M.E.; Ahsaine, H.A.; Assani, A. Recent progress on the synthesis, morphology and photocatalytic dye degradation of $BiVO_4$ photocatalysts: A review. *Catal. Rev.* **2022**, *64*, 1.
5. Wu, X.; Tan, L.; Chen, G.; Kang, J.; Wang, G. g-C_3N_4-based S-scheme heterojunction photocatalysts. *Sci. China Mater.* **2024**, *67*, 444–472. [CrossRef]
6. Fragoso, J.; Barreca, D.; Bigiani, L.; Gasparotto, A.; Sada, C.; Lebedev, O.I.; Modin, E.; Pavlovic, I.; Sánchez, L.; Maccato, C. Enhanced photocatalytic removal of NO_x gases by β-Fe_2O_3/CuO and β-Fe_2O_3/WO_3 nanoheterostructures. *Chem. Eng. J.* **2022**, *430*, 132757. [CrossRef]
7. Yue, W.; Wang, X.; Zhang, J.; Bao, J.; Yao, M. Degradation Characteristics of Nicosulfuron in Water and Soil by MnO_2 Nano-Immobilized Laccase. *Toxics* **2024**, *12*, 619. [CrossRef]
8. Shen, L.; Ye, T.; Chen, Y.; Chu, B.; Chen, H.; Hu, J.; Yu, Y. Facile Synthesis of a Novel $AgIO_3$/CTF Heterojunction and Its Adsorption-Photocatalytic Performance with Organic Pollutants. *Toxics* **2024**, *12*, 133. [CrossRef] [PubMed]
9. Kurisingal, J.F.; Kim, H.; Choe, J.H.; Hong, C.S. Covalent organic framework-based catalysts for efficient CO_2 utilization reactions. *Coord. Chem. Rev.* **2022**, *473*, 214835. [CrossRef]
10. Ding, H.; Wang, Y.-R.; Liu, M.; Shi, J.-W.; Yu, T.-Y.; Xia, Y.-S.; Lu, M.; Yang, Y.-L.; Chen, Y.; Li, S.-L.; et al. Electronic Tuning of Active Sites in Bifunctional Covalent Organic Frameworks for Photoassisted CO_2 Electrocatalytic Full Reaction. *Chem. Mater.* **2022**, *34*, 10752–10760. [CrossRef]
11. Lu, Z.; Wang, Z. Complete Photooxidation of Formaldehyde to CO_2 via Ni-Dual-Atom Decorated Crystalline Triazine Frameworks: A DFT Study. *Toxics* **2024**, *12*, 242. [CrossRef] [PubMed]
12. Nguyen, H.L.; Alzamly, A. Covalent Organic Frameworks as Emerging Platforms for CO_2 Photoreduction. *ACS Catal.* **2021**, *11*, 9809–9824. [CrossRef]
13. Kim, Y.H.; Kim, N.; Seo, J.-M.; Jeon, J.-P.; Noh, H.-J.; Kweon, D.H.; Ryu, J.; Baek, J.-B. Benzothiazole-Based Covalent Organic Frameworks with Different Symmetrical Combinations for Photocatalytic CO_2 Conversion. *Chem. Mater.* **2021**, *33*, 8705–8711. [CrossRef]
14. Hou, Y.; Lin, S.; Fan, J.; Zhang, Y.; Jing, G.; Cai, C. Enhanced Adsorption of Cadmium by a Covalent Organic Framework-Modified Biochar in Aqueous Solution. *Toxics* **2024**, *12*, 717. [CrossRef] [PubMed]
15. Chen, G.; Zheng, Z.; Zhong, W.; Wang, G.; Wu, X. Molten Intermediate Transportation-Oriented Synthesis of Amino-Rich g-C_3N_c Nanosheets for Efficient Photocatalytic H_2O_2 Production. *Acta Phys. Chim. Sin.* **2024**, *40*, 2406021. [CrossRef]
16. Lu, Z.; Lv, H.; Liu, Q.; Wang, Z. Modulating NH_2 Lewis Basicity in CTF-NH_2 through Donor-Acceptor Groups for Optimizing Photocatalytic Water Splitting. *Acta Phys. Chim. Sin.* **2024**, *40*, 2405005. [CrossRef]
17. Lu, T.; Chen, Q. Shermo: A general code for calculating molecular thermochemistry properties. *Comput. Theor. Chem.* **2021**, *1200*, 113249. [CrossRef]
18. Lu, T.; Chen, F. Multiwfn: A multifunctional wavefunction analyzer. *J. Comput. Chem.* **2012**, *33*, 580–592. [CrossRef] [PubMed]
19. Zhou, Q.; Guo, Y.; Zhu, Y. Photocatalytic sacrificial H2 evolution dominated by micropore-confined exciton transfer in hydrogen-bonded organic frameworks. *Nat. Catal.* **2023**, *6*, 574–584. [CrossRef]
20. Xiong, X.H.; Yu, Z.W.; Gong, L.L.; Tao, Y.; Gao, Z.; Wang, L.; Yin, W.H.; Yang, L.X.; Luo, F. Ammoniating Covalent Organic Framework (COF) for High-Performance and Selective Extraction of Toxic and Radioactive Uranium Ions. *Adv. Sci.* **2019**, *6*, 1900547. [CrossRef]
21. Pang, X.; Shi, B.; Liu, Y.; Li, Y.; Zhang, Y.; Wang, T.; Xu, S.; Wang, X.; Liu, Z.; Xing, N.; et al. Phosphorylated Covalent Organic Framework Membranes Toward Ultrafast Single Lithium-Ion Transport. *Adv. Mater.* **2024**, *36*, 2413022. [CrossRef] [PubMed]
22. Li, J.F.; Zhang, Y.J.; Ding, S.Y.; Panneerselvam, R.; Tian, Z.Q. Core-Shell Nanoparticle-Enhanced Raman Spectroscopy. *Chem. Rev.* **2017**, *117*, 5002–5069. [CrossRef] [PubMed]
23. Lu, Z.; Xu, X.; Wang, Z. Highly crystalline carbon nitride by covalent remedy for CO_2 photoreduction. *Sep. Purif. Technol.* **2024**, *330*, 125457. [CrossRef]

Disclaimer/Publisher's Note: The statements, opinions and data contained in all publications are solely those of the individual author(s) and contributor(s) and not of MDPI and/or the editor(s). MDPI and/or the editor(s) disclaim responsibility for any injury to people or property resulting from any ideas, methods, instructions or products referred to in the content.

MDPI AG
Grosspeteranlage 5
4052 Basel
Switzerland
Tel.: +41 61 683 77 34

Toxics Editorial Office
E-mail: toxics@mdpi.com
www.mdpi.com/journal/toxics

Disclaimer/Publisher's Note: The title and front matter of this reprint are at the discretion of the Guest Editors. The publisher is not responsible for their content or any associated concerns. The statements, opinions and data contained in all individual articles are solely those of the individual Editors and contributors and not of MDPI. MDPI disclaims responsibility for any injury to people or property resulting from any ideas, methods, instructions or products referred to in the content.

www.ingramcontent.com/pod-product-compliance
Lightning Source LLC
Chambersburg PA
CBHW061627210926
51850CB00009B/2460
9783725856893